Shaping Biology

The National Science Foundation and American Biological Research, 1945–1975

TOBY A. APPEL

THE JOHNS HOPKINS UNIVERSITY PRESS
Baltimore & London

This material is based upon work supported by the National Science Foundation under contract number LPA 88-12629. Any opinions, findings, and conclusions or recommendations expressed in this material are those of the author and do not necessarily reflect the views of the Foundation. This book has been brought to publication through the generous assistance of the National Science Foundation.

The Johns Hopkins University Press
2715 North Charles Street
Baltimore, Maryland 21218-4363
www.press.jhu.edu

Library of Congress Cataloging-in-Publication Data will be found at the end of this book.
A catalog record for this book is available from the British Library.

ISBN 0-8018-6321-X

Contents

CHAPTER 6
Funding Individuals and Institutions in the 1960s:
Opportunities and Constraints *154*

CHAPTER 7
Promoting Big Biology:
Biotrons, Boats, and National Biological Laboratories *178*

CHAPTER 8
Allocating Resources to a Divided Science:
The "New" and the "Old" in Biology *207*

CHAPTER 9
Forging New Directions after the Golden Age, 1968–1972 *235*

CHAPTER 10
End of an Era, 1972–1975 *269*

Tables

Foreword

As the story goes, President Harry Truman, an inveterate early riser, announced that he had signed the National Science Foundation Act of 1950 at six in the morning of May 10, 1950, from the back of a train in Pocatello, Idaho. This seemingly inauspicious event occurring far from Washington established NSF as an independent agency of the federal government, the first and only agency with the mandate to support basic research in all of the disciplines of science. The impetus for the legislation was the very productive partnership between the United States government and the scientific community begun during World War II. It is interesting to note that NSF's first operating budget was $3.5 million.

It is now fifty years later, and in the year 2000, NSF—now a nearly $4.0 billion agency—will celebrate and highlight the many achievements in science and engineering it has supported over the years. One of the disciplines that has made astounding progress during this period is biology.

Toby Appel's book examines the history of biological research support by NSF from 1950 through 1975. In doing so it gives us a picture of the history of biology itself during this crucial period. One of its fascinations is that it not only gives us rare insight into how the actions and policy decisions of the Division of Biological and Medical Sciences (BMS) at NSF helped to shape the field of biology, but it also provides a glimpse of how those decisions and, importantly, the people at the Foundation who made them, influenced and helped to shape all science in the United States.

From the very beginning BMS influenced the field of biology by bringing together botanists, zoologists, and ecologists to help make funding recommendations. Decisions made during those early years were hugely influential in the development of entire fields. Developmental biology, cell biology, molecular biology, plant biology, and environmental biology were born and grew up during that time. Today, new fields such as genomics and bioinformatics are emerging. All were encouraged and supported by the actions and policy decisions of NSF. Appel's book lets us see how it all began and unfolded.

Mary E. Clutter
Assistant Director
Biological Sciences
National Science Foundation

Acknowledgments

The idea of writing a history of National Science Foundation patronage of biology originated in 1988 as part of a project conceived by NSF Historian George T. Mazuzan. Dr. Mazuzan had arranged to commission histories of NSF support of several areas of science and technology, including computer sciences and engineering as well as biology. The prospect of researching the history of NSF and biology very much appealed to me because I had long been interested in the institutional development of American biology. I had previously written on the emergence of professional societies in biology at the end of the nineteenth century and the relation of such societies to the establishment of university departments. Here was a new type of biological institution—namely, the federal funding agency—one that had a profound influence on the postwar development of the biological sciences. When I began work almost nothing had been written by historians of science on the subject. Interest has since burgeoned.

I thus began research on this project under a contract in 1988. I am grateful for the early assistance of George Mazuzan and his predecessor, J. Merton England, in finding relevant records. The Historian's Files that they had amassed, and that have recently been given to the National Archives, proved to be a gold mine of information. The former Directorate for the Biological, Behavioral and Social Sciences graciously provided office space and equipment. I owe an enormous debt of gratitude to all of the former employees of NSF and biologists who shared their recollections with me. Conversations with Ron Overmann, former program director for History and Philosophy of Science, and Bob Althauser, a rotator in the Sociology program, aided me in conceptualizing issues in the early stages of research, as did visits with Robert Kohler and Scott Gilbert. A first draft was completed by the end of the contract period in 1991. Although this book began under a contract with NSF, I always felt that I had a free hand in writing whatever I chose. Needless to say, interpretations and opinions are mine, not those of the NSF.

At the end of the contract period, I felt that additional research was needed to broaden the perspective. I was privileged to obtain a grant from the NSF Visiting Professorships for Women Program. I spent two enjoyable years at the University of Florida teaching history of science and medicine and continuing my research using the facilities of a large university library that supported all of biology, from a natural history museum to medical school departments. Frederick Gregory, Bob Hatch, Betty Smocovitis, and Don Dewsbury were extremely supportive colleagues. In the summer of 1992, Deborah Warner provided space for me to work at the Smithsonian Institution.

Final polishing of the manuscript was held in abeyance for a few years while I attended library school and settled into my new responsibilities as Historical Librarian at the Cushing/Whitney Medical Library at Yale University. During this period, Dian Belanger ably assisted in editing and streamlining the manuscript. I then completed a final revision, which was submitted to the Johns Hopkins University Press. Now that I am a librarian myself, I would like to reserve a special thanks for all the librarians and archivists who helped me to locate sources, especially Janice Goldblum of the National Academy of Sciences.

Robert J. Brugger, my editor at Johns Hopkins, made a number of wise editorial suggestions, most of which I followed. Nancy Roderer and Regina Kenny Marone enabled me to use library facilities to see the manuscript through the editorial process. Joanne Hazlett of the Directorate for Biological Sciences handled communication from NSF. I found especially enjoyable the interactive nature of the copyediting by Kennie Lyman (thanks to e-mail and attachments), and I thank Juliana McCarthy for additional editing at the Johns Hopkins Press.

Throughout the long gestational process, a number of colleagues have read and commented on portions of the manuscript and supported its progress: John Beatty, Joseph Cain, Mary Clutter, J. Merton England, Robert Friedel, Gerald Geison, Elihu Gerson, Frederick Gregory, Frederic L. Holmes, Warren Kornberg, Harry Marks, Philip Pauly, Nathan Reingold, Margaret Rossiter, and Michael Sokal. I fear that I have unintentionally forgotten many others who have helped along the way.

My final note of appreciation is to George Mazuzan, who conceived the project and encouraged it throughout.

Shaping Biology

Envisioning a Federal Patron for Biology

In September 1951 Alan T. Waterman, first director of the National Science Foundation, addressed biologists assembled at the annual meeting of the American Institute of Biological Sciences on the Foundation's plan "to initiate a strong program in fundamental research in the biological sciences." He elaborated what he hoped would be NSF's unique role as a patron of biology within the federal government:

> Although the biological sciences are now enjoying a significant measure of federal support principally through the National Institutes of Health, the Department of Agriculture and the military Programs; and although a modest portion of this effort can be regarded as basic research, by and large the research in the biological sciences is directed research in the sense that it is directed toward the solution of practical problems. The novel and important aspect of the Foundation's effort will be its absolute freedom from any limitations with respect to practical application.[1]

Waterman and the biologists he brought with him in 1951–52 from the Office of Naval Research to NSF formed a coherent vision of NSF's role in funding science, and biology in particular. NSF was to be the federal government's chief patron for basic research in biology, the only agency to fund the entire spectrum of biology—from molecules to natural history museums.

How this vision of a federal patron for biology emerged and what became of it in the next two and a half decades is the subject of this book. At a descriptive level, this book is a history of the activities of NSF's Division of Biological and Medical Sciences, formed in 1952 from the merger of the Division of Biological Sciences and the Division of Medical Research specified in the NSF Act of 1950, and reorganized out of existence in 1975. But the focus of the book is science policy for research in biology at NSF and within the federal government.

Historians of science have recently begun to direct attention to the major transformation of academic science in America as a result of the entrance of the federal government as the primary patron of science in the postwar period. However, the overwhelming attention of historians of the postwar era has been given to the physical sciences, especially the relation of science to the military funding agencies.[2] Only a few historians have investigated the biological, medical, and agricultural sciences in the postwar era; most of these have focussed on large projects such as the International Biological Program, Integrated Pest Management, and the Atomic Bomb Casualty Commission.[3] This book therefore enters a vast and relatively uncharted territory, while at the same time building upon a growing literature on patronage of biology by the private foundations in the prewar period.[4]

Though never the dominant player among federal agencies supporting science, NSF is a good focal point for looking at biology and the federal patron because it was the one federal agency that funded all of biology for its own sake. In the 1950s NSF had articulate program officers with strong visions, who hoped that the agency would eventually dominate support of basic biology in universities. Moreover, in keeping with the Foundation's mission to coordinate and evaluate support of science by the federal government, NSF program officers maintained contact with and collected data from all other federal agencies funding scientific research.

The ideals concerning NSF support of biology as they were articulated in the early 1950s rested on a number of assumptions, some characteristic of NSF and some particular to biology. Primary among them was that biology would be justified and supported as a basic science, independent of any particular application. While other agencies would support basic research in the areas of their missions, NSF would be responsible for basic biology as a whole. Pluralism of funding agencies was beneficial because it provided the scientist with choices and made it less likely that politics would prevail over meritocracy.

The justification for funding biology in the early years was part of the larger argument for federal support of basic research, referred to by some historians as the "ideology of basic research."[5] Waterman, who became the nation's most eloquent advocate of basic research, declared in his address to biologists in 1951, "Basic research and the training of young scientists constitute the fundamental bases on which the gigantic superstructure of applied science and technology rests."[6] Compared to the resources expended on applied research, the nation was spending far too little on basic research.

Following the argument in Vannevar Bush's famous report to President Truman in 1945, *Science—The Endless Frontier,* Waterman and his staff advocated stocking up on basic science on the grounds that eventually at least some of it—not predictable in advance—would provide handsome dividends in terms of practical benefits to society, thus justifying support of the whole.[7]

A second major assumption, characteristic of NSF, was that the primary criterion for funding would be scientific merit. "The primary purpose of the Foundation is to develop a hard core of first-rate research by competent investigators," Waterman told biologists.[8] In biology, this came to mean supporting the best biology wherever located, though it was understood that the best biology was most often undertaken by men located in a small number of leading research universities.

A third fundamental tenet, held especially by the early biologists at NSF, was that of trust between staff and awardees. NSF would award funds to biologists in the form of "grants-in-aid" to assist them with their research programs, not to purchase research for the government. That is, grants should supplement the university's commitment to research as one of its ongoing functions. Grants as instruments of trust also implied that applications for funding and reporting requirements would be relatively simple and free of red tape. NSF could maintain congenial relations with grantees because the agency would be able to operate to a large extent insulated from politics. In fact, NSF in the 1950s made a deliberate effort not to act like a federal bureaucracy.

Particular to the founders of the biology program at NSF was the deeply held conviction that the agency could play a role in unifying biology. Through their organization of NSF's Division of Biological and Medical Sciences along functional rather than disciplinary lines, their promotion of interdisciplinary research, and their active advisory committee of academics, Foundation staff hoped to promote the ideal of one biology. Since program officers were scientists by training, many of them on a year's leave from academia, it was assumed that the interests of the staff, advisors, and grantees would largely coincide. The framework of panels for each program to evaluate proposals and advise on the program level, and above these, a divisional committee of advisors representing all of biology, was intended to create a coherent agenda for all of biology.

Because the staff saw funding of biology as a unified and cooperative endeavor of program officers and advisors at various levels, a rich body of documents was created, including annual reports at the program and division levels, revealing staff memos for the files in the form of "diary notes," reports of meetings of advisory committees, and numerous reports of ad hoc committees of staff and/or academic biologists investigating particular issues.

Central to funding biology at NSF was the strong role of the federal science administrator, or program officer, in shaping biology. The program officers in biology (even more so than those in the physical sciences) sought wide latitude to fund whatever was for the good of biology. Historical writing on the private foundations has pointed to the strong role of individuals, especially men like Warren Weaver and Alan Gregg, in shaping patronage of research. This book investigates this role of "manager of science" in the altered context of the federal government. It is suggested that there are threads of continuity from the Rockefeller Foundation through the Office of Naval Research, which funded biology in the aftermath of World War II, to the early NSF.

Led by a triumvirate of aggressive and idealistic program officers—John Wilson, William V. Consolazio, and Louis Levin—NSF's Division of Biological and Medical Sciences operated as a foundation unto itself supplying flexible aid in any form capable of furthering biology. BMS science managers stretched individual grants-in-aid to cover not only biological research but also conferences and travel, publications, educational projects, large instruments such as electron microscopes, operational support and facilities at museums and biological stations, and departmental training grants. Wilson, Consolazio, and Levin hoped their agency would become Uncle Sam's counterpart to the Rockefeller and Carnegie foundations.

Throughout this history of NSF and the patronage of biology run a number of fundamental disagreements about science policy, which are resolved differently at different periods of NSF history. The most obvious disagreement was that between basic and applied science. NSF was strongly committed to a differentiation between the two. Other funding agencies— the National Institutes of Health (NIH) for example—had less use for the distinction, in part because they feared NSF would corner the market on basic research. Even in the early 1950s, it became clear to NSF program directors in biology that other agencies were funding a great deal of basic research of very little relevance to their missions. The changing relationship of NSF to the various mission agencies funding biology, especially the National Institutes of Health, is a major theme of this book.

Although NSF wanted to fund science for its own sake, in order to obtain higher appropriations from Congress it had to emphasize the eventual applied benefits of allowing scientists freedom to pursue uncommitted research. In biology in the 1950s (before the environmental movement) the two major applications were health and agriculture, each handled by a major mission agency. Other agencies funding biology could justify their appropriations more readily on the basis of relevance. Through appeals to the possible cure of dread diseases, NIH was able to fund basic research, much of which was only remotely related to medicine. By the mid-1960s, Congress was pushing NSF in the direction of more applied research, culminating in the Daddario-Kennedy Act of 1969, which authorized NSF support of applied research, and the controversial RANN program (Research Applied to National Needs) of the 1970s. Program directors became increasingly dependent upon linking basic research to specific practical goals.

Dominance of federal patronage of biology by NIH led to a number of policy dilemmas. Should NSF maintain a comprehensive program in biology, as was envisioned in the 1950s and 1960s, or concentrate on "gap" areas left out by NIH and other agencies? Should NSF fund leading biologists who were already recipients of large sums of money from other agencies, or should it fund those who had little or no other source of support? If BMS funded the big names, as it did in the 1950s and 1960s, it could participate in the most exciting discoveries of the decade—the revolution in molecular biology. But it would get little credit for doing so as the same biologists were being amply funded by NIH and other agencies. If BMS concentrated on areas not covered elsewhere—systematics and ecology, in particular—it could make a real difference in these areas but might be frozen out of the mainstream in funding biology. These choices became more acute in the late 1960s and 1970s, when the "Golden Age" of funding had passed.

A second tension, one explored by previous historians of the physical sciences, was that between elitism and geographical distribution or support for the infrastructure of science.[9] Does one support the "best" science, which in practice tended to be of white men in the dominant research universities, or should patronage also take into consideration geographical distribution, type of institution (e.g., small colleges versus research universities), race, gender, and funding by other agencies? To what extent were such considerations influenced by political pressures? Consolazio and Wilson were strong believers in selecting grantees primarily on the basis of scientific merit. By the early 1960s, however, congressional concerns with imbalances, the rapid growth of NIH, the budget crisis of the end of the 1960s, and protests of women and other groups in the early 1970s all insured continued heated debate over this issue.

A third area of strain was the perennial debate between little science and big science, always controversial in biology, where even today most research takes place in small laboratories presided over by one faculty member. Ought NSF to fund primarily individual project grants, or should the agency fund at the level of large multidisciplinary research projects or facilities? Though academic advisors, as represented by the divisional committee, tended to favor little science, program officers in an agency that expended considerable resources on big physical science were drawn to experiment with various forms of "big biology": phytotrons and biotrons, big biological research vessels, and the International Biological Program. Such funding decisions could lead to fierce battles among staff and advisors representing different areas of biology.

Related to all these issues was the question of whether NSF should grant other than individual research awards in biology. Should the Foundation, for example, fund senior professorships, institutional facilities, support for departments, or awards for training graduate students? To what extent would the program officers in biology find leeway within the bureaucracies of NSF and the federal government to make these determinations and to compete with types of awards offered by NIH?

A fourth point of discord was the difference between a governmental gift of a grant-in-aid to an individual and federal purchasing of research from the university. The program directors and the laboratory biologists on the panels resisted siphoning funds to pay for high negotiated overheads and faculty salary. They felt that it was the responsibility of the university to support research and pay the full salaries of their professoriate. In the Johnson era, NSF became more bureaucratic and more politicized as Congress pressed for accountability, and as academic administrators on the National Science Board and on the President's Science Advisory Committee insisted on payment of the full costs of research.

A fifth cause of friction centered on the extent to which a program officer could exercise initiative. The imaginative and flexible funding of biology in the 1950s came under pressure from a number of directions: the cautious NSF hierarchy (dominated by the physical sciences), which accused program officers of overstepping the bounds of the NSF Act and of entering the jurisdictions of other arms of the Foundation; by congressional imperatives; by competition from other federal funding agencies; and finally by biologist-advisors who opposed staff-initiated funding priorities or "emphases."

Finally, the NSF's idealistic goals of establishing a community of purpose among staff and grantees and of furthering the unity of biology were both seriously challenged in the 1960s and 1970s. Biology is not and never has been a single discipline, but rather a heterogeneous and overlapping group of disciplines. The mergers and realignments of departments on campuses in the 1960s and the shift of NSF priorities toward funding ecology led to serious discord between the staff and their advisors from various areas of biology. Differences between staff and advisors were exacerbated further by the political imperative of getting congressional funding in the 1970s.

Because biology is not a single discipline, the historian wishing to study the role of the federal patron in biology must investigate a number of different fields. Practices vary greatly among fields: in some, research is carried out in laboratories with postdocs, graduate students, and expensive instrumentation, requiring constant grants for upkeep, while in others, facilities are more modest and research is carried out on a more individual basis. Each field, depending on its institutional structures, its research practices and culture, its potential practical applications, and its relations to political power, interacted differently with NSF. To address the key question of what difference NSF made to the biological sciences would therefore require a study of each separate field of biology.

Although it is beyond the scope of this book to assess the scientific results of NSF support of biology, it can contribute to the framework for such studies by comparing NSF's funding strategies and relations with the leadership of selected areas of biology. The book will focus on four areas: molecular biology; plant biology; systematic biology; and ecology, especially biological oceanography, tropical biology, and ecosystem ecology. These were chosen because their relation to NSF shifted significantly and in contrasting ways from the 1950s to the mid-1970s.

The story to be told derives from a wealth of hitherto untapped documents on biology at NSF (see discussion in Note on NSF Primary Sources) as well as on interviews by the author (or earlier oral histories) of over fifty former staff members and biologists. Interviews have been especially helpful for fleshing out personalities of key individuals, motives, reactions to events and conflicts, and reflections on changes that have taken place in the agency. These NSF sources have been supplemented by published articles and reports as well as by material from selected manuscript and archival collections in universities and other organizations, particularly the National Academy of Sciences.

The narrative concludes in 1975 with the major reorganization of the governance of NSF that brought about the demise of the Division of Biological and Medical Sciences. Its sections were transformed into divisions of the new Directorate of Biological, Behavioral and Social Sciences presided over by a presidentially appointed associate director. Biology's history at NSF was to be linked for the next fifteen years with the social sciences. To pursue the history further would require delving into the tangled politics of federal funding of the social sciences, as well as the problematic response of NSF to the rise of commercial biotechnology—both large subjects more suitable for another book.

After the 1975 reorganization, biology was no longer a unity within the Foundation. The institutional structures that underlay the ideal of "one biology" and of cooperation between staff and patrons were dismantled along with the process of developing the rich records that these structures had generated. Many of the documentary sources that served as a basis for this history were no longer created after the early 1970s, and those documents that do exist have not yet made their way into historical files. Most important, 1975 represents the end of an era that had begun with the establishment of the biology program at NSF. Although NSF continued to fund good biology, as it always had, the fundamental assumptions and expectations of the founders of the Division of Biological and Medical Sciences had undergone a thorough transformation.

CHAPTER I

Making a Place for Biology at the "Endless Frontier," 1945–1950

World War II, as is well known, marked a watershed in the organization of American science. Before the war, the federal government had supported science in its own bureaus, but federal support of academic science was limited. In the 1920s and 1930s, America's major research universities had become dependent on the great private philanthropic foundations, a source of funding that dwindled during the Depression years. The 1930s saw serious but inconclusive debate by members of the scientific elite and representatives of the government on the possibility of federal patronage for academic science. But it was the wartime experience of successful government-sponsored military and medical research in the nation's universities that produced widespread agreement, both in government and among scientists, that federal support of science, basic as well as applied, should be made permanent.[1]

As biologists in the 1940s recognized, World War II was predominantly a physicists' war and not a biologists' war. Physicists basked in public appreciation of such contributions as radar, the proximity fuse, and the atomic bomb. Medical scientists, too, won praise for achievements including the miraculous bactericide penicillin, DDT, preservation of blood plasma, improved treatment of malaria, and new devices allowing pilots to fly at high altitudes without fear of blackout. But a number of biologists felt left out of the war effort, omitted from policy-making positions and given little credit for what they regarded as their contributions to war research.[2] They feared likewise being left out of the impending postwar transformation of science.

The half-decade from 1945 to 1950 was a period of critical negotiating and intense maneuvering as various groups—among them the numerous scientific disciplinary societies, umbrella organizations such as the American

Association for the Advancement of Science, the National Academy of Sciences and its research wing, the National Research Council (NRC), and federal agencies—sought to position themselves to benefit from the expected new funding. Many policy and institutional issues had to be decided. How would the fears of scientists and university administrators that the federal government would control science be reconciled with the government's need for accountability? How would the desire of scientists to pursue basic research be reconciled with the public priority for socially useful research? Above all, what fields of science would be supported and what agency or agencies would undertake the task?

Vannevar Bush's landmark prospectus on postwar science policy, *Science—The Endless Frontier,* released in July 1945, called for support of basic research in the sciences, as well as military and medical research, by a single new agency, a National Research Foundation, which by its very name evoked its private predecessors. Deep divisions among scientists, legislators, and executive branch leaders on how to shape such an all-purpose public foundation however, delayed its creation for five years. It was in this period of extended debate that the United States' uniquely pluralistic system of funding science in universities evolved.

This postwar story has customarily been told from the vantage point of the physical sciences or individual agency histories. This chapter looks at events from the perspective of biology and medicine as their place in the proposed foundation—and in the postwar framework in general—was gradually hammered out. That biology achieved a separate identity in what was finally called the National Science Foundation (NSF), rather than being subsumed under medicine, was the result of a deliberate political campaign on the part of biologists, which was in turn part of a larger movement to create a unified "biology" from disparate biological sciences.

By the time the NSF was founded in 1950, three other major programs of federal support for the life sciences were in place—in the Office of Naval Research (ONR), the Atomic Energy Commission (AEC), and the National Institutes of Health (NIH)—and the NSF's prospects for taking a leading role in biology and medicine had altered dramatically. These other funding programs in biology and medicine were important not only because they competed with NSF but also because they, along with the prewar private foundations, served as models for the new agency's funding policies and procedures.

Biology and Its Patrons before 1945

The federal government had long sponsored research in the life sciences in its own facilities, especially the U.S. Department of Agriculture (USDA), the Public Health Service (PHS), and the Smithsonian Institution. But universities enjoyed almost no government aid for biological research except for USDA block grants, which, since the Hatch Act of 1887, funded work at agricultural experiment stations associated with land grant universities. This research was a hybrid of studies aimed at solving immediate practical problems to meet local needs and high quality "pure research."[3]

Federally funded medical research had proceeded in the laboratories of the Public Health Service. The National Institute of Health, created in 1930 from the former Hygienic Laboratory, carried on considerable in-house research centering on communicable diseases. When public attention turned to chronic diseases in the 1930s, the Institute turned its attention there as well, the PHS forming the National Cancer Institute as a separate arm in 1937. As a byproduct of the 1930s debate over federal support of science, the NCI was authorized to award grants to nonprofit institutions, which it did on a small scale for projects such as building cyclotrons for cancer research.[4] Until 1944, the NIH itself had no authority to award grants.

Except for work in the agricultural sciences, life scientists in academia before 1940 depended primarily on their own institutions for the resources to conduct research. In addition, numerous private sources offered small, supplemental "grants-in-aid"—sufficient to pay for research supplies or part-time student assistance. A few major research universities were also able to tap sometimes considerable sums from industry. Pharmaceutical companies, for example, funded selected university scientists for cooperative biomedical research projects.[5] The largest and single most important external source of support for academic biological and medical research in the interwar period, however, was the Rockefeller Foundation. Its administrative concepts can be traced through the Office of Naval Research to the National Science Foundation.

Founded in 1913, the Rockefeller Foundation (RF) awarded some $50 million for medical, public health, and nursing education in the early decades of the century. Before the 1930s, its support for the biological sciences was scattered, but among its strategic awards with long-term effects was a series of capital grants, begun in the 1920s, to such institutions as the Marine Biological Laboratory at Woods Hole, Massachusetts, Hopkins Marine Laboratory of Stanford University, and the Yerkes Primate Laboratory of Yale University. Block grants to the National Research Council's Committee on Research on Problems of Sex funded key studies in endocrinology, embryology, and related areas. In 1925, the Rockefeller Foundation helped launch the journal *Biological Abstracts,* contributing nearly $900,000 through 1948. Of particular importance to the life sciences were the RF's prestigious postdoctoral fellowships, begun in 1919 and broadened to include biology after 1923, which enabled young investigators to travel and continue their research. RF fellowships supported some one thousand biologists, both American and European, from 1923 to 1953.[6]

In the 1930s, its resources diminished by the early Depression years, the RF turned from large institutional grants to individual project grants. Its Division of Medical Sciences, formed in 1928, funded delimited areas of medical research, while the RF Division of Natural Sciences promoted a "concentrated program" of "experimental biology." The decision to focus the Division of Natural Sciences on biology to the near-exclusion of other sciences was largely the work of Warren Weaver, the former professor of applied mathematics at the University of Wisconsin who headed this division from 1932 to 1958. Weaver believed that a "new biology," ultimately beneficial to social needs, would emerge from projects entailing the transfer of concepts and techniques from physics, chemistry, and mathematics to biology. Weaver's individual project grants supported biologists in such areas as cell biology, genetics, biochemistry, biophysics, embryology, endocrinology, and microbiology. He funded work on such instruments and techniques as X-ray crystallography, the use of isotopes, the ultracentrifuge, and the electrophoresis apparatus.[7]

The Rockefeller Foundation's program was neither large nor broad; its funding strategy for biology was unapologetically elitist. It deliberately focused on only a small number of the "best" scientists, among them such notables as Thomas Hunt Morgan, H. J. Muller, Theodosius Dobzhansky, Joshua Lederberg, George Beadle, and E. L. Tatum in genetics, and E. J. Cohn, Joseph Fruton, Carl F. Cori, Fritz Lipmann, Erwin Chargaff, and Linus Pauling in biochemistry. It supported only those projects that fit into its program definition, intentionally concentrating its support where it could have the greatest impact. As Weaver later explained, the RF had a unique flexibility in this regard, for it was answerable neither to faculty, students, alumni, parents, stockholders, employees, members of Congress, nor the Bureau of the Budget.[8] Between 1933 and 1953, the RF's Division of Natural Sciences spent some $24 million on biological sciences in the United States, a sum soon dwarfed by federal research funding.[9]

Some historians have argued that Weaver's program "developed the idea of molecular biology." Weaver, himself, claimed to have been the first to use the term in print—to describe his program in the 1938 RF Annual Report. Several leading actors on the "path to the double helix" did in fact receive Rockefeller support, but Weaver's "molecular biology" of the 1930s had a much broader connotation than the molecular genetics that the term later came to signify. Like NSF program managers at a later time, Weaver deliberately sought to break down traditional boundaries through such interdisciplinary designations.[10]

Weaver developed and epitomized a new social role in American science, that of "science manager." An articulate spokesman for science, Weaver set policy, mediated between academic clienteles and the Rockefeller hierarchy, and was ever alert to locate and develop new "strategic investments." He and his staff traveled widely, consulting with trusted academic informants on promising funding opportunities. They recorded frank and often amusing personal impressions of people and projects in "diary notes," which were circulated among RF staff, a practice later imitated at the NSF. On the basis of this informal scientific intelligence, Weaver submitted his staff's funding decisions to the Rockefeller Foundation board of trustees for the expected approval.[11] The RF provided a model of an imaginative grants program in which staff initiative played a central role in making policy.

An alternative model for funding research in the life sciences emerged during World War II with the creation of the Committee on Medical Research (CMR) of the U.S. Office of Scientific Research and Development (OSRD), an agency set up to organize civilian science for the war effort. President Roosevelt appointed as its director Vannevar Bush, an engineer trained at Tufts, Harvard, and MIT, who came to the nation's capital in 1939 to head the Carnegie Institution of Washington. The OSRD pioneered in establishing contractual relations with universities to carry out specific war-related research—usually classified—using academic scientists as principal investigators. Supported handsomely by the OSRD and the military services and given a prominent place in wartime science policy, physical scientists hoped for and expected continued federal support after the war.[12]

Although the military services and the Public Health Service conducted medical research directly during the war, the chief source of funding for wartime medical projects in academia was the OSRD's Committee on Medical Research (CMR). Formed in 1941 by the same Executive Order that established OSRD, the CMR consisted of seven members: four presidentially appointed civilian scientists and one representative each from the Army, the Navy, and the PHS. Chaired by Alfred Newton Richards, a pharmacologist and vice-president for medical affairs at the University of Pennsylvania, the CMR let nearly six hundred contracts with 137 organizations, most of them universities, at an unprecedented cost of $25 million.

In the CMR model of federal patronage, leading university scientists, rather than agency staff, made the major funding decisions. The CMR relied on elite committees of outside scientists set up by the National Research Council's Division of Medical Sciences to evaluate proposals and recommend projects to the OSRD. Research supported included work on infectious and tropical diseases, nutrition, psychiatry, shock, control of wound infections, sulfa drugs and penicillin, malaria therapy, insecticides and rodenticides, and aviation physiology. The federal government provided lavish support not only for physicians' clinical investigations but also for research in such medically related academic fields as physiology, biochemistry, and pharmacology. Roosevelt and many others assumed that the government would continue its support of medical research in universities after the war.[13]

Biologists, however—as opposed to medical researchers—felt left out of OSRD and of war research in general. That they did so is related to the ambiguous nature of "biology" in this period. Central to understanding biology's relation to the federal patron is the heterogeneous nature of the life sciences. From the beginning of the era of professionalization of American science in the 1880s and 1890s, biology had been divided into separate sciences, primarily botany and zoology. Although a few departments of biology existed in 1940 (at Princeton and Stanford, for example), most universities maintained separate departments of botany and zoology. Separate long-standing national organizations—the American Society of Zoologists, formed in 1890, and the Botanical Society of America, established in 1893—represented the two fields.

Apart from botany and zoology, and often overlapping with them, a host of other biological specialties—each with its national organization—had emerged before 1940. There were national societies for genetics, ecology, bacteriology, plant physiology, ornithology, mammalogy, herpetology, ichthyology, and entomology, as well as societies representing numerous fields of applied biology such as fisheries, forestry, economic entomology, horticulture, agronomy, agricultural genetics, animal nutrition, and plant pathology. Another significant group of "biological" disciplines—anatomy, physiology, biochemistry—were located almost exclusively in medical schools. They had their own departments, doctoral programs, societies, and journals; they scarcely interacted with botany and zoology.

There was, by contrast, no unifying organization for "biology" aside from the National Research Council's Division of Biology and Agriculture and the ineffectual Union of American Biological Societies formed in 1923, which had little to its credit besides the founding in 1926 and nominal sponsorship of *Biological Abstracts*. Committees of representatives of numerous diverse biological societies governed both institutions. Biology was thus not a unified discipline, but rather a heterogeneous collection of overlapping specialties (see fig. 1).[14] Few biologists labeled themselves as such; they were first and foremost botanists or bacteriologists or geneticists. Only a handful of broad-minded leaders—like Marston Bates, a staff member of the Rockefeller Foundation, or Douglas Whitaker, professor at Stanford—identified their field as "Biology" in their biographical entries in *American Men of Science*.

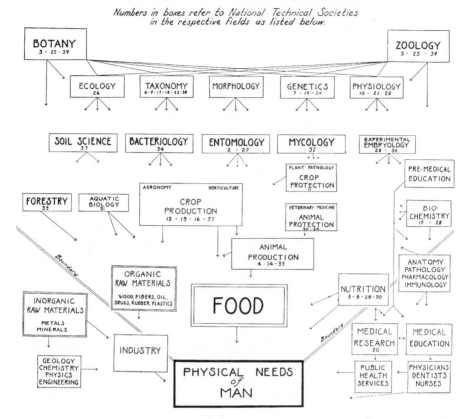

BIOLOGY and AGRICULTURE

*Numbers in boxes refer to National Technical Societies
in the respective fields as listed below.*

Fig. 1. Organization Chart. National Technical Societies: 1. American Association of Anatomists. **2.** American Association of Economic Entomologists. **3.** American Biological Society. **4.** American Dairy Science Association. **5.** American Dietetic Association. **6.** American Fern Society, Inc. **7.** American Genetic Association. **8.** American Institute of Nutrition. **9.** American Ornithologists' Union. **10.** American Physiological Society. **11.** American Phytopathological Society. **12.** American Society of Agricultural Sciences. **13.** American Society of Agronomy. **14.** American Society of Animal Production. **15.** American Society of Biological Chemists, Inc. **16.** American Society for Horticultural Science. **17.** American Society of Ichthyologists and Herpetologists. **18.** American Society of Mammalogists. **19.** American Society of Naturalists. **20.** American Society of Parasitologists. **21.** American Society of Plant Physiologists. **22.** American Society of Plant Taxonomists. **23.** American Society of Zoologists. **24.** American Veterinary Medical Association. **25.** Botanical Society of America, Inc. **26.** Ecological Society of America. **27.** Entomological Society of America. **28.** Federation of American Societies for Experimental Biology. **29.** Genetics Society of America. **30.** Institute of Food Technologists. **31.** Limnological Society of America. **32.** Mycological Society of America. **33.** Poultry Science Association. **34.** Society of American Bacteriologists. **35.** Society of American Foresters. **36.** Society for Experimental Biology and Medicine. **37.** Soil Science Society of America. **38.** Sullivant Moss Society. **39.** Union of American Biological Societies. (Robert F. Griggs, "The Organization of Biology and Agriculture," *Science* 96 [18 December 1942], 546)

In 1944, scientists could still seriously debate in the pages of *Science* the value of university courses in "biology." A committee of the U.S. Office of Education, reporting in 1943 on college curriculum during wartime, claimed to find "no objective evidence" that a general biology course had "any advantage" over well-organized general courses in botany or zoology. C. A. Shull of the University of Chicago praised the report for its courage. Biology courses, he wrote, were generally a "hodge-podge," a "fraud against the student." There is, Shull concluded, "no such thing as a science called 'biology,' any more than there is a science known as 'physical science.'" It was "no wiser to present 'general biology' instead of botany and zoology, than to present 'physical science' in lieu of mathematics, physics and chemistry."[15]

By the 1940s, though, barriers between botanists and zoologists and between life scientists on university campuses and those in medical schools were beginning to break down. New areas of biology such as cell biology, biochemical genetics, and biophysics blurred older disciplinary distinctions. In 1946, physiologists who looked for general laws of functional biology not limited to organ physiology of vertebrate animals founded the Society of General Physiologists. Biochemistry was beginning to infiltrate university biology departments. Taxonomy, primarily located in museums under curators who did not always have advanced degrees, was being transformed into the more theoretically based "new systematics." The 1940s saw the culmination of a movement to construct a synthesis linking a theory of evolution by natural selection with mathematical population genetics, paleontology, and systematics. The Society for the Study of Evolution appeared in 1946, the Society for Systematic Zoology the following year. These intellectual and institutional currents contributed to a renewed rhetoric of a unified "biology," coinciding, not accidentally, with the expectation of a new era of federal support of the life sciences.[16]

As early as 1942, spokespersons for biology publicized their perception that biologists, as distinct from medical researchers, were conspicuously absent from decision-making positions during the war. Part of the reason, as they saw it, was biology's lack of identity or unity. Botanist Robert F. Griggs, Chairman of the NRC Division of Biology and Agriculture, wrote in *Science,* "Over and over again as I endeavor to facilitate the contributions of biology and agriculture toward winning the war, I encounter the unorganized and incoherent conditions of our group of sciences," which of itself was partly responsible for their "comparatively ineffective application" to the "needs of a total war." The following year, in response to Griggs's concerns, the National Academy of Sciences commissioned a report from Stanford biologist Douglas Whitaker in an unsuccessful attempt to convince Vannevar

Bush to establish an OSRD Committee on Biological Research on the model of the CMR.[17]

These frustrations persisted into the postwar era where they became a rallying point for biologists. Columbia University geneticist L. C. Dunn, testifying before the Senate subcommittee on NSF legislation in 1945, claimed that "through lack of organization," biologists had been left out of "a great national effort." Many now realized that their absence from the "hastily improvised war agencies was bad not only for biology" and other sciences that depended on it, "but for the Nation. Their state of mind is not improved by the reflection that, by and large, the fault was their own."[18] As a result of the omission of biologists from war agencies, Griggs wrote in 1946, contributions that were biological in nature, such as penicillin, DDT, and improved hybrid crops, had been credited to medicine or chemistry. To make matters worse, many drafted biologists were not given the opportunity to use their expertise to aid in victory. He complained that at the same time that biological work had been assigned to physicians and chemists, bacteriologists were given kitchen duty. Biology was too important to the national welfare to be thus treated. If World War II had not been a biologists' war, Griggs predicted that, through the exploitation of biological warfare, World War III might well be.[19]

Transformations of content in the biological sciences, biologists' perceived neglect during the war, and their fear that they might not obtain a share of postwar government funding led, in the late 1940s, to renewed advocacy for a united "biology." Old institutions of biology, such as the venerable journal *American Naturalist,* founded in the 1860s, and the Union of American Biological Societies, made a deliberate attempt to "speak for biology." The NRC's Division of Biology and Agriculture also tried to position itself as "the united voice of the Life Sciences."[20] The short-lived American Biological Society, founded in 1941 to aid the financially desperate *Biological Abstracts* by adopting the publication, soon made common cause with the Union. By 1945 the two organizations' officers were planning an American Biological Union. An American Society for Professional Biologists, equivalent of the American Society for Professional Engineers, was also being considered in 1946.[21]

Biologists' renewed quest for solidarity culminated in 1948 in the founding of the American Institute of Biological Sciences (AIBS), modeled on the American Institute of Physics. Griggs had suggested such an organization in 1942. After the war, Robert Chambers, a cell biologist at New York University and president of the Union of American Biological Societies, and John Spangler Nicholas, Yale embryologist and president of the American

Biological Society, took up the idea. Both active in the early legislative debate over NSF, they convened concerned biologists in early 1946 at a meeting of the American Association for the Advancement of Science where
they gained endorsement for the formation of a new organization. Its purpose would be to provide a means for biologists to execute their "public responsibilities as scientists," to "safeguard" their professional interests, and to
help supply the "material means for the promotion of biological research."
In part because the NRC feared the new organization would be a competitor, Griggs and Detlev W. Bronk, new president of the Academy,
arranged in February 1948 to establish AIBS under the aegis of the NRC
Division of Biology and Agriculture. Once in operation, with an initial
membership of twelve biological societies, the AIBS acted as a focal point
for support of biology in the proposed NSF.[22]

The Rockefeller Foundation precedent, the renewed interest in a unified biology, and the frustration stemming from lack of recognition during
the war, led spokespersons for biology to monitor science legislation carefully to assure that general biology, and not just medicine and the physical
sciences, would have a place in the new postwar order.

The NSF Debate, 1945–1946

The legislative debate over the National Science Foundation is traditionally said to have begun with the publication of Vannevar Bush's *Science—
The Endless Frontier* in July 1945. Historians have interpreted Bush's classic
blueprint for federal support of science in the postwar era as a conservative
response to Senator Harley Kilgore, a New Deal Democrat from West Virginia, who introduced legislation calling for creation of a "National Science
Foundation" in 1944. Kilgore's NSF, an outgrowth of his earlier bills for
technological mobilization in wartime, was to coordinate federal science
policy, give grants and loans to advance socially relevant science, and aid in-
ventors and small businesses. To counter Kilgore's call for science to serve
national goals, Bush welcomed a way to shape postwar science according to
his own anti–New Deal views. On 17 November 1944, Roosevelt sent him
a letter, partially drafted within the OSRD, requesting his ideas on how science might serve the nation after the war. The letter made four inquiries, for
each of which Bush appointed a committee: How might war research be
best disseminated? How might medical research be organized to aid in the
"war on disease"? How could government aid scientific research in private
and public institutions? And, how might scientific talent be recognized and
developed? The resulting document included the four committee reports
plus Bush's introduction and summary.[23]

Science—The Endless Frontier elaborated a rationale for federal support of academic science that claimed science as essential to the national welfare, especially to national security, yet left scientists free to carry out research agendas of their own devising. The ideology of basic science presented in Bush's report and reiterated by scientists and NSF science administrators throughout the 1950s and most of the 1960s maintained that basic research was "the pacemaker of technological progress" in weaponry, medicine, agriculture, and industry. Investment became the preferred metaphor for justifying federal sponsorship. "Basic scientific research is scientific capital," Bush wrote. Such capital could be increased by investing public funds in undirected basic research, a fraction of which (unknowable in advance) would result in "important and highly useful discoveries." To be assured of technological progress in the future, the fund of basic research must continually be replenished. It was presumed that all science, however recondite, had the potential to prove socially and economically useful. Thus, basic research was to be supported as an end in itself, with the assumption that, as a byproduct, the national interest would be served.[24]

Despite the report's emphasis on basic research, Bush had no intention of separating basic research from its practical applications. The National Research Foundation, as Bush envisioned it, was to encompass both basic and applied research mutually supporting each other, including coordinated work on medical, military, and industrial problems in the public interest. Convinced by his OSRD experience that a large portion of military research should remain under civilian initiative and control, Bush proposed that the foundation include a Division of National Defense. And, despite the doubts of his own medical advisors, who wanted a separate agency, Bush recommended a Division of Medical Research.[25]

To insulate the foundation from the usual political process, Bush insisted that it be governed by a part-time, unpaid board of scientists. The foundation would coordinate national policy in the sciences, support research primarily in universities, and develop new scientific talent through the award of scholarships and fellowships. As successor to OSRD, it would take over OSRD's remaining contracts. Bush estimated a budget of $33.5 million in the first year increasing to $122.5 million by the fifth year.[26] The reality of NSF was to be much different.

The relation of medicine to the proposed foundation was controversial from the beginning. To Bush's consternation, his Medical Advisory Committee, chaired by Walter W. Palmer, professor of medicine at Columbia University, advocated an independent federal agency specifically for medical research. If the committee members and other leading medical consultants were unwilling to include medicine in the new foundation, they objected even more to ceding support of academic medical research to the Public Health Service of which the National Institute of Health formed part. The PHS, they claimed, was not "sufficiently free of specialization of interest to warrant assigning to it the sponsorship of a program so broad and so intimately related to civilian institutions." To "remain free from political influence and resistant to special pressures," their proposed National Foundation for Medical Research was to be controlled by a board of scientist trustees specifically not to include representatives of PHS or military medicine. The conclusions of the Palmer Report show that the shape of federal support for medical research had not yet been determined in 1945. Although Bush published the Palmer report as delivered, he claimed to have convinced the committee in the meantime to accept his own proposal for a single foundation that included a Division of Medical Research.[27]

The place of biology in Bush's foundation was also problematic. From the few references to biology in the report, it seemed likely to be subsumed under the Division of Medical Research rather than the Division of Natural Sciences. Biologists were notably absent from Bush's committees. The Medical Advisory Committee consisted primarily of clinical men with laboratory experience; the only member not from a medical school was Linus Pauling, chairman of the Division of Chemistry of the California Institute of Technology. The influential Committee on Science and Public Welfare, which under the chairmanship of Johns Hopkins president Isaiah Bowman, investigated the question of federal aid to science in universities, had no biologists at all. Some biologists felt that biology was as badly neglected by the Bush Report as the deliberately excluded social sciences. Forest R. Moulton, permanent secretary of the American Association for the Advancement of Science, who generally praised the report, noted that among its "serious gaps" were "no references to the biological sciences, except as they may be involved in medicine." He urged biologists to make their presence known at the Senate hearings on science legislation.[28]

On the same day that *Science—The Endless Frontier* was released to the public, Senator Warren G. Magnuson, Washington Democrat, introduced legislation based on the Bush Report to create a National Research Foundation. Four days later Senator Kilgore released his bill for a National Science Foundation. The different visions represented by these two bills divided both scientists and government officials and contributed to the five-year delay in creating the National Science Foundation.[29]

The most contentious issue throughout the course of the debate was governance of the proposed foundation. Would it be controlled by a board of outside scientists and thus insulated from political pressures as called for in the Magnuson bill, or would power and responsibility rest in a presidentially appointed director responsive to political and social concerns as Kilgore advocated? A related issue was whether existing federal agencies should be represented on the board, a Kilgore provision. Other contested areas included the balance between basic and applied research, ownership of patents generated from government-funded research, geographical distribution of funds (which neither original bill proposed), and support of the social sciences.[30]

Biologists were not happy with either bill, but they took especial exception to the Magnuson bill, which a majority of scientists favored. Kilgore's bill, with its greater emphasis on applied research, specified a foundation with only two subdivisions, a research committee for national defense and one for health and medicine. Magnuson provided a division of physical sciences separate from the more applied division of national defense but subordinated the biological sciences to a division of medical research, defined as "programs relating to research in biological science, including medicine and the related sciences." It thus ignored biology as a separate entity. "Biologists of every kind were astonished that the public-spirited framers of a national science bill could be so blind to the Nation's need for biological research," the NRC's Robert Griggs recalled in 1947.[31]

The first year of debate, 1945–46, was the most crucial for the future of federal patronage of biology and medicine. Almost one hundred scientists testified in twenty days of joint hearings on the Kilgore and Magnuson bills before the Senate Subcommittee on War Mobilization of the Committee on Military Affairs. Later, in November 1945, those who advocated the principles of *Science—The Endless Frontier* joined with Isaiah Bowman to form the Committee Supporting the Bush Report. Another large group of more moderate to liberal scientists, under the leadership of Harlow Shapley and Harold C. Urey, both regarded with suspicion by Bush and Bowman for their left-wing sympathies, organized a Committee for a National Science Foundation in late December.[32]

Biologists, though far less visible than physical scientists among those pressing for NSF legislation, nonetheless took an active part. Articulate spokesmen for the Bush position included CMR chair A. N. Richards; Homer W. Smith, a renal physiologist at New York University and principal author of the Palmer Report; Detlev W. Bronk, then director of the Johnson Foundation for Medical Physics at the University of Pennsylvania; and Lewis Weed, who chaired the NRC Division of Medical Sciences. During the hearings, the nongovernment medical scientists generally advocated the extreme conservative side of contested issues. Concerned about the financial plight of medical schools but fearful of federal domination of science, they followed the Palmer Committee's lead in calling for unrestricted funds to be given to schools and administered by local research committees rather than grants-in-aid for specific purposes. Smith claimed to have polled all the consultants of the Palmer Committee and found an overall preference of 226 to 2 for the Magnuson bill. He concluded, despite the biased nature of the sample, that "the overwhelming vote" showed that "these men do not want a one-man dictatorship of science, and they do not want Government domination."[33] The lead speaker for biologists at the Senate hearings, Columbia geneticist L. C. Dunn, a scientist with a strong humanitarian conscience, eloquently argued the opposing Kilgore position—that "in a democracy like ours, administration of so important a public service as the development of scientific knowledge should be responsive to public opinion and public needs."[34]

The most active spokespersons for "biology" in the critical year of 1945–46 were Robert Chambers and John Spangler Nicholas, representing respectively the Union of American Biological Societies and the American Biological Society. They claimed credit for convincing the Senate subcommittee to devote a day of the science foundation hearings to testimony of biologists. Chambers and Nicholas also encouraged individuals and biological organizations to write to the Senators. Much of this correspondence was included in the twelve-hundred-page record of the hearings.[35]

Biologists, although more divided ideologically than the medical scientists, agreed on the need for basic research in biology, graduate fellowships to compensate for the loss of training opportunities during the war, and autonomy for the scientific investigator. Several witnesses argued forcefully that basic research needed greater emphasis in the legislation under consideration. Dunn pointed out that while the Department of Agriculture supported applied agricultural research and the national health and cancer institutes were beginning to support medical research, there remained an "urgent national need" for basic research in biology, the foundation of agriculture and medicine. "Surely," he avowed, "if the fields in which fundamental facts of biology are applied, namely medicine and agriculture, are recognized as public responsibilities, the support of the basic research from which sound applications arise is a matter of even greater public importance."[36]

Adopting the capital investment metaphor of the Bush Report, biologists cited several examples of basic research in biology that had later paid off in terms of valuable application, among them Mendel's experiments on peas, Paul Bert's high altitude investigations, and Alexander Fleming's discovery of penicillin. The most persuasive example, described at greatest length and hailed as a tremendous success story, was the development of hybrid corn. Introduced commercially in 1934, hybrid corn accounted for 75 percent of all corn harvested by 1944. Lewis J. Stadler, professor of field crops at the University of Missouri, traced this achievement to the fundamental research of two Americans: George Harrison Shull and Edward M. East, who, he claimed, began their experiments around 1905 with no consideration of crop improvement. Stadler estimated that the monetary value of the increased national production of corn during the period 1942–45 alone was over $2 billion, sufficient to pay for the development of the atomic bomb. Indeed, he declared, "the return from this one application of science to the improvement of a single crop plant will amount each year to far more than the annual budget now proposed for federal support of scientific research."[37] Stadler had, in fact, minimized the researchers' agricultural ties and thereby exaggerated the distinction in this case study between fundamental and applied research. Both Shull, at the Carnegie Station at Cold Spring Harbor, and East, at the Connecticut Agricultural Station, combined interest in theoretical problems of genetics with a genuine and informed interest in improving crop yield and quality.[38]

A recurrent theme in the testimony before the Senate subcommittee was the place of biology in the proposed agency. One witness after another argued that biology ought be considered a "separate, independent discipline" and not "a subsidiary to medicine." Biology, Bronk claimed, "is the more inclusive of the two." Stadler argued, "There is no reason to expect that the needs of basic research in biology would be covered by the activities of a division of medical research, just as there is no reason to expect that the needs of basic research in the physical sciences could be covered by a division of national defense." Through their cumulative lobbying, the biologists finally convinced the lawmakers to separate biology from medicine. Chambers was pleased to report in early February 1946 that "we now have assurances from both Senator Magnuson and Senator Kilgore that in their revisions biology as distinct from the medical sciences will be considered on a par with the physical, chemical, and mathematical sciences."[39] All later bills seriously considered by Congress included a Division of Biological Sciences, a significant achievement for biologists.[40]

The legislative debate of 1945–46 ended in the failure of Congress to pass any science-funding bill, in large part due to a lack of unity among the scientists. In early 1946 hopes for successful legislation centered on a compromise bill—S. 1850, sponsored by both Kilgore and Magnuson—which contained a number of Kilgore's original controversial features. Scientists, including Bush and the Bowman Committee, at first appeared to rally behind it even though they disliked some of its provisions. But in May, when action by the full Senate was delayed, Bush and some of his supporters arranged for the introduction in the House of a rival bill that was a modified version of Magnuson's original bill.

Many scientists were disturbed by this turn of events, among them

Chambers. Although he and Nicholas had endorsed the recommendations of the Bowman Committee on behalf of their respective organizations, by February, 1946, Chambers had become a staunch advocate of S. 1850. He warned that "an uncompromising attitude, fancied or real" might "jeopardize the enactment of a measure so overwhelmingly approved of in the October hearings." In June, he deplored the divisiveness introduced with the alternative bill and issued an urgent appeal for "A United Front for S. 1850."[41] Such pleas were in vain. The compromise bill passed the Senate in July, but the House Committee on Interstate and Foreign Commerce, faced with the two bills' competing claims in the final days before adjournment, decided not to act "because the legislation was too complicated and important."[42]

Interlude: Three Alternative Federal Patrons of Biology

The demise of the NSF Act of 1945–46 was critical, as several other federal agencies, eager to expand their relations with academic scientists, took advantage of the legislative failure to fill the void left by the postwar closing of OSRD. For biology, the Atomic Energy Commission, the Office of Naval Research, and the National Institutes of Health emerged as competitors to a future NSF. These three agencies altered the perception among government officials and scientists of NSF's role in the life sciences. Moreover, they provided alternate models for managing a program in the life sciences.

The Office of Naval Research

While Congressmen and scientists were debating the form of National Science Foundation legislation, President Truman approved a bill in August 1946 creating within the United States Navy an Office of Naval Research (ONR). Viewing itself from the beginning as a surrogate for the yet-to-be-established NSF, ONR would prove to be of central importance in shaping NSF's early history.[43] ONR became the major source of federal support for general basic science, including the biological sciences, in universities from 1946 until the Sputnik era, and it led the way in establishing postwar policies for federal support of basic research.[44] Much of the conceptual and procedural framework of the NSF grants program, particularly the idea of the activist program manager, derived from ONR. In addition, ONR supplied many of the early personnel for NSF including its first director, Alan T. Waterman, and all of the permanent staff members in biology.

ONR's pioneering role as a federal patron of academic science was largely unintended by its founder, Vice Admiral Harold G. Bowen. Bowen had hoped to use the Navy's newly created Office of Research and Inventions (ORI), which he had headed since May 1945, to develop nuclear propulsion systems for naval vessels. To further that end, he had adopted a discarded plan for organizing postwar naval research suggested in 1943 by a group of naval officers known as the "bird dogs." They had proposed the creation of a permanent office, advised by a committee of civilian scientists, to coordinate weapons research within the Navy bureaus and to work with academic scientists. Bowen had hoped that by becoming the patron of scientists in universities, he would obtain access to nuclear expertise. However, by the time the wartime ORI became ONR, Bowen had lost his bid for jurisdiction over the Navy's development of nuclear propulsion. The chief functions of the ONR became those of overseeing naval research facilities such as the Naval Research Laboratory, supervising the Navy's patent program, and supporting academic research of supposed relevance to the Navy. Under the latter charge, ONR quickly mounted a sizeable program, in fiscal year (FY) 1947 spending $22 million and managing 648 contracts with 169 academic contractors.[45]

ONR modeled its system of federal patronage on the experience of the private foundations, especially the Rockefeller Foundation. Bowen, his deputy, Captain Robert Conrad, and Alan T. Waterman, a physicist from Yale who arrived in February 1946 to become Chief Scientist, established ONR's funding policies. Warren Weaver, who became first chairman of the Naval Research Advisory Committee, noted in his diary in June 1946: "O.R.I. is organized a great deal like the RF, minus our board of trustees, with 'Section Heads,' who have about the functions of directors of divisions. Their recommendations go through Alan Waterman to Conrad. When he approves (as he would expect to do in normal cases), only legal technicalities of drawing contracts, etc., remain."[46]

ONR's early success as a patron of science derived from its flexible administrative procedures. In order to attract academic scientists initially fearful of military demands and red tape, ONR adopted a liberal interpretation of the contract mechanism, its only statutorily authorized funding form. Traditionally under a contract, bids were obtained and a service purchased for which the contractor was held responsible. A grant, on the other hand, implied a gift and imposed little obligation on the grantee. OSRD had already altered the contract form to omit soliciting bids, and ONR further blurred the distinction to the point that ONR contracts functioned nearly like grants. Instead of deciding on relevant research problems and then finding scientists to do the work, ONR invited scientists to initiate their own projects. Proposals were judged on their scientific merit, the competence of the investigator, and the relevance of the project to the Navy's mission. In practice, mission relevance counted for little, especially in the period before 1950. The ONR staff acted on the belief, consonant with the Bush ideology, that if the best scientists were given free rein to carry out their fundamental research programs, eventually results of practical value would be obtained.[47] For accepted proposals, ONR drew up a contract or subagreement with the scientist's institution. (Typically a single contract sufficed for all projects in a given university.) Reporting requirements, as in the Rockefeller Foundation, were minimal, and the scientist was free to publish in standard journals.

ONR also allayed scientists' fears of government control of their research by hiring as program directors experienced civilian scientists who maintained friendly, informal relations with their clientele. Many ONR program officers came from academia and expected to return there after a stint at the agency. William Consolazio, head of the Biochemistry Branch (and future NSF program officer), spoke in 1949 of the program officer's role:

> As investigator-administrator, the establishment of a good rapport with American science should be relatively simple for isn't he one of the fraternity? The condition of mutual trust is of inestimable value. All of us are aware of the fact that we operate under a certain degree of suspicion because of our military associations. Many scientists are suspicious of federal support, and even more so of military support.[48]

Weaver, as head of the Advisory Committee, urged ONR to increase its staff and have them "visit these projects, live with them, study them firsthand," as was done in the Rockefeller Foundation.[49]

Though dominated by the physical sciences, ONR included a Medical Sciences Branch under the charge of Captain A. J. Vorwalt, M.D., to fund research in biology and medicine. By October 1946 this branch had considered some 297 proposals, of which 66 were activated, with 78 declined and 63 still in negotiation. Over $3 million had been obligated.[50] These numbers represented relative neglect. As of 1949 biology and medicine accounted for only 15 percent of ONR's budget compared to 75 percent for the physical sciences and 10 percent for the naval sciences. At that time only two biologists sat on the Advisory Committee. Thus, several biological program officers saw greater opportunity at NSF and willingly transferred to the new agency.[51]

Although the Medical Sciences Branch justified its research program as directly related to the naval mission, ONR soon supported a wide variety of biological projects of no apparent military relevance. In 1946, Vorwalt claimed that the Physiology Section funded work on "normal and pathological functions" of organic systems within organisms "as reflected in the bodily performance of Naval personnel"; the Environmental Physiology Section considered projects on "man's reaction to his environment," whether in a plane, ship compartment, or submarine, in the Arctic or tropics; and Psychology dealt with "the problems of selection, training, morals, and leadership."[52] In practice, mission relevance could be argued for nearly any project in biology. As Orr E. Reynolds, chief of the Biological Sciences Division, explained in 1952, the Navy's mission "to keep as many men at as many guns as many days as possible," encompassed "research involving such basic explorations as that of enzyme chemistry in cellular function, comparative physiology, the fundamental basis of the nerve impulse and its propagation and transmission—in short, virtually the whole spectrum of experimental biology."[53]

In its early years, ONR also supported clinical research. The Bacteriology Section in 1946 funded studies of diseases affecting naval personnel including tularemia, tuberculosis, and hepatitis. In 1949, ONR supported at the University of Pennsylvania an investigation of economic and occupational factors modifying the pathological process in tuberculosis. In 1952, the agency funded dysentery research. But by then, no doubt because of the rapid growth of NIH, medicine at ONR was either being phased out or organizationally subordinated to the biological sciences.[54]

Funding decisions at ONR were ultimately the responsibility of the staff, who could avail themselves of evaluations by outside scientists as needed. The ONR biological programs pioneered using peer advisory panels that met together to evaluate proposals, a procedure that was later adopted by the biology programs at NSF. Reynolds considered panels' advice in selecting contractors more necessary in the biological sciences than in the physical sciences, because a large number of small projects taxed the experience of any single program officer. A panel would compare all the projects and provide rough priorities as to their relative importance.[55]

Under pressure from the NRC's Lewis Weed, Reynolds first tried to organize a panel through the NRC Division of Medical Sciences, following the precedent of the OSRD Committee on Medical Research. "However," he recalled, "after trying the system for a few months, I found that the advice was frequently inappropriate and nonuseful." One may suspect, based on the NRC's experience in advising the CMR, that its panel for ONR presumed more authority than Reynolds was willing to give it. At the suggestion of Milton O. Lee, executive secretary of the American Physiological Society since 1947 and soon to hold that position in the newly formed AIBS, Reynolds then turned to AIBS to organize panels. These worked so well that several other branches of the ONR Medical Sciences Division asked AIBS to set up similar panels.[56]

ONR did not initially establish panels from a conviction that academic scientists must evaluate the projects of their peers but rather as a convenience to the ONR science managers. It was made clear to the AIBS that the panels were advisory only; program managers made the final judgment on which projects to recommend. In the physical science branches, program managers did not adopt the panel system but relied solely on ad hoc advice from individual scientists.[57] Thus the roots of the panel system later used for biology at NSF and the long-standing differences in peer review between NSF's physical and biological sciences can ultimately be derived from ONR and not from NIH as has sometimes been suggested.[58]

Intended as a stand-in for a national science foundation or not, ONR carved out its own niche as a patron of science and soon became a competitor. ONR leaders let it be known early that they had no intention of stepping aside for NSF. In fact, ONR deliberately sought during the legislative debate over NSF to limit the new agency's activities, campaigning from 1947 on for the removal of NSF's Division of National Defense. By December 1950, both Warren Weaver and Vannevar Bush were displeased with ONR ambitions. Both had strongly approved of ONR filling a vital gap, but they had assumed that once NSF was in place, the Navy would transfer all or most of its basic research to the new agency. Weaver recorded that he was "disturbed . . . that the ONR has gradually but steadily receded from that position," now seeming to have "reached the extreme point where they do not think they ought to give up anything to a National Science Foundation."[59]

The Atomic Energy Commission

A second powerful patron of biology to emerge in the immediate postwar years was the Atomic Energy Commission (AEC). In July 1946, Congress passed an Atomic Energy Act, which created a full-time, presidentially appointed civilian Atomic Energy Commission with control over the production, ownership, and use of nuclear energy and authority to sponsor by contract pure and applied research related to its mission. While the AEC's main emphasis was in the physical sciences—by design, no biologists were appointed to its General Advisory Committee—an ancillary program in biology and medicine was planned from the start.[60]

The AEC organized a Division of Biology and Medical Sciences in late 1947 under the direction of Shields Warren, professor of pathology at Harvard Medical School. It had its own advisory committee with direct access to the commission. This committee was chaired by Alan Gregg, head of the Division of Medical Sciences of the Rockefeller Foundation and counterpart to Warren Weaver, who headed the RF Division of Natural Sciences. AEC's Biology and Medicine division budgeted research in its own National Laboratories and other AEC installations and funded research in colleges and universities. Predictably, it initially focused on the biological effects of radiation, other biohazards related to the production of nuclear energy, and the peaceful uses of atomic energy in medicine and biology.

Early AEC research in the life sciences had a strong medical emphasis, especially since radioisotope therapy had roused widespread hope for a cure of cancer. Responding to public interest, Congress appropriated $5 million for FY 1948 for cancer research in the AEC. The division sponsored a program of shipping radioisotopes produced at Oak Ridge National Laboratory to universities for use in cancer therapy and biomedical research. In order to train biomedical scientists to work with radioisotopes, the AEC in 1948 funded a predoctoral and postdoctoral fellowships program, which was soon extended to the physical sciences. That year, 180 fellowships in clinical medicine, biomedical sciences, and surgery were awarded at a cost of $1 million. Expenditures for biology and medicine increased from $15.2 million in FY 1949 to $24.5 million by fiscal 1952.[61] Paul Pearson, Chief of the Biology Branch, estimated in 1953 that 20 percent of the division's budget supported some 135 projects in colleges and universities.[62]

In its contract procedures, the AEC biology program roughly followed the model of the private foundation, in which the staff had considerable authority and flexibility in making funding decisions. No panels of outside scientists met to evaluate AEC proposals. Instead the staff sometimes requested ad hoc evaluations of proposals but made the final decisions in staff meetings. The division's advisory committee reviewed the program in general terms but had no approval authority over particular projects. In its early years AEC was closely tied to ONR; until 1950 ONR actually managed AEC's biology contracts program.[63]

Like the ONR, AEC protected its prerogatives against the time when a national science agency might be created. Language in the National Science Foundation Act of 1950, first introduced in 1947, forbade the Foundation to "support any research or development activity in the field of nuclear energy," and stated that nothing in the act would "supersede or modify any provision of the Atomic Energy Act of 1946." This would restrict NSF in its support of radiation biology. AEC offered competition to NSF in other areas of biology as well. Although its initial emphasis was on applied biology, by the mid-1950s, AEC had come to have a major impact on basic research in genetics and ecology. NSF program directors in genetic biology were frustrated that AEC cornered support of a number of the leading figures in genetics (see chapter 5).[64]

The National Institutes of Health

In 1945, it was not at all clear that the National Institute of Health (then still singular) would within a few years dominate federal support of medical research and become NSF's most formidable competitor in biology and medicine. At first, NIH had few supporters for its future role, aside from officials in its parent, the Public Health Service. Bush's Palmer Committee had wanted to vest medical research in a separate National Foundation for Medical Research. President Truman supported NSF as the agency to "promote and support research in medicine, public health, and allied fields." Medical research lobbyists like the irrepressible philanthropist Mary Lasker supported first a separate foundation and then the Magnuson bill. Academic biologists and medical scientists at the 1945 hearings, all favorable to NSF's establishment, made no attempt to argue PHS prerogatives. Nor were legislators in 1945–46, despite their eagerness to spend huge sums on the "war against cancer," yet convinced that the Public Health Service was the appropriate recipient. Even the American Medical Association initially supported NSF as the agency for extramural support of medical research. To many people the PHS was an unimaginative bureaucracy that carried out in-house research that was inferior to university research. They felt it was too closely allied with public health issues and suspect New Deal social programs to serve as the chief federal patron of medical research.[65]

During the war and the immediate postwar period, PHS officials deliberately staked their claim to fund health-related investigation. They were motivated to hasten their institution-building by the fear that, at this time of great public interest, control of medical research would go elsewhere, in particular to a new science agency. As part of a major revision of the PHS statutes in 1944, Surgeon General Thomas Parran managed to include authority for PHS to award grants. At the time, Parran saw extramural grants as a secondary adjunct to an enlarged and aggressive program in public health. The Public Health Service Act of 1944 authorized NIH to award grants "for research to be performed by universities, hospitals, laboratories, and other public or private institutions." The law extended the powers of the NIH National Advisory Health Council (NAHC) to enable it to recommend to the Surgeon General research grants and policies for the support of health in fields other than cancer.[66]

At the first meeting of the NAHC under the new law in June 1945, Parran laid out the PHS view. "The Service," he stated, "does not seek to op-pose the permanent establishment of a Federal research agency such as has been proposed by Vannevar Bush to succeed the OSRD, but we look with disfavor on relinquishing to others the right to make our own decisions and develop our own programs which have already been authorized by Congress." The NAHC agreed, recommending grants-in-aid for general research by "qualified institutions and individuals." At this same meeting, the council approved the first grant under the new legislation—$92,000 to Max Wintrobe of the University of Utah for a two-year study of muscular dystrophy in a group of Mormons. The funds for this project had been specifically appropriated on the initiative of a Utah Senator but were deliberately used to begin the new grants program.[67]

Throughout the NSF debate, NIH spokesmen were both aggressive and defensive regarding their agency's prerogatives. Meeting jointly with the National Advisory Cancer Council in September 1945, the NAHC began its agenda with the Bush Report and asked each council member, "Should the PHS attempt to have delegated to it the responsibilities of administering the proposed Division of Medical Sciences of the National Research Foundation?" The council's initial response suggests doubts that the PHS program was sufficiently developed or linked to basic research to carry out this role. The members decided that taking on "this large responsibility" would require "too great an expansion of functions" and that the program should be administered by the same agency supporting basic science in fields related to medicine.[68] The joint councils adopted a resolution later in the meeting declaring the Bush Report to be "a magnificent and distinguished document," but backing PHS concerns, it also urged that the legislation be clarified to safeguard the autonomy of the PHS. The compromise bill of 1946 and succeeding bills did incorporate a provision that "the activities of the Foundation shall be construed as supplementing and not superseding, curtailing, or limiting any of the functions or activities of other Government agencies authorized to engage in scientific research or development."[69]

At the NSF hearings in 1945, Rolla E. Dyer, assistant surgeon general, director of NIH, and a key player for PHS initiatives, approved of the proposed foundation but defended the Public Health Service as "the principal federal agency engaged in health and medical research." The PHS, he claimed, already possessed "all of the authority in reference to health and medical research that is contemplated in the proposed function [of the foundation], in those fields." This authority extended not only to clinical research but also to basic research—"long-term projects without promise of immediate results." Dyer disclaimed any PHS attempt "to monopolize this field." Its "sole interest," he declared, was to be allowed to develop its programs without interference under its existing authority.[70]

The NIH grants program got fully underway when PHS officials succeeded in acquiring, as of 1 January 1946, some fifty medical research contracts managed during the war by the OSRD Committee on Medical Research. When hostilities ended in the summer of 1945, Bush had asked Richards to classify CMR contracts in terms of future handling: transfer to the military, transfer to the Public Health Service (as Dyer had suggested to Richards as early as August 1944), or hold for takeover by the proposed National Research Foundation. By assertive action, PHS got most of them and immediately replaced them with its own contracts, using funds especially appropriated by Congress.[71] Dyer brought in Cassius Van Slyke to manage these grants, placing him in charge of the Office (later Division) of Research Grants. The two worked from there to expand the grants program through increases in the NIH budget, which rose rapidly from $180,000 in FY 1945 to $850,000 in FY 1946 (reflecting the transfer of contracts), to a hefty $4 million in FY 1947.[72]

By the end of 1946, Van Slyke could report that "a large-scale, nationwide, peacetime program of support for scientific research in medical and related fields, guided by more than 250 leading scientists in 21 principal areas of medical research, is now a functioning reality." NIH had awarded 264 research grants to seventy-seven institutions at a cost of $3.9 million. New applications were coming in at a rate of over 800 a year. In addition, NIH had awarded some 100 predoctoral and postdoctoral fellowships in a program begun in 1945.[73]

By 1950, NIH had succeeded in establishing its hegemony, though not exclusive control, over medical research. (Its claims to biological research were still weak.) The 1948 appointment of Leonard Scheele as surgeon general cemented the relations of the medical research lobby with PHS and helped turn the tide in the latter's favor. That year the National Heart Institute was created, and NIH became the (plural) National Institutes of Health. Strong congressional interest emboldened the PHS leadership to ask for large budget increases, such as its $23 million request for NIH for FY 1948, which was nearly three times its budget for FY 1947. Despite the magnitude of this increase, that year marked the first of many times that Congress, led by forceful health advocates, would provide NIH with even *more* money— in this case, $3.1 million more—than the agency's official request. By FY 1951, NIH boasted eight categorical institutes and a budget of over $50 million.[74] Warren Weaver estimated that NIH was then spending some $13.7 million annually on external research in basic biology, excluding clinical research and applied science.[75]

NIH adopted procedures for evaluating grant proposals that differed significantly from ONR's. Facing some initial resistance to its role as patron (as had ONR) NIH sought to allay it, not by hiring accomplished fellow scientists as program officers, but rather by placing most of the funding determination in the hands of committees of academic scientists, or "study sections." In March 1946, the NAHC approved "the establishment of Study Sections on a broad basis" to make initial recommendations on grants which would then be referred to the council. By the end of December there were twenty-one such study sections composed of physicians and biomedical scientists.[76]

The model NIH used for organizing peer review derived from CMR; indeed, Van Slyke adopted CMR's procedures when he took over the CMR contracts.[77] Like the CMR, NIH used a two-tier system in which committees of leading outside scientists made recommendations directly to the higher level body (CMR or NAHC). While the National Academy of Sciences would have liked to furnish these committees, as it had for the CMR, NIH set up its own committees under its own auspices. Several CMR panels—the syphilis panel for example—seem to have been simply reappointed as NIH study sections.

The NIH study sections wielded considerably more authority than the biology advisory groups at ONR. As Van Slyke noted in 1946, "Each Study Section, consisting essentially of outstanding civilian scientists, constitutes a scientific group with full authority and responsibility to make expert recommendations as to whether a research project application is acceptable and can be supported by Research Grants funds."[78] Ernest Allen, of the Research Grants division, recalled his thinking that it would have been "impossible" to bring to the staff "that sort of expertise," and they "didn't try." Having chosen the best scientists to be members of the study sections, Allen claimed "that for us as staff people to try to second guess that group appeared to us fantastic."[79] Consolazio, an ONR program officer who later went to NSF, claimed that, compared to a counterpart in ONR or AEC, the "executive secretary" or staff manager of a study section had the relatively limited role of simply "executing the committee decision."[80]

The peer-review system at NIH was also unusual in that study sections functioned independently of the institutes. Congress (after 1948) funded NIH by individual institute, but grants were handled and study sections managed by the Division of Research Grants. Organized by general field of research (e.g., physiology), study sections could evaluate grants for several institutes at once. They judged proposals on the merit of the research and the capabilities of the investigator but not on how the research fit the mission of the funding institute. The institute councils provided a second tier of evaluation but rarely deviated from study section recommendations. As the number of proposals approved exceeded the amount of funds available, the study sections in 1949 formally instituted ranking procedures.[81] The councils normally activated proposals in rank order at the sums requested until the money ran out.

Most of the early NIH grants dealt with public health problems and particular pathological conditions. But the NAHC, when queried on the scope of grants, concluded that "there would be no limits in regard to the field of research so long as there was some relation to medicine." Even some of the earliest grants supported basic biological research, albeit in disciplines like physiology and biochemistry typically associated with medical schools.[82] Clearly by 1950, NIH had established itself as a premier biological grant-making agency. Although its grants were largely limited to biomedical sciences, it was soon to expand its scope to cover nearly all of biology.

In 1951 Stanford biologist Douglas Whitaker estimated for the benefit of the new Division of Biological Sciences of NSF that as of 1950 the federal government, exclusive of USDA, was spending some $10 million on "basic biological research in academic institutions." Of that sum, nearly equal parts were supplied by the PHS, ONR, AEC, and other military agencies. These amounts can be added to the Rockefeller Foundation's spending on biological research, which totaled approximately $1.2 million a year. Thus by the time NSF was founded, there were a variety of competing federal and private sources of support for biology.[83]

The NSF Debate, 1947–1950

While other agencies were organizing their programs of biology and medicine, congressional debate over NSF dragged on for another three years, revolving about many of the same issues as before. The 1946–47 campaign was an especially bitter defeat for scientists because both houses of the Republican Eightieth Congress passed a bill only to have it vetoed by the president. Based on Bush's philosophy, the bill vested the authority of the agency in a part-time board and its executive committee. Truman, in his veto message, avowed that he could not approve a foundation that "would be divorced from control of the people to an extent that implies a distinct lack of faith in the democratic process."[84]

Interestingly, the 1947 legislation provided for an even greater NSF involvement in medical research than the bill of the previous year. Senators Magnuson, Robert Taft of Ohio, and Claude Pepper of Florida proposed an amendment to direct the agency to establish special commissions on particular diseases—cancer, heart and "intravascular diseases"—and "other special commissions as the Foundation may from time to time deem necessary for the purposes of this Act." To these the House added a commission on poliomyelitis and degenerative diseases. The eleven-member commissions, appointed by the Foundation, would have six "eminent scientists" and five members of the general public. They would "make a comprehensive survey of research, both public and private" in their respective fields and recommend "an over-all research program." Although the hearings brought out the potential conflict between NSF and private foundations and the Public Health Service, the amendments were nevertheless incorporated into the final bill that year.[85]

The Steelman Report, *Science and Public Policy,* which appeared in the fall of 1947, argued that both military and medical research should be removed from the purview of the proposed National Science Foundation. This five-part study of the current status of science and the federal government, written at Truman's request by his advisor John Steelman with the aid of a staff and committee of agency representatives, maintained that NSF's primary responsibility ought to be the provision of grants for basic research in the physical and biological sciences. It also favored giving scholarships and fellowships to increase the supply of scientists. Military research, it said, belonged under the jurisdiction of the military services, while medical research in academia should be overseen by a combination of the NIH, the military agencies, and AEC. Nonetheless, NSF's role as the agency responsible for most of the federal government's expenditures on basic research was vital to the national interest. The Steelman Report recommended a budget of $50 million to start, increasing to $250 million by 1957.[86]

Congress, however, was less inclined to provide funds for basic research separated from applied ends. NSF bills in 1948 and 1949 retained some vestiges of military and medical research. In the 1948 NSF bill, worded to meet the president's objections of the previous year, the Division of National Defense was dropped, although the Foundation could still initiate and support defense-related research in consultation with the Secretary of Defense. The Division of Medical Research and the special commissions on particular diseases remained, though the latter had become increasingly controversial. Homer Smith, Walter Palmer, and a number of other scientists active in the Committee in Support of the Bush Report had issued a statement that called for the elimination of the commissions as "unnecessary." In general, medical scientists opposed the commissions as special interest legislation that would tend to prevent a balanced program for medical research. The 1948 bill passed the Senate but was held up in the House Rules Committee when the Eightieth Congress adjourned. The disease commissions disappeared in subsequent bills, although a provision for establishing special commissions on unspecified topics remained.[87]

Success finally came toward the end of the Eighty-first Congress. The 1948 election brought Democrats back in control of both houses and returned Truman to office. A new bill identical with that reported out by the Senate Committee on Labor and Public Welfare the year before, passed the Senate without amendment in March 1949, but a coalition of Republicans and Southern Democrats blocked action on passage of a variant House measure until the following March. After disagreements over controversial security provisions were ironed out in conference committee, Truman signed the bill into law in May 1950.

Floor debate in the House in February 1950 gave evidence of the power of the NIH lobby and presaged the difficulties that NSF would experience in attempting to fund medical research.[88] NIH supporters, led by Frank Keefe of Wisconsin, former Republican chairman of the House Appropriations Subcommittee who had overseen the NIH budget from 1946 to 1948, sought to prevent any future NSF encroachment on NIH. In the words of Budget Bureau representative William D. Carey, Keefe "hammered away at length on the proposition that the NSF should not under any circumstances affect the research programs of the Public Health Service." He "succeeded in getting page after page of legislative history into the record on this point."[89]

Keefe was concerned not only with the role of the proposed Division of Medical Research but also with new language on NSF's relation to other agencies. In response to the Hoover Commission Report of 1949, which promoted the creation of a national science foundation to coordinate and evaluate the mushrooming federal research programs,[90] the House made changes in the bill strengthening NSF's policy-making function. The earlier provision, supported by NIH advocates, that NSF should supplement but not supersede other agencies, was removed. Instead, NSF was "to evaluate scientific research programs undertaken by agencies of the Federal Government, and to correlate the Foundation's scientific research programs with those undertaken by individuals and by public and private research groups." When Keefe and others objected, they were told the former provision had been struck in order to assure NSF of the "the right to criticize" or point out duplication in the various agencies. The conference committee adopted the House language but specifically confirmed the prerogatives of NIH.[91] Commenting on the conference report, two Bureau of the Budget officials wrote, "Our flash reaction to this statement is that the Division of Medical Research probably should not be activated and that biological research should perhaps not be emphasized."[92]

Biology and the Long Delay in the Founding of NSF

The long campaign for a National Science Foundation had successfully ended, but much had changed in the five years during which it had been debated. The two kingpins of *Science—The Endless Frontier,* defense and medical research, were no longer prominent in the new Foundation. The Division of National Defense had been scuttled, though some expected that the Services would still transfer a sizeable portion of their basic research to NSF.[93] By 1950, there was also little need for the Foundation to sponsor medical research although the Division of Medical Research remained.

For biology, the delay meant that the Foundation would not receive a budget consonant with its mission. Without the urgency of funding defense or medical research, Congress had lost much of its incentive to support the new agency. Despite the Bush Report's recommended first year budget of $33 million and Steelman's of $50 million, the House had limited the Foundation's authorization to only a small sum for operating expenses for FY 1951 and imposed a ceiling of $15 million a year after that. Even that figure proved unattainable. NSF's first operating budget, for FY 1952, was only $3.5 million. NSF did not pass the fifteen-million mark until FY 1956, and budgets did not rise significantly until the post-Sputnik era. Thus the National Science Foundation began operations with a broad mission to fund all of biological, medical, and general basic research, but with exceedingly limited means to do so.

The years 1945–50 altered the future of NSF support of biology and medicine in another important way. During the interval precedents were set for organizing a federal grants program. Had NSF appeared in 1945 or 1946, it might have adopted procedures similar to NIH, where outside committees of scientists determined funding priorities. Instead the Foundation of 1950 modeled itself after the more flexible ONR, where staff were prominent in policy-making, a precedent that became the strength of NSF's program in biology and medicine. How that program was instituted in the critical years 1950–52 is the subject of the next chapter.

CHAPTER 2

Fashioning a New Federal Patron for Biology, 1950–1952

In 1949, when an NSF Act appeared likely, the American Institute of Biological Sciences (AIBS), at the request of the U.S. Bureau of the Budget (BOB), surveyed 360 biological departments in colleges, universities, medical schools, and agricultural schools on their research productivity and what they would do in 1950–51 if funding were available. As a result of the survey, AIBS conservatively estimated in February 1950 that the new foundation might advantageously spend $6 million a year on biology. AIBS executive secretary Milton O. Lee expected that NSF would "loom large" in the funding of biology because federal support of biology, aside from medicine and agriculture, was meager. "It is probable," Lee wrote, "that biologists will look to the National Science Foundation to a larger extent, relatively, than scientists representing other fields."[1]

Throughout its history NSF has been perceived as dominated by the physical sciences, yet in its early years, patronage of biology did indeed "loom large." The Foundation activated the Division of Biological Sciences first. NSF's first group of grants, in early 1952, went entirely to the biological sciences. Funding of the biological sciences far exceeded support of the physical sciences in the Foundation's first year of operation, and through most of the decade, the biological sciences remained on a par with the physical sciences, at least as regards individual grants.

The two years between the passage of the National Science Foundation Act in 1950 and the forming of programs and panels in late 1952 were critical in establishing the organization and policies of the new agency. The NSF Act, like any federal charter, provided only the outline of an agency with some indication of congressional priorities. Filling in the outline to create an operating federal foundation for the sciences was a major undertaking that was largely the work of the staff who joined the Foundation in 1951–52. The higher level staff members of the new agency were not merely bureaucrats carrying out the wishes of Congress and the administration, but active shapers of the structures and policies that they, themselves, considered optimum for federal support of academic science. Like Warren Weaver of the Rockefeller Foundation, they were in the fullest sense "managers of science."[2]

Among the early issues to be determined were the respective roles of the National Science Board (NSB) and staff. Who would make the major decisions, including the awarding of grants and contracts? Would the agency see its primary task as funding the "best science" wherever it could be found, or distributing funds to encourage growth more broadly? How would programs, advisory panels, and the evaluation process be set up? What would be the role of academic advisors of individual programs in making funding decisions? In the biological sciences, a chief issue was what the Foundation's role in medicine would be. The act called for divisions both of biological sciences and medical research. How would "basic research in medicine" be defined and distinguished from basic research in the biological sciences? How would NSF accommodate itself to the prior domination of medical research by NIH?

Crucial to the shaping of the new agency was the president's choice of its first director, Alan Tower Waterman, Chief Scientist of the Office of Naval Research (ONR) who, in turn, recruited a number of colleagues from ONR including the entire permanent staff for the life sciences. They brought with them ONR policy and funding traditions and an elitist and idealist vision of "basic science." Because the NSF staff was small and bureaucracy at a minimum in the early years, the assistant director and staff in the life sciences took an active role in making the basic organizational and policy decisions of the new agency. This was especially so because the biologists at NSF were the first to experiment with a process for evaluating proposals and making grants. Both at the time and in retrospect, they saw themselves as participating in a novel, exciting, and important venture. They felt a camaraderie and sense of mission that was inevitably lost over time.

The most innovative example of early policy making in biology was the decision to organize programs in the biological sciences on a functional rather than a disciplinary basis. From the beginning, NSF gave strong support to the ideal of "one biology." Though it would be several years before NSF could supply biology with the $6 million dollars estimated by AIBS— for each of the first two years, less than $1 million was available for biology—NSF developed imaginative and flexible programs, ready to take advantage of the expansion of support for science in the following years.

Setting Up: Selecting a National Science Board and Director

President Truman's announcement in November 1950 of the twenty-four members of the first National Science Board (NSB) was in many respects surprising. The selection process illustrates the tension between elite science and the New Deal goal of broad-based improvement of the nation's scientific research. The NSF Act designated that the board be appointed "solely on the basis of established records of distinguished service" but also "so selected as to provide representation of the views of scientific leaders in all areas of the Nation."[3] When the National Academy of Sciences and Vannevar Bush drew up board lists in 1949, they chose an elite group of white male scientists from major research universities. At first, the sifting of various organizations' lists was left to a group of BOB staff. Bush had full confidence that this process would yield a "sound affair," but to his dismay, the White House staff, led by science advisor John Steelman, took over the selection.[4]

Their resulting board was the outcome of a highly delicate balancing act. Although many of the expected names appeared, the president's staff made a deliberate effort to include the widest representation of scientists. There were two women, two African Americans, two representatives of Catholic schools, two members from private foundations (Rockefeller and Carnegie), two from industry, and two from small colleges. As Vernice Anderson, longtime secretary to the board put it, "It was like Noah's Ark—they came in pairs!"[5] (Best represented, however, were university administrators, who accounted for half of the members.) It is ironic that the board—the highest-level committee of scientists sharing power with the director—was more catholic in profile than any of the lower-level advisory committees.[6]

The nine board representatives from the fields of biology, medicine, and agriculture were a far more heterogeneous group than ever appeared on the divisional committee or panels of the Division of Biological and Medical Sciences. Detlev W. Bronk, Edwin Braun Fred, Elvin C. Stakman, and Robert F. Loeb were powerful, politically connected members of the National Academy of Sciences. All four (and none of the other biologists) became members of the NSB Executive Committee. Biophysicist and physiologist Bronk was the most influential biologist in America, ranking with board members James B. Conant of Harvard and Lee A. DuBridge of the California Institute of Technology as a major power broker in Washington. President of Johns Hopkins from 1949 to 1953 and then of the Rockefeller Institute for Medical Research, Bronk was also president of the National Academy of Sciences/National Research Council from 1950 to 1962. He served on the NSB for fourteen years, as chairman from 1955 to 1964. Fred, first vice-chairman of the board, had won renown for research on nitrogen-fixing bacteria. Appointed president of the University of Wisconsin in 1945, he had earlier served as director of the Wisconsin Agricultural Experiment Station. Stakman, head of the Department of Plant Pathology at the University of Minnesota and a global authority on cereal-grain diseases, was credited with increasing the world's wheat supply by hundreds of millions of bushels by developing disease-resistant strains. Loeb, the leading representative of medicine on the board, was professor of medicine at the College of Physicians and Surgeons, Columbia University. A respected bench scientist, known for his work on electrolytes in health and disease, Loeb served on a number of government committees such as NIH's National Advisory Health Council and the President's Science Advisory Committee.

The remaining biologists on the board would have been highly unlikely to appear on National Academy of Sciences lists of nominees. Gerty Theresa Cori, not a member of the inner circle because of her gender, was nevertheless the board's only Nobel laureate. Professor of biological chemistry at Washington University's School of Medicine in St. Louis, she had shared the Nobel prize in medicine and physiology with her husband Carl F. Cori and Bernardo A. Houssay in 1947 for work on biochemical pathways. Sophie Aberle, with a Stanford Ph.D. and Yale M.D., had directed research in reproductive endocrinology, human nutrition, and the anthropology of the Pueblo Indians at the University of New Mexico. She had met President Truman while associated with the National Research Council in Washington from 1945 to 1948. Despite her credentials, Aberle was little known to the postwar scientific leadership; DuBridge, for example, claimed never to have heard of her.[7] James A. Reyniers, one of the two representatives from Catholic institutions, was the controversial creator and head of the Laboratories of Bacteriology of the University of Notre Dame (LOBUND), the first center for raising "germfree" animals for use in medical research (see chapter 3).

The last two biologists, both Southerners, were especially non-elite in terms of institutional prestige or scientific achievement. Orren Williams Hyman, an able administrator, was Dean of the University of Tennessee Medical School in Memphis, with which he had been associated since 1913. With a doctorate in biology rather than medicine, Hyman had published in embryology and cytology.[8] No doubt the Board's most obscure biologist was the Reverend Patrick Henry Yancey, S.J. Since receiving his Ph.D. from St. Louis University in 1931, he had served as professor and head of the biology department at Spring Hill College, a small men's school in Mobile, Alabama. In his autobiography, *To God Through Science,* Yancey recalled that he was "surprised" and "overwhelmed" to be appointed to NSB along with men like Conant, Bronk, and DuBridge. He professed not to understand why he, "a mere biology teacher in a small Catholic college in the deep south[,] was chosen to such an August body" but surmised that what weighed heaviest was his location in the state of Senator Lister Hill, a strong supporter of the NSF Act.[9]

NSF as a whole, and biology at NSF in particular, might have turned out very differently if, rather than Waterman, one of the other candidates under serious consideration had been chosen NSF director. When, at its second meeting in January 1951, the board took a ballot on its nomination for the position, biologists figured prominently, including the first two choices. Bronk overwhelmingly headed the list, followed by A. Baird Hastings, chairman of the department of biological chemistry at Harvard Medical School and a former member of the Committee on Medical Research. The NSB resolved to recommend Bronk "as the outstanding candidate." Three nominations were sent to the President—Bronk, Hastings, and physicist Lloyd Berkner of the Carnegie Institution of Washington.[10]

Apparently the Truman administration was not happy with the Bronk nomination, in part because of Bronk's tie to the National Academy of Sciences. At the third NSB meeting in February 1951, Bronk took his name out of the running, claiming other duties in which he believed he could "more effectively serve the national interest." The board then decided to submit seven additional names, among them two more biologists, University of Wisconsin microbiologist Ira L. Baldwin and endocrinologist C. N. H. Long of the Yale University School of Medicine.[11]

Vannevar Bush had long thought that the president could make no better decision than to appoint Alan T. Waterman, chief scientist of the Office of Naval Research. "He is a quiet individual," wrote Bush in 1948, "a real scholar, and decidedly effective in his quiet way, for everyone likes him and trusts him. I would rather hope that if we had a National Science Foundation he would become Director of it . . . ."[12] Waterman had ranked seventh on the board's original list and was included on the second set of nominations sent to the president. For Truman, concerned with accountability, Waterman had the advantage of experience running a federal granting agency, and he was known to work well with the administration and Congress. Truman announced his choice of Waterman on March 9 from Key West, Florida. The board, meeting that day, expressed its unanimous approval, and Waterman, already on hand, then joined the meeting.[13]

Waterman set the tone for the new agency, first by his relaxed and open style of leadership and second by his importation of ONR concepts and procedures as well as many personnel. Had either Bronk or Hastings been chosen, NSF would probably have become a very different institution; the National Academy of Sciences might have played a far more active role in making policy, and peer review would doubtless have been more directly controlled by outside scientists as it was in the NIH. Instead, the NSF staff established its own policies, operating informally with a strong sense of esprit de corps.

Hiring Staff for the Biological Sciences

While the NSF recruited a number of staff from ONR, it was in biology that the influence of ONR was most strongly felt. All of the permanent program directors in the Division of Biological and Medical Sciences (BMS) in the 1950s came from ONR. Perhaps because biology had been relatively neglected compared to the physical sciences at ONR, staff members in biology were willing to make the move to the new agency.[14] The other BMS program directors were "rotators," scientists who came to NSF on leave from academia, federal service, or research institutions for a year or two.

From the beginning of Waterman's tenure at NSF, biologists on the board worried that biology might become subordinate to physics, as it had at ONR. The director, the newly chosen deputy director (C. E. Sunderlin, from ONR), and Board Chairman Conant, were all physical scientists. When Waterman suggested to the board in April 1951 that he hire physicist Harry C. Kelly from ONR to head the Division of Scientific Personnel and Education, biologists on the board expressed dismay at the absence of biologists among the top staff. Waterman recorded in a "diary note" that three board members had "cautioned me about having too many physicists around." Fred, in particular, insisted that "biological sciences should receive due attention." Waterman replied that the life sciences would be represented at the top level by the heads of two of the four proposed divisions.[15]

Waterman and Conant agreed to give the biological sciences priority among the three planned research divisions, since general biology appeared to them relatively neglected by other federal agencies—the Department of Defense and the Atomic Energy Commission supported the physical sciences and the National Institutes of Health funded medical research. Thus Waterman deliberately decided to fill the position of Assistant Director for Biological Sciences before seeking a counterpart for the physical or medical sciences.[16]

To head the biology division, Waterman turned in April 1951 to a former staff member at ONR, physiologist John Field II (1902–83). Field had been on leave since 1948 from his academic position at Stanford University, most recently as head of the Biology Branch of ONR. He was one of a small handful of men who combined scientific reputation and experience in managing a grants program. As physiology was on the borderland of biology and medicine, he also legitimately qualified to serve as an advisor for the proposed Division of Medical Research. Field had just accepted the chairmanship of the Department of Physiology at the new medical school of the University of California at Los Angeles when he was approached by Waterman, but on 8 June, having obtained a year's leave from UCLA, he officially joined the NSF staff.[17]

By early July, a second biologist arrived from ONR, William V. Consolazio (1910–87), who would spend the remainder of his career at the Foundation. Born to Italian immigrants in a Boston suburb, Consolazio graduated from Tufts University. He rose from technician to full member of the research staff at the celebrated Harvard Fatigue Laboratory founded by L. J. Henderson and authored or coauthored numerous papers on human physiology and biochemistry. Following wartime service with the U.S. Naval Medical Research Institute, he transferred in 1948 to ONR where he headed the Biochemistry Branch. Consolazio was the only BMS program director not to hold a doctorate. Brilliant, imaginative, and of the highest integrity, he was committed to the ideal of a non-bureaucratic, apolitical agency that would seek out and support the best and most creative scientific research. Highly emotional and excitable, he was described by former colleagues as "fiery," and more strongly as "that wild man." More than the other early program officers, Consolazio concerned himself with the internal content and the progress of science. His grantees loved him because he could almost always find some way to support a promising project.[18]

Field and Consolazio immediately set to work to help formulate Foundation policies and procedures, to defend the beleaguered 1952 budget, and to organize the first panels and grants in the biological sciences. Field made several visits to institutions around the country to talk with scientists and administrators about the NSF's patronage of biology and to informally solicit proposals.[19]

Although Waterman had hopes of obtaining eminent scientists to manage NSF programs, he found that few were willing to leave their research to come to Washington, even on a temporary basis. Instead, he filled staff positions with more recruits from ONR. Waterman at first postponed action on the recommendation of Field and Consolazio to hire John T. Wilson of ONR, because he felt that NSF already had too many ONR employees, and NSF had not yet officially entered psychology, Wilson's field. But when two prominent academics turned down NSF offers, Wilson—who eventually became NSF's central figure in biology—got the nod and joined the staff in January 1952.[20]

John Todd Wilson (1914–90) was an exceptionally talented administrator who more than any one else was responsible for NSF's policies of support for biology during the 1950s. His was an American success story. The son of a Punxsutawney, Pennsylvania, minister of modest means, he eventually became president of the University of Chicago. Wilson received his doctorate in psychology from Stanford in 1948 and, after short stints as a junior-level administrator at the American Psychological Association and assistant professor at George Washington University, was recruited to head the Personnel and Training Research branch at ONR. He headed NSF's Psychobiology Program from 1952 to 1955, and from 1956 to 1961 he served as assistant director for biological and medical sciences. After two years in administration at the University of Chicago, he returned to NSF to serve as deputy director under Waterman's successor Leland J. Haworth from 1963 to 1968. In 1968, he left NSF to return to Chicago.[21]

Former staff members recall Wilson as extremely bright and able but also as an administrator who stirred strong emotions in his colleagues. Many of the women he hired as professional assistants and secretaries adored him because he included them in division activities, offered them responsibility, and did not stand on protocol. Several men with whom he worked, however, distrusted him as a manipulator who played his cards close to his chest. By the end of the Waterman era, Wilson had major fallings out with both Consolazio and George Sprugel, another biology staff member from ONR.[22]

As prospects for inaugurating the Division of Medical Research (see below) faded by mid-1952, Field and Waterman looked once again to ONR to find a staff member who could "assist in looking after medical interests." Louis Levin (1908–81) arrived at NSF in July 1952.[23] A midwesterner, Levin had received his Ph.D. in biochemistry from St. Louis University in 1934. In his research career, primarily in biochemical endocrinology, he had closer ties with medical institutions than had either Consolazio or Wilson. Until 1948 he served as a research associate and assistant professor in the Department of Anatomy of the College of Physicians and Surgeons, Columbia University. He then joined the ONR staff, eventually succeeding Consolazio as head of the Biochemistry Branch. Recalled by former staff as gruff and demanding, Levin was nonetheless regarded as an able administrator. After serving as program director for the Regulatory Biology Program from 1952 to 1959 and deputy assistant director in the latter part of that period, he was placed in charge of the new Office of Institutional Programs in 1960. Except for a few years in the early 1960s, which he spent as an administrator at Brandeis University, he remained at NSF until 1972.[24]

Consolazio, Wilson, and Levin were to become known as "the triumvirate," and they were in essence to run NSF's program in biology for the rest of the 1950s. Although these pioneers were by no means distinguished in terms of prior research accomplishments, they were talented individuals, aware of their pivotal place in the history of managing science, and dedicated to the ideal of supporting good science with maximum freedom for the scientific investigator.

Negotiating Authority: NSF Staff and the National Science Board

At stake in 1951 was the division of authority between the new NSF staff and the National Science Board. One manifestation of this tension was the conflict surrounding the forming of the statutory divisional committees. During the NSF legislative debate, many scientists had supposed that the NSB, representing the scientific community, would formulate policies and, with the advice of scientific specialists, select grants. The NSF Act specified that the NSB appoint a committee corresponding to each of the proposed four divisions to "make recommendations to, and advise and consult with, the Board and the Director with respect to matters relating to the program of its division."[25] How, then, would these committees be chosen and how closely would they work with the board? In January 1951, the NSB voted that each committee would include at least two NSB members. In February, it formed temporary divisional committees of its own members, which began planning for the work of their respective divisions. The Temporary Committee on Medical Research, headed by Loeb, included Bronk and Cori, while Biological Sciences, chaired by Reyniers, included Fred, Hyman, Stakman, and Yancey. Some board members hoped these committees would help evaluate projects and play a major role in policy-making in biology and medicine. To their frustration, policy was, in fact, generated for the most part by the NSF staff.[26]

As the NSF staff viewed the matter of divisional committees in May 1951, there were two options. According to the first plan, the director and deputy director were to recruit the divisions' assistant directors, who would then nominate the members of divisional committees who would serve as advisors to the divisions. Alternatively, the divisional committees would be selected first and then aid the director in securing staff and formulating policies. The staff agreed on the first course of action: "Our plan follows the practices proven to be satisfactory by the Office of Naval Research, the Atomic Energy Commission and the National Advisory Committee for Aeronautics. Public Health on the other hand has developed panels that decide policy questions."[27]

In September 1951, the Temporary Committee on Biological Sciences reluctantly acceded to Waterman's recommendation that the divisional committees be advisory to the assistant directors and include no board members. The board then decided to transform its temporary committees into permanent, parallel board committees, to advise itself.[28] Although Reyniers' committee had planned to nominate the members of the Divisional Committee for Biological Sciences for full-board approval, the staff took the initiative before it could act. As Father Yancey recalled, they had "quite a hassle" regarding these appointments at the October board meeting. He had come prepared to coax Reyniers' dilatory committee into making nominations, only to discover that Field had "already drawn up a slate of his own." Yancey formally objected, especially since no Southerner was on the list. Though the protest, which Hyman backed, resulted in the addition of Donald Costello of the University of North Carolina, Yancey noted that "it was more and more apparent that the Board was becoming more or less a 'rubber stamp.'"[29] This impression was reinforced when the staff presented the first group of grants for board approval in January 1952 (see below).

Defining NSF's Role in "Basic Medical Research"

Medicine and medical research had figured prominently in the NSF Act of 1950. "To advance the national health" was among the stated goals of the agency along with national defense and economic prosperity. The act specifically authorized the Foundation "to initiate and support basic scientific research in the mathematical, physical, medical, biological, engineering, and other sciences."[30] Yet by 1950, the National Institutes of Health had come to dominate medical research and was highly suspicious of any new agency that might limit its continuing growth. What NSF's role in medical research would be and how it would define basic research in medicine, as distinguished from basic research in biology, were open questions.

Early in 1951, the board's Temporary Committee on Medical Research mapped out an expansive vision of NSF's role in medical research. It "assumed" that "basic research in medicine should include Anatomy, Physiology, Biochemistry, Microbiology, Pharmacology, Experimental Pathology, Medicine and Surgery, and possibly Experimental Psychology." That is, basic research comprised not only the preclinical sciences but also medicine and surgery. Like the Medical Advisory Committee of the Bush Report, the NSB committee strongly favored block grants to university departments for support of staff members below the rank of professor. It explicitly opposed short-term project grants, which were to become the mainstay of the NSF research program. The committee recommended scholarships for medical students, "postgraduate advanced fellowships," and a limited program of research professorships for outstanding scientists. It called for a tentative bud-

get of $3.35 million, $2 million of which would support university departments.[31] The reality was to be much different.

The Foundation's plan to organize a medical research program ran into
its first serious snag in May 1951. Frederick C. Schuldt of the Bureau of the
Budget apprised Waterman that in "marking up" the president's budget for
FY 1952, the first year in which the Foundation would award grants, the
BOB was considering eliminating the agency's $1 million request for the Di-
vision of Medical Research. Waterman recorded "a feeling on the part of
the Bureau that no new funds were really needed for medical research in
view of the large amounts of money now supporting the field." The policy
question this raised went beyond medicine. To what extent should NSF limit
its scope so as not to overlap other federal programs? In the face of continued administrative and congressional questioning of "duplication," Waterman remained firmly committed to funding basic science across the board
regardless of the programs of other agencies.

Waterman argued that it was impossible to determine whether "sufficient money" was going into medicine or any other field until the Foundation acquired a staff and studied the issue. "However," he added, "it was the
opinion of most research men familiar with both the medical and biological
side that really fundamental research was inadequate in the field of medical
research and that for the good of clinical medicine and the medical schools,
this support should be given." Moreover, touching on a function that the
BOB regarded of central importance, he argued that the Foundation could
not establish national policy "if it is excluded from operating in any area in
this fashion." When Schuldt suggested that funds could be withheld for FY
1952, since it was already late to mount a program, Waterman replied that
this "would start the Foundation off very badly indeed." Without operating
funds, the Division had little hope of being activated and certainly not with
competent personnel.[32]

At the next board meeting, the Temporary Divisional Committee on
Medical Research drew up a resolution to the Bureau stating that "to eliminate from its program any field of basic science would create a precedent
which might, at any time, be made applicable to fields in which there is current support of basic research from other governmental sources." NSF's appeal succeeded; the BOB raised no objection to including an item for medical research in the fiscal 1952 budget, but it remained skeptical of the
Foundation's role in this field.[33]

There were others in government, however, who believed that NSF
might still have a special part to play, specifically in medical education. A bill
under consideration in the House of Representatives in July 1951 sought to

increase greatly NSF's role in medicine in a manner not welcomed by Waterman or the board. Justified on the basis of national security, H.R. 3371 was intended to relieve a shortage of doctors, which had become evident during the Korean War and to aid financially strapped medical schools. The bill would authorize Congress to appropriate annually to NSF a sum not to exceed 0.25 percent of the Department of Defense appropriations of the previous year "to increase the number of well-trained doctors of the highest quality" through scholarships, fellowships, and grants—including facilities awards—to medical schools.[34]

The rationale for involving NSF, Waterman learned, was to allay fears of the American Medical Association (AMA) that the federal government would exert too much control over medicine. It was thought that the AMA might support H.R. 3371 because NSF could give "completely impartial direction to such a program." The medical school deans that Waterman consulted told him that schools were hard pressed but preferred that funding go through the Public Health Service. Indeed, an alternative Senate bill, "The Emergency Professional Health Training Act of 1951" (S. 337), placed the program where it would seem to belong—in the Federal Security Agency, then the umbrella agency for the Public Health Service.[35]

The Board agreed with Waterman that the House proposal was inappropriate for NSF. The projected figure for medical education might amount to as much as $100 million, dwarfing the rest of the Foundation's budget and taking the agency far afield from its primary responsibility for basic research in the sciences. NSF expressed these concerns to the BOB in a staff letter revised by the board, which concluded that NSF ought to administer the funds "only" if that proved to be "the most suitable means by which the critical needs of medical education can be met." But opposition from the AMA, which feared that any federal support of medical schools would lead to government interference in medicine and, eventually, to the establishment of national health insurance as advocated by Truman, assured that no action would be taken.[36] This episode reveals a certain amount of fluidity, even in 1951, over who would fund medical research.

The chief impediment to NSF support of medical research was financial. The funds available to NSF were negligible compared to the amounts the Public Health Service already allocated to extramural support of the medical and biological sciences ($18.3 million in FY 1952 by later NSF estimate). NSF's original budget of $14 million for FY 1952, approved by the BOB in May, had allotted only $1.3 million for medical research, compared to $2.6 million for Biological Sciences and $3.9 million for the Division of Mathematical, Physical and Engineering Sciences. (Another $5 million was budgeted for pre- and postdoctoral fellowships.) But this limited amount for medical research, criticized by Loeb as far too low, greatly exceeded what actually became available.[37]

Though prepared for a considerable reduction in their requested budget, NSF staff members were stunned in August when the House approved as its total budget a mere $300,000, to be used for further planning. The accompanying report stated that the "early aid" that research and fellowships could supply in the present Korean War emergency was "not very tangible." Also, it became clear that the House subcommittee judged that the new agency had not adequately planned what it would do with an operating budget.[38] Any hope of awarding grants and fellowships in 1952 depended now on convincing the Senate Appropriations Committee to restore the funds. Foundation officials had thus to demonstrate, first, that basic research was essential for the nation's security, and second, that the agency had done its homework.

At Loeb's urging, the justification NSF presented to the Senate for its medical research budget simply called for a small number of block grants to preclinical departments of medical schools, half ($400,000) to well-established departments and half to promising ones plus a few large awards (totaling $300,000) to support "research units built around outstanding research investigators." Waterman explained to the board that Loeb and other consultants felt that "this is the only way a quick presentation could be given without the danger of running into conflict with other agencies operating in the same field."[39]

In contrast to this brief plan for medical research, Field, with help from Douglas Whitaker and George Beadle, formulated a plan for the Division of Biological Sciences that was overprecise. In the crisis created by the House action, the Foundation biologists and their advisors were more willing than they were later to select specific areas of basic biology for support, justifying them on the basis of their practical applications. They divided the biological sciences into ten fields or problem areas, allotting to each $100,000 to $400,000, the total amounting to $2.2 million. They selected some biomedical fields usually found in medical schools but made no attempt to cover all of biology, leaving out systematic biology and ecology, for example. The Foundation proposed to fund research in protein structure and synthesis, enzymology, immunochemistry, biochemistry and nutrition, comparative physiology, marine biology, experimental embryology, genetics, experimental plant biology, and photosynthesis. It is likely that Waterman and the staff developed the proposed breakdown for its rhetorical value rather than as an actual means of distributing funds among fields of biology.[40]

Although Waterman and NSB Chairman James B. Conant justified the overall budget primarily on the basis of national security, the "possible applications" they presented for each area of biology were primarily to medicine or agriculture, with only a few references to military applications of physiology and medicine. For example, protein structure and synthesis would provide knowledge for the preparation of blood plasma substitutes which "would be of vital importance in an atomic bomb blast or other major catastrophe where supplies of human plasma are certain to be inadequate." Biochemistry and nutrition would throw light on such problems, important in time of war, as food acceptability, hunger, thirst, fatigue and the stress of fear.[41]

As a result of NSF's testimony before the subcommittee, the Senate voted NSF a budget for FY 1952 of $6.3 million. In October word came that the final congressional appropriation would be only $3.5 million, a serious disappointment, but Waterman agreed with Conant that "the important thing" was that it "would get us off the ground."[42] Of this limited budget, the staff allotted $1.5 million for grants in all the sciences, including medicine.

A minuscule budget and uncertain prospects for more made it difficult, as Waterman predicted, to secure a prestigious individual to serve as assistant director for medical research. He tried repeatedly but failed to convince anyone of appropriate stature to take the position for even a year. In July 1951 he solicited opinions from Milton Winternitz, professor of pathology and former dean of the Yale School of Medicine, and chairman of the National Research Council's Division of Medical Sciences, from A. Baird Hastings, and from Bronk on a list of possible candidates. Among these, ironically, was James A. Shannon, who in 1955 was to become the dynamic director of NIH, and the man perhaps more responsible than any other for limiting NSF's role in medical research. In all, some twenty-three names were mentioned.[43] In the meantime, with board approval, Waterman named Field acting director of the Division of Medical Research, and medical research proposals were considered along with proposals in the biological sciences.[44]

In early 1952, Waterman's strategy for staffing NSF's medical research division was to forge a close link with NIH through the temporary appointment of a high-level NIH staff member. This, too, ultimately failed. In mid-January, Waterman conferred with Surgeon General Leonard Scheele and Rolla Dyer and David E. Price, director and associate director of NIH, all of whom appeared supportive of the idea. Scheele highly recommended Kenneth Endicott, an M.D. and accomplished research pathologist. A longtime PHS employee, Endicott had served the previous year as scientific director of NIH's Division of Research Grants. Waterman told Scheele that he was considering a number of people "but felt the advantage, if this meets with the full approval from the PHS, of the kind of cooperation which would be most natural by this kind of tie-in with the program of NIH." Endicott came to NSF for an interview, but for reasons unknown, nothing came of the appointment.[45] In March, Waterman discussed with Bronk, Cori, and Loeb, the possibility of appointing Price to the position. Bronk felt that this "tie-in" with NIH would be "excellent," while Cori was somewhat concerned that Price was not a "research man." Whether to prevent NSF from establishing the division or for other reasons, PHS turned down the idea of loaning Price.[46]

In May, after the NIH effort fell through, Waterman offered the combined position of head of biological sciences and medical research to Hastings. A Harvard medical school professor with close ties to NIH, Hastings seemed to Waterman to have "unique qualifications" for bridging biology and medicine. When Hastings regretfully declined to come for one year, Waterman gave up trying to find a successor to Field with strong biomedical credentials.[47] Instead, he brought in Lou Levin from ONR to look after medical interests and immediately delegated to him the task of keeping track of policies at NIH.

Even before the Division of Medical Research was abandoned, Waterman's thinking concerning the relationship of biology and medicine underwent a shift, possibly as a result of preparing for the 1952 Senate budget hearings. Speaking before the annual meeting of the American Institute of Biological Sciences on 10 September 1951, Waterman acknowledged being asked why the Foundation distinguished between biological and medical research. "Biologists incline to the point that, strictly speaking, most medical research could be defined as biological research." At this time, he defended the separation as "a perfectly defensible plan, since medical research, in addition to implying the study of disease, commonly brings to bear upon its problems a point of view essentially different from that of biological research in somewhat the same way that engineering differs from physics and chemistry."[48]

A month later (after the hearings), Waterman expressed a different view in a speech at the University of Tennessee College of Medicine. Noting the "tremendous program being sponsored by the Institutes," he renounced NSF support of clinical research and suggested that its "role with relation to the medical sciences should be in the support of basic research in the pre-clinical fields of anatomy, physiology, biochemistry, micro-biology, pharmacology, and pathology." He went on to say that "It is quite likely that no distinction will be made program-wise between basic research in the medical sciences and basic research in the biological sciences."[49]

Preparation of the FY 1954 budget, to be submitted to the BOB in mid-

1952, forced the issue. BOB had never encouraged the Foundation to establish a role for medical research. For the FY 1953 presentation to Congress in early 1952, the board had agreed to submit a program for biology and medicine as a single line item, with the understanding that the programs would remain "administratively separate" and funded at the approximate ratio of two to one. That year's list of selected fields was similar to the previous one with the addition of such biomedical areas as microbiology and pharmacology.[50]

Once NSF became fully operational, however, this fictional breakdown of funds for the life sciences could no longer be maintained. In June 1952, Waterman received approval from new board chairman Chester I. Barnard, Bronk, and Loeb of a "plan for submission of the 1954 budget in which the biological and medical sciences are placed together as a group and defended as such." In late July, he proposed to his staff that the current working arrangement be formalized by the creation of a Division of Biological and Medical Sciences. This was "desirable at this time," he explained, "by the need for the Foundation's organization to correspond to its activities as described in the budget for fiscal year 1954."[51]

Seeking the board's approval in August, Waterman claimed that since basic research in biology and medicine were similar and the program had in fact operated as one, NSF's organization should be "modified temporarily" by the creation of a single division. After he assured some reluctant members that this was "an interim measure and will be reconsidered in the light of experience and study during the current year," the NSB approved, "for the time being," a Division of Biological and Medical Sciences. By August final plans were made to hire Fernandus Payne, a geneticist at Indiana University, to succeed Field as Assistant Director for the combined division.[52]

In order to satisfy his and the board's desire for some structural presence for medicine at NSF, Waterman proposed to activate a separate Divisional Committee for Medical Research "in order properly to take care of the medical research interests." In July, pathologist Ernest W. Goodpasture of Vanderbilt University, who was among those who were proposed to head the medical division, visited the Foundation at Waterman's request. Although Goodpasture declined Waterman's proposal to act as NSF's senior medical advisor, Waterman hoped he might chair the proposed Medical Research Committee.[53] Goodpasture did become a member of the committee, but anatomist and physiologist Edward W. Dempsey of Washington University in St. Louis was elected chairman. The anomalous situation of two separate committees advising a single division was not destined to last. The Medical Research Committee met only twice, in May 1953 and again that November, at which time it voted (with strong staff backing) to combine with the Committee for Biological Sciences.[54] Thereafter, the single Divisional Committee for Biological and Medical Sciences always included representatives of medical institutions.

A New Patron for Biology: First Grants

Having obtained its first operational budget in October 1951, the Foundation hurried to award its first grants and fellowships in February 1952 to demonstrate to Congress that it was underway. Because biology had been given a head start over the physical sciences (Paul E. Klopsteg, of Northwestern University's Institute of Technology, did not arrive to head the Division of Mathematical, Physical, and Engineering Sciences [MPE] until November), NSF awarded all of its first batch of twenty-eight grants in the biological sciences.

The first step of the granting process was forming the statutory Divisional Committee for Biological Sciences. In all, nine prominent senior scientists from leading research schools were invited to become members, and they chose as their first chairman Douglas Whitaker, experimental embryologist turned Stanford administrator, who had already served as Field's and Waterman's chief academic advisor for biology that summer. Compared to the biologists on the board, divisional committee members were relatively younger. All had Ph.D.'s rather than M.D.'s and only one, Wallace O. Fenn, professor of physiology at the University of Rochester, had strong medical ties. The committee was relatively balanced by area of biology, but neither by geographical distribution nor by type of school, let alone by gender or race.[55]

Field and his colleagues in other divisions had already received a number of proposals before the Foundation sent out its mimeographed four-page "Guide for the Submission of Research Proposals" to universities in December 1951. The board had approved staff-developed policies for supporting research and education during the summer and fall, including the decision to use the grant mechanism rather than contracts because grants implied an offering based on mutual trust. Despite the earlier board preference for funding individuals rather than projects, the staff decided to make grants to universities for specific projects initiated by principal investigators.[56]

In keeping with ONR precedent, grant applications were kept as simple as possible for the investigator. NSF supplied no application forms and gave little guidance for writing proposals beyond suggesting that they include a description of the research and available facilities, some biographical and bibliographical information about the investigator, a budget, and signatures of authorized officials of the institution. NSF hoped to encourage informal relationships between the staff and its grantees like those at the Rockefeller Foundation and ONR. Thus, the guide invited prospective grantees to discuss preliminary proposals with staff either in person or by letter. One bit of red tape appeared at the outset, however. Despite the free-form nature of the applications, fifteen copies were required. Some universities were ready to respond on short notice; others were not. As a result, the Foundation received a grab bag of proposals. Amounts requested ranged from $883 for two years to $343,452 for ten years![57]

Instead of asking the Divisional Committee for the Biological Sciences to evaluate the proposals, Field and Consolazio deliberately called together a separate panel of advisors—not specified in the NSF Act—who were of similar stature but entirely named by the staff. The eleven men who made up the first panel, which met on 18–19 January 1952, included ten professors or administrators associated with academic programs in biology, biochemistry, marine biology, medical physics, botany, microbiology, and zoology, and one U.S. Department of Agriculture bureaucrat.[58] While the staff attempted to obtain a representation of fields, they gave little attention to geographical or institutional distribution. In fact, two of the consultants were from Indiana University, and three were from the University of Pennsylvania.

The staff mailed the proposals to the panel members in early January along with brief informational "ground rules," asking the panelists to evaluate each proposal based on its "scientific merit," the ability and resources of the investigator and staff, and the reasonableness of the budget. They also requested the consultants to write a short comment on each proposal, to be sent to the Foundation a week before the panel meeting. The panel considered forty-eight proposals along with a special proposal for joint support with other federal agencies of the financially strapped *Biological Abstracts.*[59]

Lee Anna Embrey, an assistant to Waterman, observed the first panel meeting and recorded her impressions in a five-page memorandum to Field. She conveyed an image of noblesse oblige—self-assured men who knew each other and the successful grantees. She found it "a real pleasure to watch such highly disciplined and trained minds review each proposal as it was presented. There was a certain degree of concentration and an economy of words, yet there was an air of easy relaxation in the group which bespoke complete familiarity with both the subject matter and the investigators." She wished that representatives of the Bureau of the Budget and Congress had been there to observe, for "they could not have had a finer demonstration of critical scientific evaluation."[60]

The panel used a five-point grading system, still in use at the Foundation, with 1 as the highest and 5 the lowest score. These scores corresponded in subsequent panel evaluations to "highly meritorious," "meritorious," "acceptable," "questionable," and "decline." Scores varied widely among the consultants. No proposal received all 1's or 2's. Not uncommonly, scores for a single proposal went the gamut from 1 to 5. Comments on the rating sheets amounted to no more than a few sentences, sometimes as brief as the single word "Good." With each proposal, the chairman informed the panel of its average rating and called on members who had given high and low scores to explain their reasons, after which one panelist was asked to summarize the thinking of the group and recommend an overall rating.[61] Conflict of interest rules had yet to be established. Two of the panelists, Britton Chance, director of the Johnson Foundation for Medical Physics at the University of Pennsylvania, and Lawrence Blinks, director of Stanford's Hopkins Marine Station, submitted their own proposals; they simply did not supply a rating for them. Both received substantial grants. Ralph Cleland, head of botany and Dean of Graduate Studies at Indiana University, did not consider it improper to give the highest rating to proposals from three of his young faculty members. Nor, apparently, did anyone else.[62]

From the beginning, proposal review at NSF differed from that at NIH. At NSF, the staff's panel evaluated a set of projects to be supported with a definite budget. Although the panel recommended which proposals to fund and provided rough overall ratings and some indication of budget, it did not rank the proposals in a detailed fashion as was by then customary at NIH. Instead, the staff decided on final recommendations to the director and board and determined the level of support, which might differ substantially from the panel's opinion. In general, faculty at the top research universities (for example, California Institute of Technology, Yale, and Indiana) received grants, usually at the full requested sum, though NSF staff might cross out unacceptable budget categories such as "miscellaneous." Those from small or more obscure colleges were less likely to be selected, and if selected, often received a smaller proportion of the amounts they had requested. All proposals rated 1 or 2 were funded as well as some of the "acceptable" 3's. Overhead was an across-the-board 15 percent, which the staff added to budgets that neglected to include it. A staff summary evaluation of each proposal, based on the written and oral comments of the individual panel members, went into the project file.[63]

NSF lost no time in activating the proposals selected to be funded. After review in the director's staff meeting, the proposed awards were next reviewed by the newly formed Divisional Committee for Biological Sciences, which met for the first time on 25–26 January 1952. The committee was charged to examine grant selection procedures and discuss general policy issues such as individual versus block grants, grants to small colleges, and geographical distribution. For the final approval by the National Science Board on 1 February, the staff prepared a formal recommendation for each proposed grant, a one- or two-page document that followed the model of recommendations prepared for the Board of Trustees of the Rockefeller Foundation. It contained a paragraph recommending the project, a brief description, "collateral support" (the institution's contribution), "future implications" (NSF's likely obligations for future support), comments, and an abbreviated budget.[64]

Some board members were not at all pleased by the "rubber stamp" manner in which the "docket" of grants was presented to it for approval. Reyniers complained in a personal letter to Bronk that what occurred was indicative of the NSB's growing isolation from the operation of the agency. "The grants for the Biological Sciences Division were handed to the Board *en bloc* with a resolution that they be approved. None of us saw these grants before the meeting nor were we given any information relative to the other grants not acted upon, deferred or turned down." Although Reyniers did not object to the quality of the awards, he felt that "a little better judgement could and should have been exercised at a 'top level' relative to their distribution." He pointed out that $168,000 out of a total of $411,000 went to the West Coast, that two consultants received large grants, that one individual received two grants, and that only two went to small colleges. At the meeting, Reyniers, on behalf of the Board Committee on Biological Sciences, obtained a tabling of action on the proposals until after an executive session of the board. Later that day, the awards were approved individually.[65] In subsequent meetings, the board became gradually reconciled to considering separately only those grants that were unusual by their size or policy implications.

Given current bureaucracy in federal funding agencies, it is remarkable how rapidly the first grants were processed. By the next NSB meeting on 29 February, all twenty-eight grants had been awarded; the payments for the first installment were sent within a week of board approval. According to Waterman, recipients had reacted with "surprise as well as approval" to the short processing time, the "simplicity of the grant document, and the immediate receipt of the checks."[66]

The Foundation awarded two more batches of grants by the end of the fiscal year in June. Once the principle was established that it was not the normal function of the divisional committee to screen proposals, the staff saved time and money on the second round by combining the panel and divisional committee steps in the evaluation process. On 29 February the Board approved twenty-five grants in biology after direct divisional committee evaluation on 18–19 February. A third group of awards resulted from a panel meeting held on 24–26 April at which ninety-one proposals, some labeled biological sciences and others medical research, were evaluated. Of these, forty were approved, forty-three declined, and eight deferred. Among those approved, some were not to be activated until the next fiscal year's funds were received.[67]

In all, sixty-eight biological projects were approved for FY 1952 funding at a cost of $762,675, with twenty-one more approved for funding in FY 1953 (see table 2.1). Grants ranged in amount from $680 to $50,000 and in duration from four months to five years, with a median award of $9,000 and median period of two years.

The most substantial grants went to Berkeley coinvestigators I. Michael Lerner, professor of poultry husbandry, and his fellow geneticist Everett G. Dempster ($50,000 for five years); immunologist Manfred M. Mayer of the Johns Hopkins University School of Hygiene and Public Health ($36,000 for three years); and biophysicist Britton Chance of the University of Pennsylvania ($36,000 for three years).[68]

In later years, NSF retained few, if any, records of rejected proposals, but for the first rounds of biological sciences grants in 1952, lists of titles, investigators, and brief proposal evaluations have survived. Thus, the historian can glean some indication of why proposals were turned down. Typical among the reasons were: "lack of competence and ability of investigator" (even though the research might be judged worthy), investigator's lack of experience in the particular area of research (despite acknowledged competence in other areas), and approach not adequate to solution of the problem. Or the proposed program might have unclear objectives or not represent basic research or be "too large and diffuse," too routine, or limited to local interest (state-level taxonomic projects). Also rejected were projects judged to be "clinical medicine" or related to a "special disease."

The panels took an especially negative view of projects perceived as applied biology. As Lee Anna Embrey reported, the first panel was quick to

Table 2.1   NSF Funding of Biological and Medical Sciences, FY 1952

	Number	Amount requested ($)	Amount awarded ($)
Proposals received	296	6,908,234	—
Activated	68	1,416,090	762,765
Approved for FY 1953	21	633,412	389,100
Declined/Withdrawn	96	2,309,300	—
Pending	111	2,549,701	—

Source: Fernandus Payne to Director, "Interim Report," 27 January 1953, filed with BMS Annual Reports, NSF Historian's Files, National Archives and Records Administration.

spot "hidden jokers"—applied projects disguised as fundamental research. Such projects, the panel suggested with some derision, should be supported by agricultural, forestry, or fisheries resources, or by private industry. For example, a project to study the physiological effects of hydrocarbons on plant cells was evaluated as lacking "any fundamental promise." It "should be carried along with agricultural studies at the institution with support from oil companies, etc."[69]

Overall, the Foundation's policy, supported strongly by the staff and the panels, gave priority to the best research, which generally meant research by established investigators in leading institutions. In the male-oriented terminology of the period, the first panel characterized top researchers as "*A* men." Embrey noted that in voting to recommend one proposal, the group pointed out that it was supporting a "*B* man" at an "*A* institution." She commented: "In other words, the high standing of the institution and the quality of the work being produced there apparently assured worthwhile results, even though the investigator in question was not considered topnotch." She went on to note that panel raised questions as to "whether the Foundation would similarly support a 'B' man in a 'B' institution."[70]

In general, the staff and panels resisted supporting any project solely for the purpose of broadening the bases. They acknowledged distribution, particularly geographical distribution, as a problem but hoped to address it by identifying atypically good researchers in underdeveloped institutions and thus fanning a spark, as the typical metaphor went. Given the state of higher education in America in 1952, locating sparks was not easy. Only a small portion of the nation's colleges and universities were prepared to undertake research at all. Many states had only one doctorate-granting university, the state university; the second-level public school of higher education was generally still denominated as a state college. Even Penn State, the main state-supported school in Pennsylvania, was still Pennsylvania State College in 1952. High-quality research was far more concentrated than it came to be in the 1960s and 1970s.

NSF staff and their advisors made a sincere, albeit token, effort to fund areas of biology not receiving much federal patronage as well as young investigators, investigators at small colleges, and those with little research experience. The importance of supporting research at small colleges had been brought to the particular attention of the staff in 1951 by the "startling" findings of R. H. Knapp and H. B. Goodrich's study of the baccalaureate origins of (male) doctorates in the sciences. Knapp and Goodrich, both from Wesleyan University, a men's liberal arts college, showed that high-quality small colleges produced a higher proportion of scientists in relation to size than did leading research universities.[71]

Considerations of distribution were reflected in the summary evaluations of several of the first year's proposals. For example, Frank N. Young's
project, "Biometrical and Taxonomic Analysis of Populations of Selected
Species of Aquatic Beetles in the U.S. and Mexico" was justified as a fundamental project in the field of entomology, an area dominated by applied research. It was also "more difficult to secure funds in taxonomy than some
other field." Truman G. Yuncker's botanical survey of the Tongan Islands
gained a high rating not only for its scientific merit but also because DePauw
University was "a small but an excellent institution where research should be
encouraged." The panel was pleased to fund Irwin C. Kitchin's project in experimental amphibian embryology because "it will be carried out in a laboratory [University of Mississippi] where it will be worthwhile to have an
introduction to experimental research." In several cases, panelists pointed out
that the investigator was a young man of promise. Yet, this solicitude went
only so far.[72]

When "uneven geographic distribution, particularly insofar as California is concerned" (its institutions had won 7 of the first 28 grants) was
brought up for discussion at a staff meeting in February 1952, Waterman reiterated the NSF dictum "that merit should be the primary factor in selecting grants, and when the merit of several grants is substantially equal, geography and other factors might be taken into account."[73] The divisional
committee suggested that only at the level of proposals rated 3 should geography be a factor. Even then, as committee member Jackson Foster put it,
the purpose was not simply to fund projects in underdeveloped institutions
but to promote the growth of science there: "We should put our money
where there are sparks and make it a flame."[74]

NSF staff were more conscious of distribution of awards in 1952 than
later in the decade, tabulating all proposals received through that June and
their dispositions by state of origin and region (see table 2.2).[75] The figures
show that proposals from the Southeast and Southwest fared much more
poorly than proposals from other regions of the country.

The staff and panels gave no explicit concern to funding proposals from
women and African Americans, though the presence of women and African
Americans on the board may have offered some incentive to provide at least
token awards in FY 1952. Two women and one black scientist received
grants in 1952; several other women and at least one other black scientist had
applied. It is noteworthy that the two women—Rosalie C. Hoyt, a biophysicist at Bryn Mawr, and Grace E. Pickford of the Bingham Oceanographic Laboratory at Yale—were both associated in their research with men
who were well known to the panel, L. Joe Berry and Alfred Wilhelmi, respectively.[76]

Table 2.2   Geographical Distribution of BMS Proposals and Awards, FY 1952

Region[a]	Received	Activated	Approved for FY 1953	Declined/ Withdrawn	Pending	Percentage[b]
Northeast	32	8	6	8	10	64
Middle Atlantic	50	10	6	17	17	48
Southeast	46	8	0	19	19	30
Great Lakes	56	17	3	17	19	54
Plains	30	8	2	11	9	48
Southwest	26	4	0	9	13	31
Rocky Mts	3	1	0	1	1	50
Far West	42	10	4	14	14	50
National	9	2	0	0	7	
Foreign	2	0	0	0	2	
Total	296	68	21	96	111	

Source: "Quarterly Summary of Proposals Received in Biological and Medical Sciences Division through 30 June 1952 (Geographic Distribution)," Biological Sciences Divisional Committee Meetings, Box 20, 70A-2191, National Archives and Records Administration.
[a]The data have been retabulated for modern regional designations. (The "South" in 1952 included Maryland, while the District of Columbia was placed in the Middle Atlantic.)
[b]The column "Percentage" has been calculated by dividing the number of proposals activated or approved for later activation by the number of proposals received minus those pending.

A grant to plant physiologist James H. M. Henderson of the Carver Foundation, Tuskegee Institute, raised controversy. Though Henderson had impressive credentials (a Ph.D. from the University of Wisconsin and two years' research at the Caltech Kerckhoff Biological Laboratories), the panel justified the grant largely as a means of encouraging science in a lesser school. When one divisional committee member questioned the award on the ground that the recipient was a "second rate investigator," Foster replied that it was a "high risk investment" but perhaps NSF "could do a real service here as a sort of sociological experiment in the aid of a weak institution where facilities and time for research may now be pretty inadequate." He suggested evaluating the project not by its scientific results but rather by "its usefulness in drawing research to the attention of students who otherwise would never have any contact with research. I think we should face the issue here."[77]

Among those rejected or deferred in the first year were a few prominent (or rising) biologists: M. C. Chang was a well-known reproductive endocrinologist at the private Worcester Foundation for Experimental Biology, which was developing the birth control pill. NSF deferred (and never funded) Chang's request for support of a metabolic study of the mammalian ova because the panel doubted his competence in biochemistry and found the proposal too vague about the biochemical collaboration Chang hoped to obtain.[78]

When C. D. Michener and Robert R. Sokal of the University of Kansas proposed to study a growing environmental problem—the increasing resistance of insects to DDT—they were soundly dismissed with a rating of 5. The panel "questioned" the "adequacy" of the investigator's genetics training. Michener, a leading entomologist and systematic biologist who later served on the systematic biology panel, had little experience in genetics, but his young assistant Sokal had trained with Sewall Wright, a founder of population genetics, and later acquired fame for his development of numerical taxonomy. The panel might also have disliked the applied aspect of the project. The following year, NSF funded Michener for work on the "caste behavior" of bees, which was more clearly related to his previous research.[79]

Recent immigrant Knut Schmidt-Nielsen, later recognized as a premiere comparative physiologist, faced rejection of his proposal in an area in which he was to excel: "Role of Body Surface Insulation in Animals in Hot Environments, and Effect of Body Size of Animals upon their Oxygen Consumption." Panel members judged the work in narrow terms as "not of sufficient importance to warrant a higher rating."

Waterman's and the NSB's intention in 1951 had been to fund the biological and physical sciences equally, but since the Division of Biological Sciences started earlier than other divisions, it gave two-thirds of the NSF's ninety-seven first-year research grants. In that year, the Foundation spent more than twice as much ($762,675) on biology and medicine than on the mathematical, physical, and engineering sciences. In December 1952, MPE Assistant Director Paul Klopsteg protested both the skewed distribution in 1952 and the presumption that the biological and physical sciences would continue to be funded evenly. His argument was that the physical sciences deserved a greater share of funding because there was less societal demand for biologists than for physical scientists. Moreover, both the Defense Department and AEC were likely to reduce their commitment to basic research, while NIH, claiming mission relevance, was unlikely to turn over any basic research to the Foundation. Waterman saw some justice to these claims, but was not yet willing to alter the balance between the biological and physical sciences. The even split in funds for individual research grants was retained in presentations to Congress through most of the decade, primarily to avoid having to justify an inequality.[80] Thus, though NSF was later to give increasing priority to the physical sciences, in the early years the biological sciences held a prominent place.

Delineating Programs: "The One Biology Concept"

One of the hallmarks of BMS was its organization. It was not organized along traditional disciplinary lines like the programs of other agencies, but according to an innovative functional scheme decided upon during the summer of 1952. In May 1952 the staff was already considering a functional division but still planning a more traditional organization. For the biological sciences, Field initially projected eight programs in "disciplinary areas": systematic biology, genetics, biochemistry, microbiology, experimental botany, experimental zoology, physiology, and experimental psychology. For the medical sciences, he proposed four programs: anatomy, medical physiology, medical biochemistry, and related therapeutic areas. Field selected disciplinary labels, he said, "because it seems likely that both the public and non-biological scientists will understand such labels better than more functional ones."[81]

But once the Foundation decided to create a combined Division of Biological and Medical Sciences, the staff faced the challenge of supporting "basic medical research" in the context of biology. Consolazio, Wilson, and Levin seized upon a new functional reordering of biology that avoided disciplinary labels. The idea had originated with embryologist Paul Weiss, who, at the time, chaired the Division of Biology and Agriculture of the National Research Council. Weiss hoped to position the division to advise federal agencies in the postwar era by creating a set of committees to cover all of biology. In an internal 1951 memorandum, Weiss divided biology into six areas: molecular biology, cellular biology, genetic biology, developmental biology, regulatory biology, and group and environmental biology. "You will note," Weiss wrote, "that there is no longer any reference in this scheme to particular forms of life, but that every biological phenomenon can be defined in terms of one or more of these six categories." Weiss saw his ordering as a means of reintegrating the field.[82]

By July, the NSF staff had adopted Weiss's scheme in modified form. Field had left in May, and his replacement was not due to arrive until September. With no money left for grants until the new appropriation and plenty of time at hand, Consolazio, Wilson, and Levin brainstormed and argued over organizational concepts that would "make the Foundation the cornerstone of federal government support" for biology, although as Wilson recalled, none of the three had ever taken a course in "biology." They—especially Consolazio—were impressed with Weiss's conceptual scheme.[83] Their new plan, based on Weiss, divided biology into seven areas: molecular, regulatory, developmental, genetic, environmental, and systematic biology and psychobiology. They also initially proposed a program for microbiology, but Levin pointed out (and the divisional committee agreed) that the category was morphological rather than functional and should be abandoned.[84]

Organizing along functional rather than disciplinary lines had many advantages for the fledgling NSF Biology Program. As Consolazio explained, their approach did not just "grow like Topsy" but came about because of a "realization that biology was becoming more and more splintered," that the "acknowledged discipline breakdown failed to describe the state of biology today," and that a "real need existed for a reinforcement of the one biology concept." They looked for "an approach that made no distinction for medical sciences, biological sciences and agricultural sciences, microbiology, plant biology or animal biology. We wanted a horizontal approach in the overall field of life sciences, and this is what I think our system accomplished."[85]

The functional program designations allowed NSF to cut across traditional divisions of the life sciences. Except for psychobiology, which was limited to animals, each of the functional programs could encompass research in botany, zoology, microbiology, or the biomedical sciences. Most radical was the discarding of the boundary between botany and zoology but, in Wilson's words, "we figured, as Paul Weiss argued, biological processes are the same in plants and animals." To Waterman, an advantage of the functional scheme was "that interdisciplinary relationships in the life sciences may be more easily handled from both an administrative and a scientific information point of view." Besides, NSF could support its own definition of "medical research" without being vulnerable to the charge of "duplicating" the work of NIH, since biomedical categories were effectively hidden under biological rubrics.[86] Finally, this system allowed the NSF program managers, who distrusted the National Academy of Sciences, to coopt Weiss's plan to adapt the functional scheme to making policy for biology through the National Research Council (see chapter 4).

Since these programs persisted for decades at NSF, they deserve some comment. The program in molecular biology, begun the year before Watson and Crick discovered the double-helix structure of DNA, represents perhaps the first formal institutional use of the term. Before 1953, "molecular biology" tended to be defined broadly, in contrast to the post-double-helix narrowing of the term to molecular genetics. Like Warren Weaver, who had first used it in 1938 to characterize a core Rockefeller Foundation program, the NSF staff conceived the term in a broad sense. Consolazio defined molecular biology in January 1953 as "a study of structure, synthesis and molecular properties of biologically important molecules and the kinetic properties of their reactions." It included protein structure and synthesis, photosynthesis, and immunochemistry.[87]

Of all the programs, the one in regulatory biology was the most diverse and difficult to delimit. According to Levin, it dealt with "those aspects of biological and medical science which are concerned with or related to the control and regulation of vital processes," and included plant, animal (vertebrate and invertebrate), and microbial physiology, biophysical mechanisms, metabolism, nutrition, immunology, and "inter-organismal regulatory mechanisms" such as symbiosis.[88]

Genetic biology, closer to a discipline than most of the categories, had yet to be transformed by the molecular revolution. Consolazio defined it as "the study of heredity and variations of the resemblances and differences between organisms." It compassed classical genetics, biochemical genetics, population genetics, cytoplasmic inheritance, and medical genetics.[89]

Developmental biology included all aspects of growth and regeneration, thus signifying a broader area than traditional embryology. Weiss, a founder of the Society for the Study of Growth and Development in 1939, now the Society for Developmental Biology, was largely responsible for successfully promoting the term among biologists.[90]

Environmental biology dealt with the "interrelationships between living cells or complex organisms and their external or sometimes internal environment."[91] It thus included ecology but also any study of interrelationships of particular organisms and the environment, as, for example, parasitology and migration.

Systematic biology, not present in Weiss's NRC scheme, was a term recently given new significance. Museum departments and professional societies of taxonomists had traditionally been organized by class of organism—mammalogy, ornithology, herpetology, and so on. In the 1920s to 1940s, older methods of classification by external morphology and comparative anatomy were supplemented by newer methods of "experimental taxonomy" based on genetics, cytology, and ecology. As part of the modern evolutionary synthesis, a broader overview of the principles of taxonomy in relation to evolution emerged. The so-called new systematics was reflected in the founding of the Society for the Study of Evolution in 1946 and the Society for Systematic Zoology in 1947. Since other agencies generally neglected systematic biology, NSF staff were especially concerned to support it; after molecular and regulatory biology, it received the next largest share of funds.[92]

Finally, psychobiology, a term variously used by the Rockefeller Foundation, Yale psychologist and primatologist Robert Yerkes, and others to convey an interdisciplinary approach, encompassed physiological psychology including neurophysiology, learning, and animal behavior. (Social psychology was still outside the scope of the Foundation.)[93]

In 1958, the functional system was enlarged by metabolic biology, which took over portions of molecular and regulatory biology. On the whole the system was remarkably resilient. It lasted intact for well over a decade; remnants of it persisted into the 1980s.

As there were not enough staff to go around in the fall of 1952, programs were grouped with the expectation that eventually each would acquire a separate program director and separate panel. Consolazio handled molecular and genetic biology; Levin, regulatory biology; and Wilson, psychobiology. In September Frank H. Johnson of Princeton University, the division's first "rotator," arrived to take charge of developmental, environmental, and systematic biology, even though these areas were far from his own research in bioluminescence.[94]

NSF's 1952 *Annual Report,* which appeared in November, listed the first year's biological grants according to the new functional system, although they had all been approved by a single panel before the system was established. Members of the four newly created advisory panels, each corresponding to a program director, were also listed. Thus, by the second year of grants, NSF's biology program was substantially in place.

A Flexible Organization

It is ironic that biologists had campaigned in 1945 to separate biology from medicine in NSF only to have them recombined by 1952—this time with medicine subordinated to biology. Waterman had not abandoned NSF's charge to fund medical research, but he had redefined it to fit realities. Forced in part by the success and opposition of NIH, he and his staff renounced clinical research and research related to particular diseases as applied science and therefore a priori outside the scope of NSF while they interpreted basic medical research as a subset of basic research in biology. As BMS Assistant Director H. Burr Steinbach expressed it in 1953, "basic biological research and medical research are really not distinguishable. A real distinction appears only with regard to applied research in these areas."[95] Thus, NSF funded biomedical research in medical schools through interdisciplinary programs which also funded biology in universities and colleges. NSF consequently became less vulnerable to any political charge that it was "duplicating" the work of NIH.

From the beginning, the advantage of the NSF program in biology was its flexibility to fund widely varying projects. John Wilson liked to boast that while NIH had the money, NSF had the ideas.[96] Though it had awarded less than a million dollars in grants by the end of fiscal 1952, the NSF Division of Biological and Medical Sciences was organized with an eye to the future. It was prepared to take advantage of increased budgets later in the decade to run—despite NIH competition—a broad, imaginative, and significant program in biology.

CHAPTER 3

Expanding and Experimenting
in the 1950s

When John Wilson and Bill Consolazio looked back on the 1950s they recalled an exhilarating time when the NSF staff was small, communication was open, and opportunities to advance biology seemed to be everywhere. Far from acting bureaucratically, program directors in the Division of Biological and Medical Sciences were able to engage in a highly personal and experimental style of grant-making in a manner reminiscent of their private foundation precursors. As Wilson recalled it, lack of money in the early years of NSF, rather than limiting options, gave program directors the advantage of plenty of time to think and plan.[1] The NSF staff in biology were talented managers of science, inventing and shaping programs, actively identifying opportunities to invest in biology, and continually negotiating with academic constituencies, top NSF management, the National Science Board, and, indirectly, with the Bureau of the Budget and Congress, who held the purse strings. Compared to their counterparts in later decades, program directors maintained close informal contacts with their academic advisors and grantees and had a relatively free hand to shape their programs and to use their available funds to benefit their areas of biology.

Though growing, NSF remained relatively small until the Soviet Union launched the artificial satellite Sputnik in October 1957. Then, with public clamor to meet the Russian challenge, Foundation budgets shot up rapidly: from $16 million in FY 1956 to $133 million in 1959 to $320 million in 1963.[2] Though funding for education rose relatively faster than that for research, the BMS budget burgeoned from about $5 million in 1956 to $20.5 million in 1959 (more than doubled from 1958) to $39 million by the end of the Waterman era in 1963 (see tables 3.1 and 3.2).

Table 3.1   National Science Foundation Obligations, FY 1951–FY 1976
(in Millions of Current-Year Dollars)

Fiscal Year	Research and Related Activities	Total Obligations
1951	0.2	0.2
1952	1.9	3.5
1953	3.0	4.4
1954	6.1	8.0
1955	10.4	12.5
1956	12.5	16.0
1957	24.3	38.6
1958	30.3	49.5
1959	69.3	132.9
1960	88.7	158.6
1961	106.1	175.0
1962	175.1	260.8
1963	223.4	320.8
1964	244.9	354.6
1965	287.9	416.0
1966	333.4	466.0
1967	334.2	465.1
1968	358.0	500.3
1969	303.0	432.5
1970	328.7	462.5
1971	383.4	496.1
1972	480.0	600.7
1973	504.0	610.3
1974	540.2	645.7
1975	593.2	693.1
1976	613.4	724.4

Source: National Science Foundation, *Report on Funding Trends and Balance of Activities, National Science Foundation, 1951–1988* (Washington, D.C.: NSF, 1988), 18.
Note: Research and Related Activities includes individual awards, groups, institutional support, other research resources, facilities, centers, and Research Applied to National Needs (RANN). The difference between this and Total Obligations is the funds spent on Science and Engineering Education and on the U.S. Antarctic Program.

BMS patronage of biology in the first decade illustrates three tensions that have beset NSF throughout its history: (1) funding the best science versus improving the nation's science infrastructure more broadly by wider distribution of support, (2) supporting merit-based individual project

Table 3.2   Division of Biological and Medical Sciences, Grants Awarded
by Year, FY 1952–FY 1968

Fiscal Year	Proposals Received	Amount Requested (in Millions of $)	Number of Awards	Amount Awarded (in Millions of $)	Percentage of Proposals Receiving Awards
1952	294	6.84	68	0.76	23.1
1953	266	5.17	72	0.82	27.1
1954	448	8.97	177	1.95	39.5
1955	727	13.04	275	3.50	37.8
1956	827	16.93	426	5.08	51.5
1957	1,137	28.19	561	7.88	49.3
1958	1,362	40.20	605	8.88	44.4
1959	1,549	52.02	967	20.46	62.4
1960	1,586	60.25	908	24.87	57.3
1961	1,922	91.35	981	27.28	51.0
1962	1,934	105.11	1,058	32.86	54.7
1963	2,158	126.10	1,143	39.10	53.0
1964	2,047	123.61	1,253	41.35	61.2
1965	2,351	166.74	1,179	43.29	50.1
1966	2,421	193.13	1,363	51.82	56.3
1967	2,433	212.59	1,390	54.02	57.1
1968	2,462	226.15	1,331	53.59	54.1

Source: Data from BMS annual reports, NSF Historian's Files, RG 307, National Archives and Records Administration; *NSF Annual Reports;* John T. Wilson, "Preliminary Program Plans and Budget Estimates—FY62," 11 March 1960, in folder "Budget—1958, 1959, 1960, 1961," Box 6, 70A-3915, RG 307, National Archives and Records Administration.

grants versus providing a variety of modes of support, including departmental and institutional awards, which were more subject to political considerations, and (3) claiming jurisdiction for BMS of all aspects of research and research training in biology, as Wilson aggressively and unapologetically attempted, versus the perennially competing interests of NSF's education division.[3]

By the end of the Waterman era, several impediments had arisen to BMS's informal and flexible style of science management. Some were external, posed by Congress, the Bureau of the Budget, university administrators, the growing federal science advisory establishment, and competition from other federal granting agencies (see chapters 4–6), while others were internal, the inevitable result of rising budgets and increasing bureaucratization. Despite the obstacles, the division's position both within the Foundation and among academic scientists remained strong throughout the first decade. Staff and grantees looked back to these halcyon days when all things seemed possible, including indefinite growth.

Programs and Program Directors

Much of the Foundation's flexibility in funding biology in the 1950s stemmed from the organization of BMS into functional programs headed by knowledgeable program directors with broad discretionary authority. As budgets increased, each of the seven programs gradually gained its own director and advisory panel. Molecular and regulatory biology overshadowed the others, accounting for nearly half of the dollars expended. Because they were so large, a new program in metabolic biology, was created in 1958 from parts of each to cover research in metabolic processes. After the split, molecular biology remained the largest and most prestigious program by far, accounting for 22 percent of BMS funds in FY 1960. The smallest programs that year were developmental and genetic biology, with respective shares of 6.9 percent and 8.6 percent of the total funds (see table 3.3).[4]

Among fields of biology, the chief beneficiary of the functional system was biochemistry, the specialty of both William Consolazio and Louis Levin. Biochemists were to be found on practically all the program panels. Physiologist Wallace Fenn, who chaired the initial Divisional Committee for Biological Sciences, observed to Assistant Director Fernandus Payne in 1952

Table 3.3   Division of Biological and Medical Sciences, Grants Awarded by Program, FY 1954–FY 1967 (in Thousands of Dollars)

Program	FY 1954	FY 1957	FY 1960	FY 1963	FY 1967
Developmental	111	390	1,706	3,937	5,067
Environmental	43	770	2,592	4,657	5,884
Genetic	157	674	2,136	3,774	5,214
Metabolic	—	—	3,096	4,794	4,949
Molecular	458	1,788	5,560	8,224	11,862
Psychobiology	307	850	2,161	3,328	4,642
Regulatory	460	1,888	3,575	5,150	6,226
Systematic	249	760	2,771	3,939	5,589
NCE/Special Programs[a]	171	760	1,289	1,640	4,584
Total[b]	$1,955	$7,880	$24,887	$39,443	$54,017

Source: Data from BMS annual reports, FY 1957, FY 1960, FY 1963, FY 1967, NSF Historian's Files, RG 307, National Archives and Records Administration.
[a]These figures do not include Facilities.
[b]Columns do not add up exactly due to rounding.

that the roster of BMS panel members was "evidently selected with exaggerated emphasis on the biochemical aspects of proposals." Two of the panels, he claimed, were "composed almost entirely of biochemists." In contrast, there was only a single representative of physiology, H. Burr Steinbach, on any of the panels. Agreeing, Payne explained that the panels had been formed before his arrival at NSF.[5] Wilson later recalled that "biochemistry was the cornerstone of everything, of course. You could have had one great big biochemistry division, and that would have encompassed everything."[6]

The areas that were probably most neglected or discouraged by the functional system were anatomy, a biomedical discipline, and morphology, the traditional mainstay of university zoology. In 1955, the staff considered forming a program in structural biology but decided two years later that none was needed. Wilson contended that there was little call for research "on structural problems for the sake of knowledge of structure per se" and that most morphology was done "with reference to some reasonably immediate functional problem." The Developmental Biology Program handled whatever "structural" proposals were received, possibly not giving them much consideration.[7]

In order to fund other than research projects, BMS staff established the category NCE, for Not Classified Elsewhere. Each year they set aside divisional monies for NCE projects such as equipment awards, support of stock collections, or support of summer training at field stations. Typically evaluated by ad hoc mail reviews, these grants were administered by the closest program. As Wilson wrote in 1959, "The importance of the '*Not Classified Elsewhere*' program cannot be overemphasized because it insures a degree of flexibility within the Division as a whole that cannot be met by strict adherence to individual area programs."[8]

Heading the division was an assistant director who reported directly to Waterman. For the first four years, Waterman was able to obtain the services of four distinguished scientists on leave from university positions. With the aid of each incumbent, Waterman conducted the search for their successors himself, visiting candidates in their university settings. Influential board members not only suggested names but applied pressure on reluctant candidates to accept. It was not easy for Waterman to find candidates of the stature he sought. Each year he had to offer the position to several men before he could find someone suitable who was willing to give up his research for a year to come to Washington.[9]

John Field's immediate successor, Fernandus Payne (1881–1977) of Indiana University, was an early product of the Thomas Hunt Morgan school of genetics at Columbia. In his twenty-year tenure as chairman of Indiana's zoology department and dean of the graduate school, he had worked with

Indiana's president Herman B Wells to stimulate growth of the university's research. When he came to NSF in September 1952, he was 71 years old and had recently returned to full-time research. Highly recommended by Detlev Bronk, he impressed Waterman as "an able administrator, having a quiet direct way in speaking and dealing with people."[10]

Payne was followed in 1953–54 by the more dynamic H. Burr Steinbach (1905–81), a general physiologist well known on the Washington circuit as a spokesperson for biology and appreciated by colleagues as a cheerful, open, and "uniquely successful administrator." A professor of zoology at the University of Minnesota when he came to NSF, he was later a professor at the University of Chicago and director of both the Marine Biological Laboratory in Woods Hole, Massachusetts, and the Woods Hole Oceanographic Institution. Steinbach was perhaps BMS's most astute administrator for he recognized that BMS needed to forge closer ties with the higher echelons of other federal agencies and with clinical medical, dental, veterinary, and agricultural scientists. However, his advice to develop such a constituency was little heeded.[11]

Lawrence Blinks (1900–89), although the most academically distinguished of Waterman's recruits, was by contrast a reluctant administrator who recalled hating his year in Washington. A leading plant physiologist and biophysicist and a member of the National Academy of Sciences, Blinks had spent most of his career at Stanford University as professor of biology and director of Stanford's Hopkins Marine Station in Pacific Grove, California. At NSF, Blinks found it especially frustrating having to administer a budget prepared by his predecessor while spending much of his time working on the budget for the following year when he would no longer be there.[12]

In BMS's first years, the presence at the helm of eminent members of the biological fraternity was no doubt important in establishing good relations with the academic community. But by 1956, continuity, knowledge of government, and administrative ability had become more important than stature as a biologist. According to Wilson, Waterman had come to recognize that "rotators" were far more effective at the program level than at the divisional level where budgets were formulated. When he could not readily find another replacement, Waterman turned to the capable Wilson, who had served as acting assistant director as early as the summer of 1952. On his own admission, Wilson, a psychologist, had little training as a biologist, nor had he significant research experience.[13] He served as assistant director through FY 1961, while still retaining charge of the program in psychobiology through 1959.

Although the assistant directors for BMS took the lead in making policy for NSF patronage of biology, much authority and autonomy was left to

the individual program directors. The concept of the program director, derived from ONR, was central to the functioning of NSF. As Field explained
in 1952, the program director was the "fundamental unit" of the division's
organization, "the chief authority in the Federal Government on research
activities in his area," who would "spearhead the Foundation's activities relating to evaluation of other Federal programs in his field." Field optimistically envisioned the position as "one equivalent to that of full professor in a
major university."[14] Besides managing their own programs, deciding what
types and amounts of grants to give, and whom to place on advisory panels,
BMS staff worked together to define (in as nonthreatening ways as possible)
"national policy" for biology as a whole (see chapter 4). Not limiting their
activities to biology, program directors like Wilson, Consolazio, and Levin
concerned themselves with the overall effect of federal funding on universities and on the scientific community.[15]

    "The BMS Division is characterized by our effort to promote extensive
personal relations between the staff and the scientific community," Wilson
wrote in 1960. From the beginning prospective grantees were encouraged
to discuss their developing proposals with appropriate program directors.[16]
Grantees kept the staff informed of progress, future plans, and the state of
their departments and disciplines in general. As soon as rising budgets permitted, program directors went on rounds of site visits during which they
gave presentations and conferred informally with biologists in their laboratories. Like their predecessors in the Rockefeller Foundation, BMS staff circulated "diary notes" of their formal and informal contacts with scientists
and others.[17] As Consolazio, wrote to Waterman in 1963: "To me it is the
individual contact between the operation groups of NSF and the recipient
scientists that develop the ideal Government-university relationship."[18]

    In the 1950s, program directors of the research divisions not only had
good access to the scientific community but also to the top hierarchy at NSF.
Informality ruled; the small staff held frequent meetings at all levels and any
staff member had good access to the director.[19] Program directors viewed
themselves as direct links between their constituencies and the director and
the board. The NSF staff, especially program directors in the research divisions, tried their best to avoid the image of a bureaucratic government
agency.

    The BMS program directors in the 1950s were far more autonomous
than their NIH counterparts, the executive secretaries of study sections,
who were less likely to maintain personal ties with grantees and had much
less influence over the granting process, since study sections rank-ordered
the proposals and institute councils usually funded them in order at the requested sums. Moreover, BMS staff resisted any attempts by outsiders to re-

duce their role. For example, some members of the short-lived Divisional Committee on Medical Research, familiar with NIH procedures and disturbed by the way BMS program directors took the initiative in awarding grants, grilled Consolazio on the evaluation of proposals, especially those in the NCE category, where staff exercised "executive judgment" based on proposal evaluations mailed in by expert reviewers they selected. (These reviews were known in the agency as "mail reviews.") Consolazio reported to Steinbach, "The Committee kept harping that panel advice was necessary, its understanding being that panels were more than advisory, and that staff served merely as executive secretary."[20]

Much of the power in BMS throughout the 1950s lay with the permanent program directors—Consolazio (Molecular Biology), Levin (Regulatory Biology), and Wilson (Psychobiology) plus George Sprugel Jr., a recent Ph.D. in economic zoology, brought over from ONR in 1953 and soon given charge of Environmental Biology. All other BMS program directorships were reserved for rotators, whom the permanent staff considered important because they brought an infusion of new ideas and support for the agency through their networks of professional colleagues.[21] It was not unusual to attribute growth in proposals to the tenure of a rotator with extensive academic contacts. In the Waterman era, these positions attracted several biologists of wide reputation through their research or textbooks (see appendix A).

Early program directors, with one exception, were male and, also with one exception, from Ph.D.-granting universities or museums. The one woman to serve, albeit in an acting capacity, was geneticist Margaret C. Green, who was hired initially in 1953 to aid Consolazio in surveying federal extramural grants in the life sciences and stayed on for another two years as head of genetic biology.[22] The second exception, a controversial appointment, was that of Hubert B. Goodrich of Wesleyan University, coauthor of the celebrated Knapp and Goodrich report on the undergraduate origins of Ph.D. scientists (see chapter 2), to be head of the combined Developmental, Environmental, and Systematic Biology Program in 1953–54. Payne deliberately hired him as an advocate of research support for undergraduate institutions.[23]

As funds became available, beginning in fiscal 1954, BMS gradually engaged "professional assistants" for each of the programs. These second-level positions were reserved for women. While some, like Consolazio's assistant, Helen Jeffrey, had doctorates, most of them were bright young women with bachelors degrees in biology. It was claimed that Wilson liked to hire women with laboratory experience (so they could communicate comfortably with scientists) but without Ph.D.s (so they would not aspire to be program directors). Several assistants, like Estelle ("Kepie") Engel, who came to work with Consolazio in 1960, were brought over from intramural laboratories in NIH. Like the higher ranking positions, none of the professional assistant-ships was openly advertised; they were filled by word of mouth.[24]

Assistants played a vital role in the division. They regularly attended panel and divisional committee meetings and participated in policy making. Long-time assistants like Engel in molecular biology/biochemistry or Josephine Doherty in environmental biology were well known and appreci-ated by their respective biological communities. They held their programs together during frequent changes of rotating program directors. Wilson re-called that his assistant, Mary Parramore, saved him from embarrassing situ-ations created by his lack of training in classical botany or zoology. Once during a speech at Virginia Polytechnic Institute, a member of the audience asked Wilson if NSF funded research on arachnids (spiders). Having no idea whether arachnids were plant or animal, he muddled through a positive re-sponse. Thereafter, "it was one of Mary's jobs to go through every piece of paper I had to deal with, and wherever there was a Latin taxonomic name she wrote in English what it was."[25] After 1962, "professional assistants" be-came assistant or associate program directors, depending on whether or not they held a Ph.D., and a few men began to be hired at the associate level. Despite the gender typing, the women found these positions professionally rewarding and well paid. Even the program secretaries (all female) who per-formed the clerical work felt they were part of an exciting venture.[26]

In addition to direct contacts with grantees, the staff communicated with the scientific community through advisory panels and the divisional committee. Borrowing from the practice at ONR, the BMS program direc-tors set up advisory panels, each consisting of approximately nine leading re-searchers, that met to consider proposals. Although the physical science di-vision also established "panels" in the 1950s, theirs did not usually meet together. As Wilson explained to the incoming Payne in September 1952, "You may be interested in knowing that the MPE Division handles all of their evaluation by mail, but we have felt rather strongly that the round-table discussion of proposals by Panel members is much more satisfactory."[27] In the 1960s, when program funds became tighter, BMS staff staunchly de-fended panel meetings against the efforts of physical scientists in the upper NSF hierarchy to eliminate them as an unnecessary expense. The panels pro-vided, in varying ways, a rough evaluation of each proposal and sometimes offered general advice. Program directors, guided but not bound by the panel and also by mail reviews, made the final funding decisions. For pro-posals other than typical investigator-initiated research, program directors could bypass the panels altogether and select scientists to review them by mail. When a grant was made, the program director negotiated the budget with the investigator, often providing less than the full amount requested.[28]

The staff also took much of the initiative in interactions with the divisional committee (see appendix B). After 1953, the divisional committee no longer passed judgment on dockets of recommended research projects. Rather, the staff arranged meetings two or three times a year and presented policy issues and new ideas for funding in the form of reports by staff or ad hoc committees of academic scientists. In the Waterman era these were on such topics as electron microscopy, systematic biology, basic research in agricultural experiment stations, controlled environment facilities, forestry, biological oceanography, and tropical biology. The divisional committee almost always supported the staff's current designs and future visions and, moreover, championed them with the board and the academic community. When, for example, Wilson sought to enlarge the scope of the project grant to include funds for training graduate students, the committee prepared resolutions to the director and NSB upholding Wilson in his resulting conflicts with the NSF education division and the NSF general counsel.[29]

The close and informal personal relations between staff, panels, and divisional committee operated in effect as a well-intentioned elite male network, albeit one that was more democratic in funding research than the private foundations that preceded them. Except for Libbie Henrietta Hyman on the systematic biology panel, no women or blacks served in BMS advisory groups from 1952 through 1969.[30] It is not surprising that, though some lip service was paid to the infrastructure of science, NSF patronage of biology in the 1950s emphasized funding the most "meritorious" science, which was generally presumed to be that carried out by white male scientists in major research universities, the predictable source of NSF's biology advisors.

Funding the Best Science: The Project Grant

The backbone of NSF support of biology was the investigator-initiated "project grant" for carrying out specified research. Each year the number of grants increased; by 1961, BMS was making about a thousand a year (see table 3.2). Awards varied greatly in amount and duration. Some researchers, especially those in small or obscure institutions, received only a few thousand dollars; by the end of the 1950s, leading biologists in major universities might receive $20,000 a year, occasionally over $30,000. The duration of awards ranged from one to five years; program directors tried to give the best investigators three- to five-year grants. Awards also differed significantly from program to program. In 1960, the average multiyear award in molecular biology was $46,400 while it was only $18,275 in systematic biology. In the Waterman era, the entire sum for a multiyear grant was reserved from the current year's budget.[31]

Although program directors typically held some grants over to be funded in the next fiscal year, they did not, like NIH officials, point to large numbers of "approved" grants they could not fund in order to seek higher budgets. In fact, they usually did not supply statistics on ratings of proposals in their annual program reports.[32] Typically they funded all proposals in the "highly meritorious" category and a portion of those rated "meritorious." Investigators with ratings of "acceptable" and above but nevertheless unlikely to be funded were invited to withdraw their proposals, presumably to save face with university administrators, rather than have their proposals declined.[33] Statistics usually combined withdrawals and declinations.

Compared to the 1970s and 1980s, the proportion of proposals funded to proposals received was high; in the post–Sputnik era, this figure was consistently over 50 percent, and as high as 62 percent in 1959 (see table 3.2). Program directors measured success by the proportion of funds granted to the total funds requested in proposals, the proportion of requested funds awarded to those receiving grants, and by the average length and size of grants. In perhaps the best year, FY 1959, Wilson reported that 40 percent of the total requested funds were awarded as well as 70 percent of funds requested by those selected for funding. But no matter how large their budget increases, program directors each year projected an increased "demand" for the following year and expressed concern that the funds available would not allow them to keep pace. They fueled that rising demand by encouraging many new types of grants—for field stations, instruments, facilities, research education. BMS staff assumed that the nation had an obligation to meet the proliferating needs of its biologists.[34]

Through the Waterman era, most academic biologists of note, except for those in medical schools and oriented toward medical research, were likely to have received NSF grants. It was not difficult for an established researcher to obtain continuous support from NSF. Although renewal requests were processed as if they were new, previous grantees were almost always successful (86% in 1959, for example). Each year renewal requests accounted for a greater percentage of BMS proposals, rising from 12 percent in 1956 to 45 percent in 1963.[35] NSF did not consider the amount of a grantee's funding from other sources in making its decisions; many good investigators in major universities, especially those in cellular and molecular biology, received simultaneous support from a number of federal agencies (see chapter 5).

Basic research grants were likely to pay for some permanent equipment, supplies, the support of a predoctoral or postdoctoral student, publication costs, and perhaps some travel. For FY 1956, Wilson noted that 20 percent of BMS grants paid the salaries of postdoctoral students, 75 percent funded predoctoral students, and 35 percent supported technicians (paid support staff who were not students).[36] Though some advisors and staff objected, NSF began early on to pay faculty salaries during the summer months so that, it was argued, impoverished scientists, especially those at small schools, would not have to teach or take another summer job to make ends meet. The same privilege, however, was also extended to better-paid scientists. BMS rarely paid any salary to principal investigators except during the summer, since staff and advisors generally felt that investigators' institutions were responsible for paying their faculty and conceived of an award as a "grant-in-aid" to assist universities in their research function, not to pay the full costs of research. Thus, some 30 percent of BMS awards in FY 1956 included summer salaries for principal investigators, but only 4 percent allowed for any portion of academic year support.[37]

NSF staff were proud of the implied mutual trust and simplicity of the Foundation's grant instrument—a short letter signed by Waterman—and the freedom it accorded to the investigator to alter the plan of research or shift funds among the various budget categories. The university business office provided accounting in semiannual one-page reports which simply listed the funds expended by broad category: salaries, permanent equipment, expendable supplies, travel. As long as all the money was eventually accounted for, no further questions need be asked. It was a dictum that the investigator knew best how to use his or her grant.[38]

In the post-Sputnik era, the size of the largest grants for project research in biology increased dramatically. When it appeared likely that the BMS budget would double in fiscal 1959, the divisional committee and staff agreed that priority should be given "to lengthening the period of grants and reducing the amount of budget trimming" rather than to increasing the number of grants and lowering the "qualitative standards of research."[39] In 1958, only one award was larger than $100,000; Nobel prize winner Fritz Lipmann of the Rockefeller Institute for Medical Research received a four-year award totaling $120,000. The following year, BMS made eight such awards. By 1963, BMS was awarding some forty six-figure research grants.

The largest grants for individual research were the "coherent area" grants, stupendous awards to subsidize an entire laboratory, which Wilson inaugurated in 1960 partially in response to Waterman's limitations on funding graduate training through departmental grants. The first went to University of Pennsylvania biophysicist Britton Chance, who received $700,000 for a five-year period, an astounding sum for an NSF grant, even in the 1990s. It supported his research "on the nature of the enzyme substrate complex and further exploration of techniques for studying energy transfer systems." The next year Fritz Lipmann, a long-time grantee of the Molecular Biology Program whom Consolazio especially admired, received a similar five-year grant for $750,000.[40] These two awards represented the apogee of the project grant, but there were numerous other large awards during the closing years of the Waterman era. Molecular biology made by far the greatest number of six-figure awards, followed by the Metabolic and Genetic Biology Programs. The fewest such awards were found in systematic biology.[41]

Those who received coherent area grants were likely to have worked them out through a long process of informal negotiation rather than asking for such sums out of the blue. As Wilson recalled, men like Chance "came in and quite rightly complained that they were running laboratories," not projects. They had "10 or 20 postdoctoral people, 15 or 20 Ph.D. candidates, and a few undergraduates, I suppose, and you couldn't write projects for each one of these guys, so why couldn't we underwrite the laboratory?" To find the right legal language, Wilson and BMS created the new "coherent area grant" concept and selected Chance, a long time friend of Waterman and Bronk's associate and successor at the Johnson Foundation, as an unimpeachable first recipient. Waterman, Wilson observed with appreciation, had a "great tolerance for ambiguity and would allow you to stretch limits."[42]

Beginning in 1959, NSF awarded a few grants to outstanding foreign scientists. The criteria, laid out by Waterman after several such awards had already been made, were far more stringent than for Americans. NSF was interested not only in the caliber of the researcher but also whether the institution provided unique facilities or scientific training to Americans. Among foreign scientists funded by BMS in the period 1959–63 were future Nobel laureates Jacques Monod and François Jacob, biochemists at the Institut Pasteur in Paris; future Nobelist R. R. Porter of St. Mary's Hospital, London; Marianne Grunberg-Manago of the Institut de Biologie Physico-Chimique of Paris; and Swedish biochemist Arne Tiselius.[43]

Although Waterman, the NSB, and staff agreed that the major test for any NSF award had to be scientific merit, they were under some pressure to consider other criteria. The issue of "distributive funding" was usually raised in the 1950s in two differently treated contexts: (1) the funding of "small colleges," and (2) the funding of "underdeveloped" or "developing" universities, especially those in states or regions where science was weak.

The BMS staff were generally sympathetic to funding biological research in high-quality small colleges, convinced by such studies as the Knapp-Goodrich report that these schools sent an inordinate number of students on to graduate school. They also realized that these institutions could not compete well with major universities for research funds. At its first meeting in January 1952, and from time to time thereafter, the divisional committee suggested that money should be set aside for small colleges. Without establishing a formal policy, BMS regularly funded researchers at high quality small schools, usually at low sums compared to those for large research universities. During Goodrich's term as a BMS program director, he helped organize a conference at Bryn Mawr "on the role of research in small college departments of biology."[44] In addition, Wilson sought means to give departmental awards to small colleges through his Psychobiology Program in order to train future experimental psychologists (see below).

The issue of "underdeveloped schools" was more controversial. Those who defended awarding grants strictly on merit argued in narrow terms that if funding were based on any other premise, NSF would have to decline excellent research in favor of funding second-rate projects. Though staff were inclined toward the merit criterion, they always paid some lip-service to what became familiarly known as the "G-factor" (for geography). Staff and panels were delighted to identify a good project in an out-of-the-way place. Program directors liked to say that it was policy to fund good research wherever it was located or, in the divisional committee's words, to "go along with the man, and not with the school."[45]

The NSF Act had directed the agency to avoid "undue concentration of funds," yet everyone realized that NSF practice further contributed to the concentration of resources in a relatively small number of states and institutions. It was common knowledge that most good research was located in the major research universities. As Levin explained in 1953, "The established institutions attract the best men who in turn submit the best proposals, attract about themselves the best young men, etc., in a vicious circle." He believed it important for "the National interest" to determine "to what extent it is desirable to deliberately stimulate research activity, even though perhaps it will be of second grade quality for a time, in institutions in order to establish new research centers where now only 'backward' ones exist."[46] Levin, who was later to supervise the Science Development Program, was more receptive to spreading support than either Consolazio or Wilson.

Although the staff calculated geographical distribution for grants awarded through December 1952, most program directors showed little concern for geography in their reporting for the remainder of the Waterman era. In fact, funds were becoming ever more concentrated. Of a total of $3.5 million in BMS awards in fiscal years 1952–54, the top ten states accounted for 62 percent, the top fifteen institutions 44 percent. By 1957, the number of institutions receiving some support from BMS had risen from 134 in 1952–54 to 177, and some funds went to all but one state. Yet, the top ten states received over 70 percent of the individual-project funds, and the top fifteen schools accounted for 49 percent of the total (see table 3.4).[47]

Table 3.4   Concentration of BMS Support by State and Institution,
FY 1952–FY 1954, FY 1957, and FY 1962

FY 1952–FY 1954	FY 1957	FY 1962
TOTAL OBLIGATIONS		
$3,500,000	$6,585,918[a]	$31,968,412[a]
STATES RECEIVING THE MOST FUNDING		
1. California	California	California
2. Massachusetts	New York	New York
3. Pennsylvania	Massachusetts	Massachusetts
4. New York	Wisconsin	Pennsylvania
5. Connecticut	Illinois	Illinois
6. Illinois	Pennsylvania	Indiana
7. Wisconsin	Connecticut	Wisconsin
8. Indiana	Michigan	Maryland
9. Missouri	Missouri	Connecticut
10. New Jersey	Indiana	Michigan
TOTAL OBLIGATIONS FOR LEADING STATES		
$2,158,130 (62%)	$4,686,543 (71%)	$20,406,657 (64%)
INSTITUTIONS RECEIVING THE MOST FUNDING		
1. Yale	Harvard	Harvard
2. UC Berkeley	U. of Wisconsin	UC Berkeley
3. Harvard	UC Berkeley	UCLA
4. U. of Wisconsin	Yale	Purdue
5. U. of Pennsylvania	Caltech	U. of Wisconsin
6. Caltech	UCLA	Johns Hopkins
7. Princeton	Cornell	Yale
8. Indiana U.	U. of Michigan	Columbia
9. U. of Iowa	Washington U.	U. of Illinois
10. U. of Illinois	Columbia	Western Reserve
11. NAS	NYU	U. of Michigan
12. Stanford	Duke	Dartmouth
13. U. of Chicago	U. of Pennsylvania	U. of Chicago
14. Duke	WHOI	Duke
15. U. of Michigan	Northwestern	Brandeis
TOTAL OBLIGATIONS FOR LEADING INSTITUTIONS		
$1,542,730 (44%)	$3,210,900 (49%)	$12,564,612 (39%)

Source: Calculated from lists of awards in *NSF Annual Reports,* FY 1952–FY 1954, FY 1957,
FY 1962.
[a]Excludes "General" (NCE grants)

Not until about 1960 was the issue of small and developing colleges forcefully raised again. Wilson defended his division's support as adequate although, in fact, grants to such schools represented a small portion of his budget. Of a sample of 900 grants, he pointed with satisfaction to 55 awards to small colleges and 112 to "developing universities" like those of Wyoming, Vermont, and Mississippi. He concluded, "To the extent that research funds are useful in assisting both small colleges and developing universities to enhance their status, we feel that the Division's research support program is making a reasonable and equitable contribution as compared to funds going for the same purpose to the larger, research-oriented institutions of higher learning."[48]

In contrast to their concern about small colleges and geographical distribution, no one in BMS openly worried about funding women or minorities. Although the number of women scientists grew during this period, women were even more marginalized than before 1945, as universities, in a time of expanding federal funds, sought to enlarge and upgrade their research capacity. Few administrators regarded women scientists as contributing to their relentless pursuit of prestige.[49] BMS did little or nothing to alter these prejudices. Women received only a tiny percentage of research awards. Even grantees at women's colleges were as likely to be men as women. African American scientists, just about all of whom were still segregated in black institutions, were only sporadically funded in this period. Here, too, some of the largest awards went to white male faculty.[50]

In one set of controversial experimental educational grants to undergraduate departments of psychology, however, it appears that Wilson deliberately supported female and black colleges. Justified on the rationale that such schools sent students on to graduate study, his awards included Fisk, Howard, and Bryn Mawr as well as several "underdeveloped" universities (see below). His panel summary sheet for a Sarah Lawrence grant awarded to two female psychologists in 1955 contained a rare written acknowledgment of negative attitudes toward women scientists. In this confidential internal document, Wilson recorded that "certain Panel members admit extremely strong bias against graduate training for female students."[51]

Thus, the tenor of research funding reflected cultural assumptions about where scientific merit would be found. From the hindsight of the more egalitarian 1970s, Wilson recalled NSF's first ten years as an exciting time with a strong sense of mission for supporting the best science. He allowed that "there was an element of elitism in the Foundation, which you now wouldn't admit. I would but the Foundation wouldn't. But in a sense it was in many ways the government counterpart of the Rockefeller Foundation."[52]

Stretching the Project Grant

Individual projects represented only one of a wide variety of BMS grants. Throughout the Waterman era, BMS staff and the divisional committee sought legal ways to stretch the project grant concept to cover other types of awards, including departmental and institutional support, most of the time with Waterman's strong encouragement. In the 1950s BMS funded conferences, travel, journals, instrumentation, operations of special facilities and agencies, facilities construction and renovation, surveys and data collection, and a host of educational projects aimed at everyone from high school students to seasoned investigators.

From its first meeting in January 1952, the divisional committee advocated more flexible forms of science support than could be accommodated in individual project grants. "Fluid research funds" administered "by responsible local committees or individuals in reputable institutions," the committee claimed, would assist in "maintaining a healthy balance between institutional financing of basic research and individual financing of the usual type."[53] Later that year the committee sent a formal resolution to the National Science Board "that a series of special five-year block grants be made to selected institutions in selected geographical areas as an experimental method of stimulating research in scientifically underdeveloped areas."[54] Here, as in other examples, the staff's strategy for introducing departmental or institutional awards was to appeal to distribution.

Waterman and the board discouraged "block grants" as likely to make the Foundation vulnerable to "political influence." Moreover, the general counsel of NSF questioned the legality of grants other than project grants. Despite these difficulties, the staff and divisional committee argued forcefully that the progress of biological research required a broader approach and found an informal means out of the dilemma. While opposing the awarding of "fluid research funds" as such, Waterman encouraged BMS to stretch the project grant, legitimating such awards on the grounds that they supported experimental or pilot projects rather than officially announced programs open to all institutions.

As program directors in biology expanded the scope of their funding, they ran into increasing opposition from other arms of the Foundation. Conflicts increased as budgets grew and NSF's organizational chart became more complex. The chief issue was whether BMS, advised at the panel and divisional levels by leading research biologists, ought to take charge of all support of biology, or whether different kinds of projects should be funded by NSF offices devoted to, say, publications, data collection, or education. According to BMS, these other offices lacked biological expertise and were not sufficiently attuned to the needs of the field. Wilson, Consolazio, and the divisional committee wanted to fund, or at least retain advisory status over, all aspects of biology and stubbornly fought limitations on what kind of projects they could consider. The wide variety of grants made in the category "Not Classified Elsewhere," Wilson claimed, "illustrate the Division's feeling that it should be able to support any legitimate and worthy need, however unorthodox."[55]

Conferences

One of the least problematic expanded forms of support was that of conferences. Especially desirable were conferences on an actively pursued but limited topic, with a small number of invited attendees, followed by publication of the proceedings. In its first few years, BMS funded conferences on such topics as photosynthesis, methods of determination of steroids, specificity in development, the role of proteins in the transport of ions across membranes, luminescent systems in biology, glutathione, the physiological development of the mammalian fetus, and fundamental problems of perception. In the 1950s, the division regularly subsidized such well-known series as the annual Growth Symposia of the Society for the Study of Development and Growth, the annual conference on quantitative biology at Cold Spring Harbor, New York, and the annual symposium on systematic biology at the Missouri Botanical Garden in St. Louis.[56]

Conferences could play a seminal role in giving momentum to preselected areas of emphasis. In 1951, before BMS adopted its functional organization plan, staff chose for special attention protein synthesis, enzymology, photosynthesis, and marine biology. To encourage research on photosynthesis, NSF began in 1952 to fund the National Academy of Sciences Committee on Photobiology, which in turn organized, with NSF assistance, two conferences on photosynthesis and one on bioluminescence.[57]

Sometimes program directors initiated conferences, seeking out biologists who would apply for grants to organize them. When Lou Levin attended the Laurentian Hormone Conference in 1952, he was struck by the weakness of methods for determination of steroids. He therefore took a convenient opportunity to discuss the matter with Gregory Pincus of the Worcester Foundation for Experimental Biology, a member of the regulatory biology panel. In effect, Levin solicited a proposal for a small invitational conference on methods for determination of steroids in blood and urine, which was held in 1953 under the auspices of the Laurentian Hormone Conference at the Worcester Foundation in Shrewsbury, Massachusetts.[58]

Occasionally conferences initiated by BMS led to the founding of new biological organizations. For example, in systematics, an area where NSF support was critical, staff brought together in 1957 and 1958 the directors of the major systematic collections. Out of these meetings developed the Conference of Directors of Systematic Collections, from which was founded in 1972 the more broadly based Association of Systematic Collections.[59] Through similar support of conferences, BMS played a major role in founding the Association for Tropical Biology in 1962 (see chapter 7).

BMS funded travel grants for a wide variety of international meetings, among the earliest being the second International Congress of Biochemistry in 1952 and international congresses on microbiology and genetics in 1953. Program directors resisted the attempt of the National Research Council to act as intermediary in making these awards, preferring to work directly with representatives of biological societies.[60] BMS also provided miscellaneous travel awards for research purposes; societies and individuals simply negotiated their needs with BMS program directors.

Publications

In the 1950s, BMS gave grants to a number of biological publications. *Biological Abstracts,* after surviving the withdrawal of its initial sponsor, the Rockefeller Foundation, in the late 1930s, was on its way to solvency when the war created a new financial crisis. As one of its first awards, BMS, relishing NSF's coordinating role, administered, on behalf of a group of federal agencies, an emergency grant of $65,000. The Foundation continued to fund *Biological Abstracts* until the publication became self-supporting in the mid-1960s.[61] BMS also aided a variety of primary journals, awarding, for example, startup funds for the *Journal of Limnology and Oceanography* and money to reduce publications lags in the *Journal of Experimental Psychology* and *Journal of Comparative and Physiological Psychology.* In addition, it funded proceedings of conferences and multivolume monographs, particularly in systematic biology.[62]

Operational Expenses, Equipment, Construction

Wilson was especially eager to stretch the project grant in the direction of funding laboratory and other facilities. Well before BMS began a formal facilities program in 1957, it supported scientific equipment purchases, collections of biological stocks, operational expenses, and even construction costs in the guise of project grants. Equipment was purchased for the Marine Biological Laboratory and other biological stations and was included in

Wilson's grants to undergraduate psychology departments. In 1955, the division awarded its first electron microscope, to the Department of Bacteriology at Indiana University. From 1957 on, BMS grants purchased a number of electron microscopes and such other instruments as an analytical ultracentrifuge, an infrared spectrophotometer, a mass spectrometer, and an amino acid analyzer.

BMS funded collections of biological stocks, conceived as national resources for biologists, as early as 1952 when Paul Burkholder won $10,000 for the development of a "National Culture Collection of Algae." The American Type Culture Collection, which maintained bacterial cultures for distribution for research, received its first of many NSF grants in 1954 to collect bacteriophages. In 1957, the division formed a Committee on Genetic Stocks under the aegis of the American Institute of Biological Sciences to study the general problem of forming national collections of organisms with known genetic characteristics for use in research. From 1958 on BMS provided continuous funds to such collections as Drosophila at the California Institute of Technology and mice at the Jackson Memorial Laboratory in Bar Harbor, Maine.[63]

Research monies on several early occasions supported operational expenses of biological organizations. In 1952, BMS funded the National Academy of Science/National Research Council's Pacific Science Board and Committee on Photobiology. BMS contributed along with other agencies to the Biological Sciences Information Exchange, which collected abstracts and data on grants awarded by private and public agencies. And provision of interim operational support for AIBS in 1955 enabled the institute to sever its ties to the NRC and become an independent organization (see chapter 4).[64]

With its 1950s grants for temporary operational expenses NSF played a crucial role in helping selected biological facilities shift to large-scale funding by other agencies in the 1960s. In 1955 the BMS Psychobiology Program awarded $120,000, then a very large sum, for three years' general support of the Yerkes Laboratory of Primate Biology in Orange Park, Florida, later the Yerkes Regional Primate Center. Founded in 1930 by Yale psychologist Robert M. Yerkes, the laboratory was owned by Yale and supported by the Rockefeller and Carnegie Foundations. By 1955, Yerkes had retired and Yale had lost interest in the laboratory, then under the charge of Henry W. Nissen. In 1958, though disappointed with research progress made under the first grant, BMS provided another $40,000 to Emory University to aid a transition to new sponsorship, and the following year Emory formally acquired the facility from Yale for one dollar. Once it survived this difficult period, the laboratory flourished again as one of the regional primate centers established in 1961 and supported by NIH. In 1964, the laboratory moved to Atlanta with the aid of a large institutional grant from the National Heart Institute.[65]

In collaboration with the Office of Naval Research and the Rockefeller Foundation, NSF began in 1952 to fund operational expenses of the White Mountain Research Station of the University of California. Located near Bishop, California, the station was formally established in 1950 under the directorship of Nello Pace of Berkeley as a site devoted primarily to studies of high-altitude physiology. Its initial support had come from Consolazio at ONR who, from his Harvard Fatigue Laboratory experience, was thoroughly familiar with the concerns of high-altitude physiology. After a site visit, he approved in 1950 a proposal to build the station's Barcroft Laboratory at an altitude of 12,470 feet. Consolazio's transfer to NSF in 1951 seems to have led to Foundation involvement. In May 1952, despite the somewhat skeptical divisional committee, he developed a grant of $32,800 as NSF's contribution to operational support. Later that month, he and Levin, who was then still at ONR, convinced the Rockefeller Foundation's Warren Weaver to join ONR and NSF in funding the station, which Consolazio hoped would eventually become a national biological facility. In 1955 a second BMS grant of $55,000 allowed for road extension and construction of the small stone Summit Laboratory at 14,250 feet. A third grant of $132,000 in 1957 paid for an electric power line to the laboratory sites. NSF continued to subsidize operations until the mid-1960s, by which time the National Aeronautics and Space Administration had assumed the major share of support of this unique and productive facility.[66]

Similarly NSF support of the Barro Colorado Island Biological Laboratory aided that facility in a period of transition from a marginal institution to a major center for tropical biology and evolutionary ecology. Founded in 1924 by an umbrella organization known as the Institute for Research in Tropical America, the laboratory was located on an island in Gatun Lake in the Panama Canal Zone. A relatively primitive facility, it operated on a shoestring budget, even after the Smithsonian Institution took it over just after World War II. A BMS grant of $29,000 for operational expenses in 1954 represented a substantial increase in the laboratory's budget. In addition, BMS awarded research grants to two regular users, Theodore Schnierla of the American Museum of Natural History and Robert K. Enders of Swarthmore College. Just after the period of the BMS operational grant, which coincided with a key change of laboratory leadership, public and scientific interest in tropical biology burgeoned and the Smithsonian, whose total budget was rising rapidly, was finally able to provide large-scale support.[67]

By way of contrast, the LOBUND Institute provides a revealing example of staff initiative applied to *not* funding a controversial facility in the face of considerable outside advocacy. The Laboratories of Bacteriology of the University of Notre Dame (LOBUND) was a unique, controlled-environment facility that produced "germfree" (completely sterile) animals (mammals, especially rats, and chickens) for use in biological and medical research by staff and visiting scientists. Since the 1930s, LOBUND had been under the charge of James A. Reyniers, who had developed its techniques and instrumentation and who served on the first National Science Board. In an ultimately unsuccessful attempt to prevent the establishment of competing facilities, Reyniers generated ambitious plans for expansion, including construction of a new laboratory building, which he hoped would be funded by the federal government. The excitable Reyniers and the indomitable Father Theodore Hesburgh, president of Notre Dame, who replaced Reyniers on the NSB in 1954, leaned on BMS staff and also communicated directly with Waterman and members of Congress in support of LOBUND. Moreover, ONR and NIH, at interagency meetings held to discuss the future of germfree research and LOBUND, also urged BMS to take major responsibility for operational costs and new construction.[68]

The BMS staff, especially Assistant Director Blinks, Levin, and Wilson, exercised considerable skill in parrying the pressure to provide funds. They were skeptical of LOBUND from the beginning because its research productivity did not seem to match its ambitions. Matters came to a head in the mid-1950s. In February 1955, Hesburgh laid out a grandiose plan for the enlargement of the LOBUND Institute and invited federal agencies to "collaborate" in supporting it. Blinks, who with Levin and other government representatives visited Notre Dame, decided to delve behind the formal presentation and made discreet inquiries of former employees and others familiar with operations. In June 1955, he detailed to Waterman a long series of problems ranging from high staff turnover to serious doubt as to whether the animals were truly germfree. Most disturbing to Blinks, however, was the lack of an academic atmosphere, especially the absence of students being trained. It appeared to Blinks as though Reyniers were trying to protect proprietary rights as long as possible.

To buttress their position, the BMS staff arranged for the backing of leading microbiologists, convening in late 1955 an Ad Hoc Advisory Committee on Microbiological Facilities chaired by Conrad Elvehjem of the University of Wisconsin. After four meetings, including a visit to Notre Dame, the blue ribbon committee submitted a damaging critique of LOBUND which was distributed to board members, the House Appropriations Committee, and NIH.[69] Levin asked the divisional committee for a formal endorsement of the report. Reyniers and Hesburgh raised objections in a barrage of communications. Undeterred, in 1956, Hesburgh submitted a new proposal for a laboratory building at a cost of $987,000. The staff was well prepared, giving the same divisional committee, now acting as panel for special facilities proposals, the task of evaluating the new request. Unconvinced by Hesburgh's plans to reorganize LOBUND and inaugurate graduate training, the committee firmly declined LOBUND, using the same criticisms made in the Elvehjem report. The following year, at Hesburgh's request, Reyniers left Notre Dame.[70]

By the time BMS formally announced a facilities program in FY 1957, individual program directors had already informally given some twenty-five facilities-type awards.[71] In 1957, BMS received a budget line item of $1 million for facilities which was, in effect, a compensation to biologists for the "big science" facilities NSF was building for the physical sciences, notably the National Radio Astronomy Observatory at Green Bank, West Virginia, and the Kitt Peak National Observatory in Arizona. For the next two years, the BMS Divisional Committee examined facility proposals as a panel, at least one member participating in the required site visits of staff and scientist teams.

Major facilities grants in this period included BMS's first grant for curatorial support of systematic biology collections awarded to the Academy of Natural Sciences of Philadelphia, a five-year grant of $60,000 a year for curatorial support and improvement of facilities at Harvard's Museum of Comparative Zoology, and two major grants to the Marine Biological Laboratory (MBL). The first, for $415,000, funded the complete modernization of Crane Wing, one of MBL's two main brick laboratory buildings built in the early twentieth century. The second, for $544,250, funded a quarter of the cost of erecting the new Whitman Laboratory. (The Rockefeller Foundation contributed half and NIH the remainder.) It also paid the entire cost of the MBL's first group of 25 rental cottages. This latter controversial expense, which Wilson justified by the need to eliminate barriers for young researchers with families who could not afford "resort housing prices," illustrated the division's strong conviction that any type of award that advanced biology was legitimate.[72]

In 1959, Wilson set up a separate program and advisory panel for facilities and recruited Harve J. Carlson, former scientific director of ONR's San Francisco Office, to run it. After several years of lobbying by the BMS staff and divisional committee, the Foundation began in FY 1960 a program for the building and renovation of graduate research laboratories to complement the Special Facilities Program (see chapter 6).[73]

Data Collection

Through the 1950s BMS organized its own surveys of the state of biology and biological funding. In the early years, as part of an attempt to contribute to "national policy," it sponsored extensive projects by the American Physiological Society and the American Psychological Association to survey the status of these disciplines (see chapter 4). Many of the staff reports presented to the divisional committee, such as that on agricultural experiment stations, were based on extended travel, interviews, and data collection (see chapter 5). But the largest data collection effort in the 1950s was the division's own annual survey of federal funding in the life sciences.

Although the Foundation was publishing general reports on federal research activities and BMS was contributing to the support of the Biological Sciences Information Exchange (a clearinghouse located at the Smithsonian for information on individual grants), BMS staff felt the need for data according to its own format. Each year, from 1952 through 1958, the division compiled and published a list of all extramural grants in the life sciences by federal agencies broken down into categories (the BMS programs plus others for applied biology). Designed for limited distribution, the reports, published as *Federal Grants and Contracts for Unclassified Research in the Life Sciences,* enabled science managers throughout the government to determine funding patterns in biology by agency, institution, or area of biology and also sources of funding for any given scientist. A separate series handled grants and contracts in psychology.[74]

Jurisdictional Conflicts: Funding "Research Education"

It was in science education that Wilson and the BMS staff ran into the greatest resistance in their quest to expand the project grant. Science education was supposedly the province of the Division of Scientific Personnel and Education (SPE), created by the NSF Act to manage fellowship programs. Yet because BMS staff believed that education should not be separated from research, they were unwilling to leave education in the life sciences in SPE's hands. They extensively funded educational projects in biology in the 1950s, claiming jurisdiction over "research education."

SPE impinged most directly on BMS's sphere of concern through its predoctoral and postdoctoral fellowship programs, begun in FY 1952. Awarded on merit, NSF's fellowships were evaluated by the National Research Council, which had administered fellowships for the Rockefeller Foundation since 1919. Although NSF funded the research divisions on a par with each other in the early years, the fellowship program was from the beginning heavily weighted toward the physical sciences, which provided a much larger number of applicants. In FY 1953, 813 life scientists applied for predoctoral fellowships versus 2117 physical scientists, of which 150 (18%) and 365 (17%) were selected respectively. For postdoctoral fellowships, of 191 applicants in the life sciences and 177 in the physical sciences, 18 and 24 were selected respectively.[75] The disparity continued to widen in the 1950s. By 1959, NSF was awarding 216 predoctoral and 113 postdoctoral fellowships in the life sciences, compared to 866 and 158 in the physical sciences.[76]

Although BMS program directors could not award fellowships, they, too, supported graduate and postdoctoral students through the medium of project grants. Almost all research awards paid salaries of postdoctoral research associates and research assistants, usually graduate students, and it was assumed that a sizeable portion of regular grant funds would support thesis research undertaken in the faculty advisor/principal investigator's laboratory. By 1961, research grants were supporting 985 predoctoral students and 213 postdoctorates. These figures, always a matter for boasting in annual reports, were, as Wilson had earlier pointed out, much higher than those representing the education division's fellowship program. Consolazio predicted as early as 1954 that over the years the research divisions would play a far greater role in training scientists through project grants than the more "bureaucratic" and, in his opinion, ineffective SPE.[77]

Relations with SPE over funding any other forms of educational projects were sensitive. As early as 1954, H. Burr Steinbach tried to set out respective boundaries based on spirited discussions with SPE. Competitive fellowships and strictly educational projects such as conferences on teaching were the province of SPE, he wrote, while proposals for advanced training limited to biology "gave a presumption of primary interest by B&M." Agreeing that SPE's general program should "not be infringed upon," Steinbach concluded, "it is equally important that both MP&E and B&M retain primary cognizance over all matters relating to the special fields within their respective Divisions."[78]

One BMS vision that never materialized due to SPE opposition was that of funding "a type of senior investigatorship." Wilson first suggested this means of broadening the grants program in 1956, stimulated by an unusual proposal from physicist-turned-biologist Leo Szilard. This request for $106,000 for five years was submitted jointly by Caltech, Chicago, the University of Colorado School of Medicine, the New York University College of Medicine, and the Rockefeller Institute for Medical Research to cover salary and travel expenses for Szilard to conduct collaborative research in theoretical and quantitative biology at each institution. Consolazio had hoped to fund the innovative proposal but, to his disappointment, the divisional committee, which disapproved of multiple weak institutional affiliations, did not think such an award would increase Szilard's productivity. In Consolazio's interpretation, the committee "seemed to lack faith in an individual left to his own resources and to feel that theoretical biology had no place yet in the biological sciences." As a result of the deliberations, however, the committee placed itself on record as favoring career investigatorships for up to five years as a possible means of encouraging "theoretical biology."[79] But, in part because the idea competed with SPE's one-year, senior postdoctoral fellowships, BMS was never able to make this form of award.

Over the next several years, BMS experimented with a number of types of educational grants, often not publicly announcing them or, as in the cases of summer training at biological field stations and medical student grants, declaring them only several years after giving the first money. Here, as with other experimental awards, grants were hammered out by informal negotiations between program officers and grantees. Since many initially fell under NCE, they were evaluated not by panels but solely by mail reviewers whom program directors chose for the purpose.

Some of the earliest and least controversial of BMS education awards went to marine laboratories and inland field stations. Such institutions, Lawrence Blinks wrote, were "the equivalent of observatories, cyclotrons, atomic piles, computers, and such large facilities needed by the physical sciences."[80] In contrast to investigator-initiated research grants, those to biological stations, begun in 1952, were "block grants" for support of graduate students and independent investigators using the facilities. The stations chose the scholars, sometimes by running their own advertised competitions. In its first two years of grants, BMS supported tables (places for researchers) at the Naples Zoological Station and summer research and education at the Marine Biological Laboratory and the Hopkins Marine Station. These laboratories were special cases because their predominant research was physiological, biochemical, and biophysical rather than ecological. The Environmental Biology Program handled other stations. By the summer of 1959, this program was sponsoring over 200 participants in summer work at nine biological stations located from Puget Sound to Bermuda.[81]

Among other educational projects BMS supported were several sponsored by the American Physiological Society, one of the few biological societies of the 1950s to have a paid executive secretary backing an ambitious Education Committee. In 1955, NSF funded the conference "Teaching of College Physiology" at Storrs, Connecticut, and the following year awarded a three-year grant for summer workshops. Yet another grant to APS in 1957 enabled teachers of physiology in small colleges to undertake summer research at established laboratories.[82] BMS also funded special courses in advanced research techniques. In 1956 BMS began to support the Tissue Culture Association's annual intensive summer course in techniques of tissue culture for "mature investigators" at the University of Denver. The following year, on the strength of a staff study on ways to promote electron microscopy, BMS began funding a course in electron microscopy for research biologists to be given at Cornell every other summer.[83]

Given the presence of NIH, it is somewhat surprising to find NSF creating a special program in 1957 to fund summer research by medical students—its first and only series of awards directed solely at medical schools. The impetus came from the recently merged Divisional Committee for Biological and Medical Sciences which still included a number of members of the former medical research committee. These grants, too, began as an experiment. One had already been made in FY 1954 to Washington University and two more went to Wisconsin and Minnesota in FY 1955. That year E. W. Goodpasture of the BMS Divisional Committee carried out a preliminary study of this form of award, and Levin prepared a staff study based on visits to medical schools. After a year of discussion, George Wald, on behalf of the divisional committee, officially wrote to Waterman in favor of establishing such awards. He argued that, with few exceptions, "the most able persons now being trained in biology" went on to medical school, where, after their preclinical training, they constituted "perhaps the most valuable potential for biological research that the nation possesses." Yet, "almost the whole of this training, oriented as it is toward medical practice, is lost for the advancement of science." The purpose of the NSF program was to give medical students research experience so that they might later pursue research careers. Awards were to be given in the form of block grants to special committees in the medical schools who would then select the students to receive stipends.[84]

Six more awards were made in 1956 before BMS sent a formal announcement to medical school deans and organized an advisory panel in 1957. The "demand" dwarfed the supply. BMS received proposals totaling $1.8 million, when the funds that had been allocated were only $150,000. Most grants were for three years and ranged from $1,380 to $10,800 annually. Because no overhead was allowed, a few schools declined the awards. From 1954 to 1961, 81 of the 85 U.S. medical schools applied for grants and 70 received them. In all, BMS gave 98 grants, providing an estimated 2,150 stipends for a total of over $1.2 million (see table 3.5).[85] While SPE did not openly contest these grants, Waterman questioned from the beginning whether they were not more appropriate for NIH. For its part, NIH objected to NSF's running a program directed solely at medical schools.[86]

Table 3.5   BMS Support of Short-term
Research by Medical Students, Grants
to Medical Schools, FY 1954–FY 1961

Fiscal Year	Number of Grants	Total Amount
1954	1	$ 6,900
1955	2	20,700
1956	5	43,950
1957	16	188,400
1958	17	200,400
1959	19	207,000
1960	18	295,560
1961	20	247,680
Total	98	$1,210,590

Source: Calculated from lists of awards in *NSF Annual Reports,* FY 1954–FY 1961.

Waterman and the board viewed awards to field stations and medical school deans, though clearly block grants, as special cases. Wilson reached a roadblock, however, when he tried to provide funds to undergraduate and graduate departments expressly for training students. Though BMS made a number of such grants—all justified as experimental awards—no permanent programs could be established. Wilson's earliest effort in this direction was his initiative to fund training and research in undergraduate departments of psychology. Already, in FY 1954, a five-year BMS grant to the department of psychology at Swarthmore provided graduate research assistants to faculty members. That fall, Wilson's assistant Marguerite Young conducted a staff study of faculty qualifications and research productivity of small college psychology departments, in response to which the psychobiology panel recommended "that proposals be encouraged from departments in smaller institutions as well as from those in under-supported institutions." The awards would aid departments that were attracting good students for "sub-doctoral" work and successfully "feeding" them to larger graduate schools. BMS sent out no formal announcement, then or ever, but by March 1955 had received four proposals and selected three for funding.[87]

In the next three years, Wilson funded such schools as Bryn Mawr, Fisk, Howard, Wichita State, and the University of Nevada, to purchase equipment and laboratory assistance. For example, a 1955 grant to Sarah Lawrence was intended to introduce more experimental psychology into a curriculum that had hitherto emphasized personality, child development, and teacher training. The grant paid for an undergraduate assistant, remodeling an office and building an animal room, and purchasing equipment including a Skinner box, control box, and recorder. As Wilson recalled, he had made a number of similar grants before NSF's general counsel "blew the whistle, and said they were not within the legal authority of the agency." They were, "pure and simple department grants," Wilson admitted. "We didn't call them that, but we were stopped."[88]

In 1956, the BMS staff experimentally extended the concept of short-term research for medical students to undergraduate biology students. In late 1955 they made awards to Harvard and Reed College to support summer research by undergraduates.[89] The following year, they made similar grants to Johns Hopkins and Carleton College, this time also involving high school students. The grant to Hopkins, for example, included funds for high-school students and teachers to discuss science-fair projects with university faculty. Waterman, who had no doubt that experimental grants such as these would "succeed," given the quality of the recipient institutions, nevertheless worried that such grants were "dangerous," and set a precedent unwarranted until the board and Congress had given approval of direct aid to departments and institutions.[90]

Perhaps most controversial of all Wilson's educational grants was his attempt in 1957 to fund departmental "training grants" for graduate students, similar to those that NIH had so successfully awarded. BMS made its pilot award to George Beadle, chairman of the Division of Biology at the California Institute of Technology, providing stipends for training graduate students in biology. Wilson argued that biology at Caltech was a special case because Beadle strongly disapproved of students doing thesis research on faculty research grants.[91]

Harry C. Kelly and Bowen C. Dees of NSF's education division vigorously opposed the Caltech award, claiming that despite deliberate changes in the proposal language, "the 'research stipends' requested are essentially locally administered graduate fellowships, differing from those now awarded in our national competitive program in only one respect—the fact that they are awarded at the discretion of the Biology Division of Caltech." Moreover, they argued, such grants, if made only to schools of the caliber of Caltech, would open the Foundation to further criticism of concentrating support in a small number of schools.[92] Although the board approved the Caltech grant, Waterman decided to lay down the law in 1957 and prohibit Wilson from offering further departmental training grants. He ruled as outside the scope of Foundation operating policies any "block grants" not for specific research as well any research training grants except for medical student and field station awards. Moreover, if proposals were "primarily intended to support education or training," they were henceforth to be under SPE jurisdiction.[93]

Apart from the legal issues, the general counsel and Waterman were wary of departmental grants because they could invite political interference. Waterman told the BMS Divisional Committee in 1957 that he and the board felt that any extension of research support "could best take the direction of broadening research grants in contrast to providing support to departments or other subdivisions of institutions." Departments, he pointed out, were difficult to evaluate because members varied in competence. To depart from the practice of peer review for individual projects would "put the Foundation in a less tenable position by introducing factors other than the merit of the work being proposed."[94] When the divisional committee questioned General Counsel William Hoff as to why medical student grants could not be extended to graduate students, he offered the egalitarian argument that that "might involve some 1700 different requests coming in from various schools and colleges." The divisional committee gave the elitist response that if proper screening procedures were used to select according to the competence of the scientific staff, the number of eligible graduate departments would be no larger than the number of medical schools.[95]

Wilson, with the backing of the divisional committee, resisted Waterman's decision. SPE had not renewed a number of BMS grants transferred to it, including the Caltech training grant. Wilson wrote in his 1958 annual report to the director, "As we have stated many times, we are vigorous in our belief that this type of grant is almost foolproof in providing a profitable return to science for the small amount of funds invested. We feel that the Foundation is losing a precious opportunity to exert a most beneficial influence on the future of the biological sciences in the United States by its current policy of mitigating against this type of grant."[96] The divisional committee, at its meeting in March 1958, formulated a resolution to the director and board recommending institutional grants for graduate training, "to be used for the stipends and research expenses of grad students selected by the institutions for these awards." They further proposed that "since graduate work is intimately and inextricably related to research," these awards should be handled by the research divisions. Wilson reported to Waterman the committee's concern that the grants docket had become too narrow in scope as a result of policies inhibiting the division's characteristic "flexible trial-and-error experimentation with a wide variety of grants."[97]

When SPE presented its own plan for "training grants" in 1958, the BMS Divisional Committee voiced vigorous opposition. The new awards were intended to counter the criticism that the Foundation's fellowship program tended to increase the concentration of graduate students in the top universities. Although NSF had tried to select fellows from home states throughout the union, the awardees were then free to choose their graduate schools and overwhelmingly selected from among a small number of institutions. Kelly's post-Sputnik plan for a "cooperative fellowship program," was designed to bring about a greater geographical distribution of graduate schools in which NSF fellows were trained. Students were to apply through the graduate institution of their choice, and after an initial screening by the institution, the Foundation was to select the final awardees, with consideration to the distribution of schools selected.

On the basis of experience with NIH, in which training grants were made to departments which then selected their graduate students, several members of the divisional committee avowed that Kelly's proposed "type of administration would be highly unacceptable to the biologists in the country and might lead to a considerable loss of prestige of the whole Foundation program." Such awards, they claimed, abrogated the rights of institutions to select their own graduate students "without having to feed information through a committee sitting in Washington." Steinbach, as chairman of the divisional committee, voiced these objections to the National Science Board.[98] Nevertheless, the program was begun on a large scale in FY 1959, when the Foundation made 1,050 awards, about a quarter of them (238) in the life sciences.[99]

As a result of pressure from Wilson and the divisional committee, Waterman relented somewhat in 1959 and allowed BMS to make grants strictly for thesis research or training grants in a specific or "coherent area of science," especially in new and critical areas not represented by fellowships. Such grants could be considered extensions of project grants and judged on the merit of the principal investigator's research. Grants to departments or institutions, however, were to be avoided, since departments typically aimed at diversification instead of a single coherent research program.[100] While considering these options a "partial substitute," the divisional committee still contended that NSF "in effect has lost face in comparison with the National Institutes of Health with respect to training grant programs."[101] Wilson immediately proposed a block grant for thesis research under University of Chicago population ecologist Thomas Park. Dees and Kelly contested this form of award also, not only because it provided "locally administered fellowships," but also because they disapproved of students working as paid employees on directed research instead of carrying out independent dissertation research. Waterman allowed the award to Park, but only after Wilson omitted tuition payments for the students, so that the awards would be less like fellowships.[102] Another response to Waterman's restrictions on grant-making was Wilson's inauguration in 1960 of the "coherent area" grant—a large enough "project grant" to support all the graduate students and postdoctorates in a sizeable laboratory.

The Constraints of Growth

As NSF grew in size and complexity, BMS was gradually forced to relinquish its broad range of projects. By Waterman's decisions from 1957 to 1960, surveys were transferred to the General Survey Office (thus terminating the surveys of biological grants), BMS grants for departmental laboratory improvement went, after 1960, to the new Office of Institutional Programs, journals and monographs were to be funded by the Office of Scientific Information Service, and most educational projects in biology were placed under the control of SPE.[103] Moreover, a new NSF-sponsored Antarctic Biological Research Laboratory, opened in 1959, was entirely outside of BMS jurisdiction; a separate Antarctic Research Program handled those research projects, with advice from the NRC's Committee on Polar Research.[104]

Wilson complained vehemently against NSF's tendency to establish new offices with "increasingly specialized functions" that bore "little or no relation to substantive content." As he made the case, the Foundation's most important task in its "support of science" responsibilities was that of "maintaining effective and helpful relationships with the various science communities." To do that, NSF should strengthen and consolidate the role of the scientific program director, "who is viewed by the working scientist as the individual to approach in connection with any question in this field," not have it "progressively weakened by splintering responsibility between numerous offices within the Foundation."[105] As Consolazio put it, "There are more activities, more pigeon holes and more barriers."[106]

Despite the setbacks to his experimentation with types of grants, Wilson continued to generate plans, based on the 1950s best-science ideology, to increase program director discretion, and grant flexibility. In 1960, in order to "streamline" operations, he encouraged program directors to take direct action (or perhaps telephone a few panel members) "in cases of small grants and in cases of scientists whose stature is unquestioned," rather than submit the proposals to panel review. He experimented with "logistic" awards to combine funding of operational support of a specialized laboratory and its research projects. The following year he unsuccessfully proposed "writing base grants or contracts with the 25 or so major research institutions" such that individual grants could be charged against these funds, so money could be available when needed.[107]

By the end of the Waterman era, BMS had forfeited a large share of its coveted freedom to fund any worthy project related to biology. Yet what BMS lost in breadth in the 1960s, it could make up in size, with million-dollar institutional grants and major international biological ventures in oceanography and ecology. There remained ample scope for imaginative and expansive thinking (see chapters 6–8).

The opportunities and limitations of the biological program at NSF can only be understood within the context of a complex institutional environment. NSF and BMS made their decisions within a framework of other federal agencies and private institutions in Washington, most notably the Bureau of the Budget, the White House, the U.S. Congress, competing federal funding agencies, and the National Academy of Sciences. NSF was only one of several agencies supporting biology in the 1950s; it was never the largest, and after 1959, its share of funding continued to decline as that of NIH continued to grow. Other agencies not only provided models for emulation or criticism but also increasingly challenged NSF's ability to function effectively as a patron of biology. NSF's role as one small part of a pluralistic system of federal support of science is the subject of the next two chapters.

CHAPTER 4

Government Relations and Policy-making in the Cold War Era

The post–World War II federal system for support of science in America was characterized by pluralism. A multitude of agencies, their appropriations overseen by a multitude of congressional subcommittees, offered the government advice on scientific matters and funded basic and applied scientific research. Despite determined efforts, the Bureau of the Budget (BOB) managed little effective coordination of federal grant-making to American colleges and universities, as agencies, jealous of their prerogatives, resisted all attempts to delimit their scope. Scientists, for their part, preferred this looseness because it meant more funding choices and fewer opportunities for federal control of science. And despite recurrent legislative proposals for a Department of Science or other measures to centralize science funding, Congress, too, ultimately favored the pluralistic structure.[1]

NSF was only one player—and a small one—in a league of powerful agencies supporting science. Claiming a unique place as the only agency supporting basic research in all the sciences, NSF was, paradoxically, primary in only a few areas. Historically, other granting agencies could extract funds from Congress for basic research by invoking the critical nature of their applied missions more easily than NSF could make the case for basic research on its merits alone. Though BMS—and NSF—officially approved of mission agencies supporting relevant basic research, they hoped, in vain, that NSF would gradually assume a larger share of it. NSF's biology program, although it grew rapidly as part of the post-Sputnik across-the-board expansion of NSF, remained only a small part of the federal commitment to the biological, biomedical, and agricultural sciences.

The BOB, in its campaign to coordinate and rationalize the programs of federal agencies supporting science, expected NSF to actively assist the presidential administration by exercising its statutory authority to make policy for science and evaluate programs of other granting agencies. Lou Levin, whose task it was to survey the biological research of other agencies, urged NSF to take steps "to correct the present uncoordinated and chaotic sponsorship of scientific research" in which "each agency operates, as it were, in a 'vacuum['] and considers only its own real and fancied needs."[2] But Waterman, fearing repercussions from much larger and stronger agencies, understandably avoided a substantive effort to coordinate federally supported science. BMS staff led in interpreting NSF's evaluative role as supporting self-studies of scientific disciplines, holding conferences, and collecting data on federal research. These innocuous measures raised few antagonisms.

This chapter also explores three episodes of more controversial NSF policy-making related to biology: NSF's one forced foray into evaluating the operations of another agency, NIH in 1955, did not succeed in controlling the power of NIH lobbyists, as the BOB hoped, nor in preventing NIH's expansion into general biology. However, NSF, coaxed by BMS program directors, did make a significant policy decision in the 1950s when it belatedly but publicly fought the McCarthy era loyalty restrictions that had resulted in blacklisting of biologists by NIH. If BMS program directors were usually reluctant to make policy decisions concerning federal support of biology, they were jealous of their right to do so when challenged by a rival source of authority. Thus they sabotaged the attempt of the National Research Council's Division of Biology and Agriculture to create a central policy-making Biology Council. This episode reveals both the strengths and the weaknesses of the pluralistic system of federal funding of biology in the 1950s.

Promoting NSF Biology before Congress

Effective programs in biology grew not just from their managers' and advisors' imaginations or ability but from the resources available to them. What the NSF Division of Biological and Medical Sciences could accomplish as a federal patron depended on the drawn-out annual budget process, which provided NSF's primary relationship with the administration and Congress in the 1950s. Each year the National Science Board approved a budget for submission to the Bureau of the Budget, which oversaw and attempted to coordinate the president's overall budget. The Bureau "marked up," and typically reduced, the agency's request by line item. Through sharp administrators like William D. Carey and Hugh Loweth, BOB wielded considerable power over funding science.[3]

After Congress received the president's budget, NSF faced its most difficult hurdles: the House and Senate hearings. The budget went first to the Independent Offices Subcommittee of the House Appropriations Committee, chaired by Albert Thomas of Texas. Hearings at which Waterman, the NSB chairman, and other staff and board members made statements and answered questions were held early in the calendar year. Usually the assistant director for BMS or his deputy was present, but Waterman handled most of the testimony. "I always felt like you were going through your Ph.D. orals when you went up before Mr. Thomas," John Wilson recalled.[4] In most years the House reduced the NSF budget significantly. After another, briefer hearing before the more sympathetic Senate subcommittee, chaired for most of the period by NSF's old friend Warren G. Magnuson, the Senate restored some of the funds. The final amount, determined by the conference committee, was usually somewhere between the House and Senate figures, but inevitably lower than the president's budget had requested. For FY 1956, for example, the Foundation submitted to the BOB a budget of $24.7 million, of which $20 million appeared in the president's budget. The House subcommittee recommended only $12.3 million. Waterman convinced the Senate to restore the full $20 million. The conference committee agreed on a final appropriation of $16 million, only 65 percent of NSF's initial request.[5]

Waterman was perhaps the federal government's most eloquent spokesman for basic research. Yet, ever the realist, he recognized as early as 1951 that "support of very fundamental science with no reference to its possible application would probably result in an ever diminishing budget."[6] So Waterman argued that there was an unhealthy imbalance between the amount spent on basic research and that spent on applied research and development. Like Bush, he insisted that basic science was an investment in a stockpile of knowledge that would ultimately lead to practical benefits in military security, health, and agricultural and industrial progress. Neglecting this investment jeopardized future dividends. No one could predict in advance where the next breakthrough would occur. This ideology of basic research, which presumed a one-way causal chain from basic science to applied science to technological advances, appeared year after year in both hearings and published reports directed to Congress and the public.[7]

Waterman typically measured the adequacy of NSF's support of basic research not by the number of grants the Foundation made or the proportion of proposals that resulted in grants, and not in relation to other agencies, but rather by the ratio of dollars awarded to dollars requested for "meritorious" proposals. His ambitious goal, he claimed, was to raise this ratio to 60 percent or more. Waterman's original request of $24.7 million to BOB for fiscal 1956, for example, would have provided as much as 80 percent of the funds requested for meritorious proposals. When questioned by members of Congress as to what limits could be set on funding increases, Waterman inevitably replied that the limit was determined by the number of competent scientists who could be supported.[8] Throughout the 1950s, NSF representatives repeatedly argued that the federal government must do all in its power to identify and nurture scientific talent in order to relieve a claimed critical shortage of scientists.

The relevance of basic research to maintaining America's lead in the Cold War competition with the Soviet Union for scientific hegemony became NSF's strongest selling point. Well before the Sputnik crisis, Foundation representatives alluded to Russia's scientific growth as an argument for garnering more support. "There is every reason to believe that the U.S.S.R. has already surpassed us in the production of engineers and is on the way to doing so in the production of scientists," Waterman told the House subcommittee in early 1955.[9] Later that year, Nicholas DeWitt's *Soviet Professional Manpower,* a much cited study funded and published by NSF, described in impressive detail the Russian system for educating scientists. Congressman Thomas, who was initially skeptical of Russian "propaganda" on production of scientists, was converted by this book. "In another 5 to 6 years they are going to be ahead of us," he declared during the 1956 hearings.[10] That year, for the first time, the Foundation received almost all of its official request. But, to Waterman's consternation, Thomas insisted on earmarking considerably more funds than requested for high-school-teacher training—better science teaching would recruit more scientists—without increasing the overall appropriation.[11]

With the exception of a discussion of biological facilities later in the decade, biology was not talked about much in the House and Senate hearings. The committees devoted more attention to fellowships, science education, facilities for the physical sciences, and more mundane matters such as staff numbers, travel, and consultation funds. Agency officials made no specific effort to justify basic research in biology or to delineate NSF's comparative role in biology. They usually mentioned health in passing as one of the areas basic research would ultimately benefit. Occasionally they referred to the basic research background of such well-known medical achievements as penicillin or, in 1955, to the recent successful testing of the Salk vaccine for poliomyelitis.[12] But few specifics were provided.

If asked during hearings for results, Waterman would respond with brief descriptions of sample NSF projects that were considered to be significant contributions to knowledge. For example, in his FY 1956 testimony, he highlighted Stanley L. Miller's experiments at the University of Chicago on the production of amino acids by discharging lightning in a simulated primitive atmosphere of water vapor, methane, ammonia, and hydrogen. A *Fortune* magazine article had listed this work as one of ten major discoveries in basic research in the past year. The same testimony described the beneficial work of microbiologist Robert E. Hungate of Washington State College who, in studying the life cycle of bacteria in cows' stomachs, discovered that microbial fermentation was a factor in causing cattle bloat. With other investigators, Hungate subsequently identified a chemical agent that could prevent the malady.[13]

In most cases, however, the payoff for basic research was implied and distant. In another 1956 example, NSF staff reported that Wolf Vishniac at Yale had succeeded in creating a "cell-free" laboratory model of significant features of the process of photosynthesis. After noting new lines of investigation in photosynthesis that the research opened, the staff offered a highly speculative statement on possible future applications: "It is now possible to visualize production line or continuous flow processes in which high energy materials useful for food and fuel are created through the action of sunlight." Similar accounts of exemplary awards in biology appeared each year in published NSF annual reports, sometimes with investigators and institutions identified, sometimes not. Waterman asked program officers to collect files of "research highlights" or "gold nuggets" for the purpose.[14] It is unlikely that such dry and impersonal descriptions of esoteric biological research made much of an impression on either lay readers or members of Congress.

If, however, a congressman asked what NSF was doing about a particular practical problem, Waterman was at a loss, since NSF research support was not organized to that end. In one revealing interchange in 1957, Thomas queried Waterman about biological research: "In what fields do you want to assist the applied sciences through this basic research? What fields ultimately will need attention? Is it cancer and blood chemistry, or what?" Waterman replied cautiously that NIH would be expected to "take the lead" in such areas but "at the same time we support pure biological and medical research that is brought to our attention." Thomas, pressing for the "two big fields that are undeveloped so far as the medical world is concerned," again asked if NSF's biology was going to "supply the fundamental research . . . that is ultimately going to solve the problems of mental health and cancer." Board chairman Bronk, more sanguine than Waterman, interjected, "I think that is the one sure thing you can bank on. It may be that somebody will come up with a lucky idea, but that will be a chance." Later, Thomas concluded the discussion with reference to the good work being done at the National Institute of Mental Health.[15] The underlying message to NSF was that the best way to invest in basic research that might lead to advances in medical treatment was to provide more money to NIH. The NSF approach was too scattershot to satisfy those who were anxious to solve specific problems of public interest.

The prickly issue of overlap or duplication was one that might be raised at any time. When members of Congress questioned NSF about duplication with NIH in the 1950s, they were concerned with the extent to which NSF funded medical research, not biology. Waterman either ducked the issue or assured his questioner that NSF-supported medical research was limited to "the fundamental variety, the biological sciences."[16] Most congressional probing of NSF on duplication, however, involved Waterman's attempt to support construction of nuclear reactors and accelerators on university campuses when the Atomic Energy Commission, whose budget was overseen by the same subcommittee, was already providing substantial sums for such facilities. Underlying the duplication issue was whether NSF should support basic research in all the sciences or concentrate on filling in gaps left by other agencies. Waterman adamantly claimed full coverage of the sciences, even if it meant significant overlap with, say, AEC or NIH. "Duplication," like "basic" and "applied," was an ambiguous term, defined differently by representatives of Capitol Hill, who usually meant overlap in areas of science, and NSF officials, who construed the term narrowly to mean the same research receiving duplicate support. Since NSF staff always knew what projects other agencies were funding, there was no possibility of this occuring.[17]

In addition to answering questions at the hearings, the BMS staff replied to requests from Congress for commentary on specific health-related bills. Generally the Foundation argued against any measure calling for more money for research on specific diseases. In its first few years, NSF opposed additional funds for cancer and for the establishment of a National Institute for epilepsy. In both cases, NSF spokesmen maintained that what was most needed for medical progress was not more funding for diseases but more basic research on normal function.[18] Although they balked at making policy themselves, they opposed any bills or executive orders that would challenge NSF's statutory authority in that area. In 1955, for example, when the administration considered creating a Federal Advisory Council on Health, Waterman suggested that medical research policy be specifically excluded from the jurisdiction of this body.[19] BMS staff sometimes expressed hope that NSF might benefit from congressional zeal for health-related research but, especially after Sputnik, the argument for the primacy of basic research in medicine was more likely to serve biology at NIH than NSF (see chapter 5).[20]

Occasionally the BMS staff would respond to requests for information from members of Congress concerning the status of particular proposals submitted to NSF by friends or constituents. Albert Thomas, for example, made several inquiries to Waterman about the application in 1958 of Rubin H. Flocks, a urological surgeon at the University of Iowa, who had operated on him. Although Levin was not much impressed by Flocks's proposal, "Antigenic Properties of Urogenital Organs," which was more clinically oriented than most NSF projects, he submitted it to the usual evaluative procedures. The panel rated it too low to fund, but Waterman, after consulting with staff and biological members of the board, arranged for Flocks to be given a two-year award. When Flocks mentioned to Levin that Thomas had urged him to apply, Levin replied "by telling him that I wish that Mr. Thomas would appropriate the funds we seek for research without juggling our budget rather than soliciting proposals for us."[21]

By 1960, if not earlier, members of Congress began inquiring about grants that later became known as "target titles"—projects that, on the basis of their titles, various muckrakers could exhibit to the public as a waste of taxpayers' money. Such grants were more likely to fall in areas of organismal biology or the social sciences, where titles seemed comprehensible to the layman, than in molecular biology or the physical sciences. To give one example, in 1960, Senators Hugh Scott and Joseph S. Clark each wrote to Waterman concerning an award that their constituents had inquired about. The unidentified news article targeted a three-year psychobiology grant to two investigators at Cornell University for a project titled "Ethological Investigation of Bird Sounds." The source reported that NSF "—a Government-supported organism of cloudy purpose—has announced it is giving away $2,900,000 in outright grants to various colleges and universities with no strings attached" and "has just given $50,000 of your dough and mine to Cornell for, God help us, a study of bird calls." Waterman took the standard NSF defense in a lengthy and patient letter, drafted by Wilson, explaining once again the ideology of basic research.[22] In later decades, with the zealous combing of titles by publications such as the supermarket tabloid the *National Enquirer* and by Senator William Proxmire for his "Golden Fleece awards," the problem of "target titles" in the biological and social sciences continued to plague the agency.

The Failure to Centralize Basic Research

In the early 1950s, the BMS staff hoped that NSF would take over an increasing portion of federal support of biology. From roughly 1952 to 1954, it appeared that with prodding from the Bureau of the Budget, other agencies, particularly the Department of Defense (DOD), might relinquish at least part of their basic research and fellowship programs to the NSF. BOB encouraged reductions in the budgets of other agencies and corresponding increases in NSF's. The Eisenhower administration codified its view that NSF ought to be the main federal agency for support of general basic research in Executive Order 10521, dated 17 March 1954, which stated that "the Foundation shall be increasingly responsible for providing support by the Federal Government for general-purpose basic research through contracts and grants. The conduct and support by other Federal agencies of basic research in areas which are closely related to their missions is recognized as important and desirable, especially in response to current national needs, and shall continue." Other agencies were to ensure "that the Foundation is consulted on policies concerning the support of basic research."[23]

The bureau's intention of transferring at least some basic research programs to NSF placed mission agencies, especially NIH and AEC, on the defensive. As Bill Consolazio wrote in early 1953: "We have had little effect on other agencies except that great fear of the Foundation exists," fear that NSF would move into "especially those fundamental areas" they now supported. "By the fact that we were born, we created fears and antagonisms . . . ." It was, he thought, "a little like an elephant fearing a flea."[24] Though Waterman insisted that other agencies ought to support basic research closely related to their missions, defining "closely related" was always a contentious issue.

For a few years, Waterman tried without success to obtain much larger appropriations on the grounds that NSF should command a larger share of basic research. For FY 1954, BOB agreed to a budget of $15 million, over three times the previous year's budget. Although Waterman defended the increase on the basis of the executive order, the House approved only $5.7 million. In an attempt to restore the cuts, Waterman explained to the Senate committee that the Foundation had established "a very clear policy" that "since basic research cuts across all the agencies and all the subjects from every point of view, we believe that general purpose basic research should be centralized in one agency." That could not go forward, he warned, unless the Foundation's appropriation was large enough to compensate for reductions; otherwise, "the national interest may be prejudiced by a reduction in the overall level of needed basic research activity."[25] Though the Senate sympathetically approved $10 million, the final budget was set at $8 million.

The hope of centralizing basic research in NSF proved illusory. NSF was unable to take over programs of other agencies, first because the latter resisted giving them up; second, because NSF was unable to win the budgets necessary to increase significantly its share of basic research; and third, because NSF was unwilling to channel resources into the specific areas to be relinquished by other agencies. Moreover, NSF staff received little support from their scientist-advisors, who wanted to protect the programs of DOD,

AEC, and NIH from curtailment so that the amount and variety of federal support would continue undiminished.[26] In the end, the executive order only sanctioned the role of other agencies in supporting basic research. It had little effect in limiting their scope or giving BOB control over mushrooming budgets and overlap among agencies.[27]

The difficulty of "transferring" basic research programs to NSF is well illustrated by the case of NIH fellowships. In 1953, under pressure from the Budget Bureau to avoid "duplication," the AEC and NIH were slated to terminate their "general purpose" fellowship programs. BOB cut NIH's predoctoral fellowship program by $390,000. At the same time, the president's budget allowed for a substantial increase in NSF fellowship funds, from $1.4 million in FY 1953 to $4.3 million in FY 1954. Waterman told the House committee, "From now on the plan is for the NSF to administer the general purpose fellowship program of the Federal Government." But the House approved only $1.9 million, the amount of the previous year's program plus the $390,000 of NIH's proposed reduction. Despite Waterman's pleas, the Senate did not restore the funds.[28] Moreover, NSF did not set aside the additional $390,000 for biology but simply added it to the overall program. Of a total increase of 179 fellowships, NSF gave only 27 more predoctoral fellowships in the biological sciences in 1954 than in 1953.[29] Thus, NIH could easily argue that NSF did not in fact assume its program.

In late 1953 and in the hearings for the FY 1955 budget, NSF continued to press through the BOB and Congress for further concentration of fellowships in NSF. When the decision was made to transfer all predoctoral fellowships to NSF, NIH shifted emphasis to postdoctoral fellowships, awarding about 350 in fiscal 1954.[30] Waterman and the NSF fellowship staff held two meetings with Director William H. Sebrell Jr. and other officers of NIH in October 1953 to discuss possible transfer of postdoctoral fellowships to the Foundation. NSF claimed that a single agency handling all fellowships could obtain a better balance among scientific fields. Waterman proposed that NSF take over all "general purpose" postdoctoral fellowships, leaving to NIH special-purpose training programs directly related to NIH's mission.

NIH countered that "general purpose" fellowships could not be separated from those of "special purpose." Sebrell, according to NSF staff member Bowen C. Dees, replied that many fellows would be hard to classify, offering as an example "the case of a geneticist who was doing work on sweet-peas but whose ultimate goal was to do research in the field of hereditary cancer." Ernest Allen, head of the NIH Division of Research Grants, claimed that "practically 100%" of postdoctoral fellows were disease-oriented and warned of a likely unfavorable reaction by scientists if the program were transferred. In order to prevent further disagreement, NIH agreed to present its postdoctoral program to BOB as "categorical," that is, restricted to special-purpose fellowships.[31]

Through direct intervention with congressional committees, NIH soon thwarted NSF and BOB. In May 1954 Harry Kelly, head of NSF's education division, heard a rumor that NIH had requested through the House Subcommittee on Appropriations the reestablishment of its predoctoral program at the previous level. Kelly told Waterman the request was made "without prejudice to the NSF program, and implied that NIH would like to see the NSF program increased, but because of the low order of magnitude of the NSF program they felt they should make a special play for fellowships in the biological and medical field." BOB apparently had no knowledge of NIH's request to the House.[32] The subcommittee, headed by John Fogarty, partisan for the cause of medical research, readily included the supposedly transferred $390,000 in NIH's appropriation. The House Appropriations Committee report recommended that NIH's predoctoral fellowship program be restored since the transfer to NSF had resulted in a great decrease in the number of fellowships in medically related fields.[33]

After negotiating with the BOB, NIH agreed to a short-lived compromise whereby it could maintain its program on the understanding that it be limited to categorical areas. Sebrell told Waterman in late 1954 that the $390,000 would be used "for special training of individuals going into public health or related fields." Thus NSF could still claim to have the only general-purpose predoctoral program. Assistant Director Steinbach noted that the net result of the entire transaction meant that there would be fewer awards in general biology.[34] Within a few years, with the aid of congressional supporters, NIH was able to reinstate and greatly expand its general fellowship program and eventually dominate federal fellowship support in biology.

Making Science Policy: Studies of Biological Fields

A second means by which the administration sought to gain control over competing federal programs was by attempting to force NSF to exercise its policy-making authority. Listed first among NSF's functions in the act of 1950 was "to develop and encourage the pursuit of a national policy for the promotion of basic research and education in the sciences." A related function was "to evaluate scientific research programs undertaken by agencies of the Federal government, and to correlate the Foundation's scientific research programs with those undertaken by individuals and by public and private research groups."[35] These requirements represented a tall order that Waterman and the board would have been happy to have ducked. As a latecomer to federal support of research, NSF could not afford to alienate larger and more powerful agencies by interfering with their programs. It thus found itself in a delicate position. Shortly before his confirmation as NSF director, Waterman noted in conversations with board members Bronk and Conant, "This is a matter which must be handled with extreme care in order to avoid (a) the actual evaluation of research programs because of the difficulty and danger of doing so, and (b) failure to provide the President, the Bureau of the Budget and Congress with information which is evidently called for by this clause. If the Board's effort in this direction should at any time be regarded as inadequate the Foundation would be in trouble."[36]

The Division of Biological and Medical Sciences led the way in formulating a response to Waterman's dilemma. NSF needed to give the appearance of establishing a national science policy without alienating scientists fearful of federal control or encroaching upon the prerogatives of other agencies. One way to evaluate the national effort in science was through information gathering without interpretation or recommendations. John Wilson was instrumental in creating NSF's Program Analysis Office, which produced the first volumes of *Federal Funds for Science,* a compilation of the amounts of money federal agencies spent on basic and applied research in various areas. To avoid offense, NSF allowed each agency to categorize its own research support.[37]

BMS also initiated the sponsorship of surveys of various scientific disciplines by leading practitioners, which became a chief way for NSF to claim publicly to be making national policy. Such self-surveys of fields of science conformed to the board's insistence that science policy must emanate from scientists. As Fernandus Payne, assistant director for biological sciences, expressed this dictum in 1952, "They are the ones who know best past history and present developments and they should be the ones who are best able to vision future needs, insofar as any one is able to look into the future."[38]

Pressed by the need to satisfy the administration's requests for making policy, BMS was receptive when the American Physiological Society (APS) informally approached the Foundation in late 1951 to support an ambitious survey of the physiological sciences. This study and an even broader one organized by the American Psychological Association the following year represented two of BMS's largest early awards. While the program directors for the physical sciences were less active in sponsoring surveying of fields, the Foundation funded a smaller study in the area of applied mathematics. NSF highlighted these surveys, while they were underway, in annual reports, hearings, and press releases as evidence that it was evaluating the needs of science. Yet, it is not clear that the surveys did in fact contribute either to national science policy or to the Foundation's own granting policies. Nothing was publicly said about the surveys after the reports appeared in print.[39]

The American Physiological Society proposed to investigate all aspects of "physiology": research trends and support of research; recruitment and motivation of physiologists; career paths and career satisfaction; teaching at the high school, college, and graduate levels; the role of publications and societies in the profession; and public attitudes toward "physiology." The society had its own reasons for carrying out such a study. Before World War II, APS was almost exclusively a research organization, dominated by medical school interests and limiting its activities to holding annual meetings and publishing two highly successful journals. After the war, some of its more reflective members, led by E. F. Adolph of the University of Rochester and Ralph Gerard of the University of Chicago, began to look at the status of physiology in a broader sense, especially the role of teaching physiology and the recruitment of physiologists. They held symposia at annual meetings on these issues, where they argued that the APS ought to become more representative of the discipline of physiology conceived not as human or mammalian physiology in the service of medicine but as functional biology.

The genesis of the survey came from Gerard, president of the society, who arranged a meeting in November 1951 with Orr Reynolds and Emmanuel Piore of ONR and John Field, assistant director of BMS.[40] As the APS's needs complemented those of the Foundation, another meeting of APS and NSF representatives was held in February, a short proposal was written, and in March 1952 a contract was approved for $120,000 to be spent in a period of two years and three months. The contract was in some ways a safe and cozy arrangement in that Reynolds, an APS member and well known at NSF as chief of ONR's biosciences division, was willing to take a leave of absence to direct the pilot phase of the project. Moreover, Field was also an active member of the society.

The resulting volume, written by Gerard and published in 1958 with an additional NSF award, contained an impressive amount of data, a large part of it based on a questionnaire answered by over four thousand "physiologists." While the survey committee members were disposed to take a broad view of physiology, Lou Levin, the contract administrator, encouraged them to encompass all the physiology in his Regulatory Biology Program. Not only were invertebrate physiologists included in the survey, but also plant and bacterial physiologists. As the training and careers of medical school physiologists, nonmedical zoological physiologists, plant physiologists, and bacterial physiologists proceeded along four entirely separate tracks, some of the aggregate statistics were not especially meaningful. Unlike later National Research Council surveys of sciences, which were aimed primarily at funding agencies, the diffuse recommendations of the physiology report were directed to all segments of the profession. In fact, they made no plea at all for more government support. Gerard reported, "Funds for physiological research seem, for most investigators, adequately underwritten."[41]

While the report was a disappointment to some, including Reynolds, because it was hastily written and did not fully address the issues initially set forth, the survey project did encourage the APS to expand its functions, and in particular to take an active role in promoting college teaching of physiology and recruiting students to become physiologists. Its education committee, formed in 1953, sponsored a wide range of innovative projects, funded by NSF and NIH, and placed itself at the forefront of education committees of biological disciplinary societies.[42]

The Survey of the Physiological Sciences was soon followed by the even more elaborate Study of the Development and Status of Psychology. In October 1952 NSF approved a contract with the American Psychological Association (APA) giving them $40,000 for the first year of the study. Wilson, who was a member and former assistant executive secretary of the APA, managed the project for NSF.[43] Here again, the impetus for the survey came from the disciplinary society, and the chief concern was not financial resources but a postwar identity crisis. Before the war, psychology and the APA were predominantly academic, but World War II had given a tremendous boost to applied human psychology, especially to the practice of clinical psychology. In the postwar decade, experimental psychology, the academic wing of the field, was beginning to feel overshadowed by the clinical practice wing, both of which were represented in the APA. Moreover, academic psychologists debated whether psychology was a legitimate science and whether it had any unifying principles amidst the welter of conflicting theories. The reports resulting from the two-part survey were written primarily from an academic point of view.[44]

The first part of the survey, "Project A," resulted in six volumes edited by Sigmund Koch of Duke University and published by McGraw-Hill. The series treated the substantive content of psychology and was widely used in teaching. Chapters by various authors explored the state of psychological theories, methodology, and empirical knowledge in problem areas of the discipline.[45] "Project B," based on surveys of psychologists, dealt with such matters as training and employment of psychologists, factors motivating students to enter psychology, and factors influencing research eminence. Its outcome, a single volume, *America's Psychologists,* was edited by Kenneth Clark of the University of Minnesota.[46]

For a time, NSF hoped to expand the survey concept to other fields of science, but it soon ran into resistance from scientists. BMS planned in 1952 to sponsor a survey of biochemistry, to be handled from the NSF side by Consolazio. There was also discussion of a survey of ecology to be managed by George Sprugel. The problems encountered in setting up the biochemistry survey, however, marked the death knell of "status of fields" studies as a means of making policy. The biochemistry section of the American Chemical Society, which first broached the idea of the survey, was enthusiastic, but after protracted negotiations with the NSF staff, the Council of the American Society of Biological Chemists (ASBC) refused to cooperate. The chief opponent of the project was Carl F. Cori, a member of the ASBC council, who was influenced by his wife, NSB member Gerty Cori. The Coris claimed that the money would be better spent on research projects, although Levin and Consolazio explained that policy funds could not be transferred to research. A deeper issue was that even seemingly harmless scientist-directed surveys raised the specter of central control of science. In their postmortem of the abortive biochemistry survey, Levin and Consolazio told Waterman, "One point which needs real emphasis" was that a major reason scientists opposed such surveys was fear "that it will lead to the formulation of National Science Policy which, in turn, is interpreted as being synonymous with control, regulation and direction of science."[47]

BMS's pioneer studies of physiology and psychology were forerunners of the studies of various sciences carried out by the Committee on Science and Public Policy (COSPUP) of the National Academy of Sciences in the 1960s. Yet they differed in an important respect. The early surveys were internally generated by real needs within professional communities to understand the intellectual development and social dynamics of their disciplines, while the later National Research Council–sponsored studies were directed primarily to those outside the disciplines and focused on public relations and resource needs.

Making Science Policy: The "Long Report"

NSF's one foray into evaluating the research program of another agency, the NIH, did nothing to clarify the relations between the two. In fact, the so-called Long Report, completed in 1955, paved the way for the rapid expansion of NIH into general biology.[48] Because NIH was so successful in obtaining money directly from Congress through the intervention of disease-related lobbies, the Bureau of the Budget found itself with little authority over the NIH appropriations process and hence unable to coordinate federal support of science. Each year BOB would negotiate a budget with NIH officials only to have congressional appropriations committees overturn it by voting millions of dollars more than the agency had officially requested in the president's budget. NIH officials, rather than defending the figure negotiated with BOB—the amount appearing in the president's budget—were more than willing to reveal what higher figure they had originally proposed to the BOB. By 1954, the bureau considered the situation completely out of hand. The president's budget called for $56.3 million for NIH, but Congress appropriated $71 million, 26 percent more than the administration requested. The drama was repeated the following year when the administration asked for $71.1 million and the appropriation came to $81.3 million.[49]

Thus, BOB, which had been pressing NSF to use its statutory evaluative authority, asked the Foundation in late 1954 to undertake a study of the internal (intramural) and external (extramural) medical research programs of the Department of Health, Education and Welfare (HEW). Waterman, treating the matter as "private and confidential," informed the board on 5 November that BOB director Rowland Hughes had suggested the study "to solve the problem of pressure groups exerting sufficient influence with Congress to increase the appropriations requested in the President's Budget." BOB had proposed, with the (perhaps reluctant) approval of the secretary of HEW, Olveta Culp Hobby, "a careful and authoritative study" of the relative emphasis that should be placed on research related to specific disease categories and to basic research as well as intramural versus extramural research.[50]

NSF was understandably hesitant. The "special committee" of the board, chaired by Detlev Bronk, which was appointed to consider the task, at first tried to duck BOB's invitation by recommending to Hobby and Hughes that the HEW secretary set up her own advisory commission to conduct the study. But this was acceptable neither to the BOB nor to Nelson Rockefeller, undersecretary of HEW, who was aware of and concerned over the lack of executive branch control over medical research. An in-house commission, Rockefeller thought, would not be "entirely adequate in objectivity" to consider NIH's proper level of support in relation to the total national medical research effort.[51]

At a December meeting with Hughes and other BOB officials, Waterman and board members Barnard and Bronk continued to argue against the feasibility or appropriateness of the task. Waterman suggested such an effort could "stir up hostilities which might seriously damage the whole program of support of the medical sciences." Moreover, he claimed, NSF was in the process of preparing a comprehensive study of the role of federal support of scientific research. But BOB officials insisted that the Foundation had both the authority and the responsibility to act. Therefore, at Hughes's request, Waterman drafted a letter from the secretary of HEW to himself, asking that

NSF, in view of its responsibilities in evaluating the scientific research programs of federal agencies, undertake "an early interim appraisal" of the medical research programs of HEW.[52] Although "misgivings were expressed by numerous Board Members with the questionable value of such a study and the fact that it set an undesirable precedent," the board agreed on 21 January to undertake the review.[53]

Waterman took six months to appoint the Special Committee on Medical Research and secure a suitable chairman, choosing finally C. N. H. Long, chairman of the Department of Physiology and former dean of the Yale University School of Medicine.[54] In correspondence with committee members, Waterman (presumably at BOB's request) omitted any reference to the Budget Bureau and attributed the genesis of the report solely to the secretary of HEW. As promised, the committee completed its work within six months of its first meeting in July. The report, which Long emphasized to the NSB was unanimous, was "released" to the public in mimeographed form on 28 February 1956 but never published.[55]

Formally titled *Medical Research Aspects of the Department of Health, Education and Welfare,* the Long Report was a controversial document. Although it made recommendations concerning all of HEW's medical research programs, it focussed on the National Institutes of Health. It predictably criticized the process whereby "special pleaders, including citizens committees" were able to convince congressional committees to raise NIH budgets well beyond the ceiling proposed by BOB.[56] The report's main thrust, however, was its critique of the categorical disease approach to medical research. Concerned that an "unbalanced situation" had developed "to an alarming degree" in educational institutions, the committee expressed fear that public pressure would lead to the further creation or expansion of categorical institutes and programs, when what was most needed for the understanding of disease was more "knowledge of the normal physiological and chemical functions of living organisms."[57] Thus, the report advocated no new institutes and no significant increase in intramural research. Rather, it recommended "non-categorical" academic research grants, "unrestricted institutional grants," expansion of teaching and training grants from categorical areas to "departments and disciplines in medical and graduate schools whose contributions to the over-all pattern of medical research . . . are equally important to the nation's welfare," and expansion of noncategorical pre- and postdoctoral fellowships for "substantially all qualified applicants." The report's most controversial recommendation was to separate NIH's extramural program from its disease-oriented intramural program and, in fact, remove it from NIH altogether into a new HEW agency with the suggested title of Office of Medical Research and Training.[58]

While the recommendations, if enacted, might have restored to the BOB some measure of leverage over NIH's budget by reducing the power of the disease lobbies over the extramural program, they would have done nothing to further the aims of Eisenhower's executive order or to delineate the respective jurisdictions of NSF and NIH. The report did not discuss the issue of jurisdiction except to suggest that the new extramural agency "establish such liaison with the National Science Foundation . . . that it can act with full knowledge of pertinent NSF policies and activities." Indeed, the recommendations would have greatly increased the areas of overlap. However, the recommendations were never acted upon, and the Long Report failed to bring about a reorganization of medical research in HEW. Even prior to its release, Public Health Service officials were said to be apprehensive about its effect on their "legislative program." Although there was talk in 1956 of setting up a special HEW committee to consider the recommendations, nothing was done.[59]

Science policy analyst Philip Smith, who in 1980 reviewed for the board the history of its policy-making activities, pointed to the Long Report as an example of the Foundation's failure to influence policy. "The recommendations stepped on both institutional and political toes in a large department with strong patrons in the Congress," he wrote.[60] However, despite Smith's conclusion, the report was significant for advocating the expansion of grants for "non-categorical research." (The report avoided the term "basic research" in favor of "non-categorical research," perhaps to make it clear that such work should remain in HEW.) It was the first in a series of commissioned reports that paved the way for NIH to support a broader spectrum of biological sciences (see chapter 5).

Making Science Policy: The Loyalty Issue

Perhaps NSF's greatest success in making national science policy was its handling of the security crisis of the early 1950s. As NSF's stand on this issue was occasioned by the policies of NIH and was a result of pressure on the part of the Division of Biological and Medical Sciences, the loyalty episode serves as a significant case study of BMS's relation to other organizations.

Although members of Congress had raised security issues since the debate over the founding of NSF, by late 1953 and early 1954 McCarthyism and the investigation of Communist infiltration into academia had reached a peak. A number of professors lost their jobs because congressional investigating committees found them to be disloyal or simply because they took the Fifth Amendment when questioned. About the same time, in the spring of

1954, the Atomic Energy Commission held widely publicized hearings that resulted in the removal of the security clearance of J. Robert Oppenheimer, the physicist who had been in charge of the Manhattan Project during World War II.

BMS's protest in this climate of fear and suspicion was modest but highly significant. When HEW Secretary Hobby announced in a press release that the Public Health Service would terminate grants of scientists *suspected* of disloyalty, BMS deliberately funded blacklisted biologists and urged NSF to establish a general policy opposing security checks of grantees. By its eventual action on behalf of scientific autonomy, NSF created considerable good will among biologists and scientists generally.[61]

BMS first confronted the security issue in the fall of 1953 when the division convened a conference of federal agency program directors in the life sciences, chaired by Lou Levin. High on the agenda was discussion of rumors that the HEW counsel would impose security checks on all NIH grantees. NIH grants managers greatly feared this security initiative. Ernest Allen, director of the Division of Research Grants, expressed his concern at the conference that investigations of some 5,000 scientists would not only consume an enormous amount of time and resources but would also seriously jeopardize the good relationships that NIH had built up with the scientific community.[62]

Bill Consolazio goaded NSF into taking a public stand on the issue of scientific freedom. In a memorandum to Waterman in October 1953, he warned that the problem of security was now beginning to seep into federal programs for unclassified research in academic institutions and was reaching the stage "where all the hopes and ideals entailed in a freedom of science are being jeopardized." He called attention to rumors that "already some members of the various Study Sections and Councils of the Institutes of Health had been asked to resign, and of those who had resigned, also had their research programs terminated." Those present at the program directors' meeting, he reported, felt "that the NSF had the responsibility under its policy-making functions to reopen the whole subject of security and its effect on the Extramural Basic Research Program." The BMS staff believed, however, that the problem went beyond the Foundation and had to be brought to the attention of the Interagency Committee on Scientific Research and Development and, Consolazio added, to the president of the United States.[63] In November, the Divisional Committee for Biological Sciences meeting jointly with the short-lived Divisional Committee for Medical Research spent a half day discussing the issue. While making no recommendations, they "hoped" NSF was looking into "all possible actions in protecting the scientists from undeniable ramifications in security programs."[64]

By the spring of 1954 the situation had worsened as rumors circulated of NIH's termination of grants of biologists, several of them of eminent stature. As a result of resolutions by the American Society of Biological Chemists and the American Physiological Society, Detlev Bronk, president of the National Academy of Sciences (and also chairman of the NSB), wrote to Secretary Hobby to request a clear statement of HEW policy. Hobby replied, in a letter subsequently released to the press, that HEW did not require security checks but when "information of a substantial nature reflecting on the loyalty of an individual is brought to our attention, it becomes our duty to give it most serious consideration. In those instances where it is established to the satisfaction of this Department that the individual has engaged or is engaging in subversive activities or that there is serious question of his loyalty to the United States, it is the practice of the Department to deny support." She stated that there had been fewer than thirty cases of denial of support since the policy began in 1952.[65]

Among the biologists who had their grants terminated in late 1953 and 1954 for having engaged in left-leaning political activities were such prominent figures as Linus Pauling, Elvin Kabat, and Martin D. Kamen.[66] Consolazio's Molecular Biology Program readily funded these scientists and probably others who remained unidentified. Kamen, a biochemist working on photosynthetic bacteria at Washington University in St. Louis, wrote in his autobiography that he was informed by a curt note from the chief of the Division of Research Grants that his present grant, supporting the salary of his collaborator, biochemist Leo Vernon, fell in a "category which can not be supported by the Public Health Service." He was rescued first by the private Kettering Foundation, and then, as he recalled, "I had the good fortune to meet . . . Bill Consolazio," to whom he attributed the award of a five-year grant for $32,500. "It appeared that what was treason in one area of government-supported science was patriotism in another," he wrote.[67] Kabat recalled that when he lost his NIH support, Consolazio, who at ONR had practically foisted a grant on him, now suggested that he also apply for an NSF grant. The Foundation gave him a three-year award for $60,000 for "Immunological Studies on Polysaccharides." Kabat boycotted NIH for many years, refusing to allow anyone in his laboratory to accept NIH funds. NSF remained his chief source of support into the 1980s.[68]

As a matter of principle, several other biologists switched their source of funding from NIH to NSF. Consolazio noted in his 1954 annual report that his grantees included several formerly supported by the Public Health Service.[69] Harvard biochemist John T. Edsall, also a molecular biology grantee, declared that he felt free to speak out in the pages of *Science* against the grave threat to scientific freedom posed by HEW "because I derive my research support from other agencies that have maintained the tradition of freedom."[70] Decades later, scientists such as Howard K. Schachman of Berkeley recalled with admiration Consolazio's heroic actions to save the research careers of talented biologists who were under a political cloud.[71]

Meanwhile, the Divisional Committee for Biological and Medical Sciences pressured the National Science Board to take a stand. Its resolution, drafted by George Wald and Jackson Foster, urged Waterman to request the NSB "to announce at an early date the general principles and procedures of the National Science Foundation involving security clearance of applicants for research grants and fellowships."[72] Waterman asked NSF General Counsel William J. Hoff to draw up a position document, in which Hoff concluded, "It is my opinion that, since consideration of national defense and security are not involved in the making of grants in support of unclassified research carried on by persons not employees of the U.S. Government, the laws and executive orders relating to loyalty and security are not applicable." Consequently, NSF had no legal obligation "to inquire into or consider matters relating to the 'loyalty or security' of persons (other than U.S. Government employees) who would participate in work under a proposed grant."[73]

The NSB adopted on 21 May a statement based on Hoff's position paper that declared that while NSF would not knowingly fund avowed Communists or those who were determined by judicial or other official process to be disloyal, it would not take into account rumors or allegations. The board deliberately approved Consolazio's grant to the blacklisted Linus Pauling.[74] However, it did not publish its statement, because Waterman, ever cautious, had decided that the current climate created by the McCarthy and Oppenheimer hearings was a poor one in which to make a stand.[75] The more idealistic Consolazio continued to plead for a public announcement. He wrote in his annual report for 1954:

> The security problem is probably the major problem the Foundation faces, and to my mind it is the stand that NSF takes on this issue that will determine whether the Foundation will become a potent force on the American scene or just another Federal agency . . . . We must support individuals strictly on the basis of scientific merit and never allow a political consideration to enter the picture. This is the only way to demonstrate to the world that NSF is an organization for scientists and for the support of basic science. I believe that in our support of research we should be on an equal basis with the Rockefeller, Carnegie, and Ford Foundation. The fact that our funds come from the taxpayers should be acknowledged, but this fact should not allow NSF or the Federal Government to disturb the free principles of science.[76]

NSF did not publicize its position until 1955 when McCarthyism was clearly on the wane. By early 1956, the Public Health Service had quietly shifted to the NSF policy, but the issue was not completely resolved until later that year.[77] Sherman Adams, assistant to President Eisenhower, formed an interagency committee to consider the loyalty problem and asked the National Academy of Sciences to submit a report. The Academy committee, chaired by Julius Stratton of MIT, adopted a modified version of the NSB statement. These principles, Adams announced on 14 August 1956, were satisfactory to the other agencies and were "essentially those which support the policy of the National Science Foundation." NSF hastened to include a self-congratulatory account of its role in the security crisis in its sixth *Annual Report* released in November.[78]

Consolazio hoped the Foundation's stand on the security issue had demonstrated the superiority of NSF awards. He reported in 1955 that "the basically oriented scientists are looking particularly to the NSF for support, and less and less to DOD, AEC, and HEW." The recent developments, he claimed, "have convinced American scientists of the desirability of association with an organization whose interests are those of promoting the basic principles of science and the development of a policy commensurate with the freedom of scientific and intellectual inquiry."[79] Although NIH may have temporarily lost prestige, most biologists had short memories when they discovered they could get larger and longer-term grants from NIH than from NSF.

Making Science Policy: Sabotaging the NRC's Biology Council

Although BMS staff did not want to assume an active role in making national policy for biology, they were jealous of any rival source of policy—in particular, the National Academy of Sciences and its research wing, the National Research Council. Before the war the Academy had stood at the pinnacle of private science organizations. Its role challenged by the advent of large-scale federal funding of science, the Academy maneuvered to remain a central force in the postwar period. In 1945, Robert F. Griggs, chairman of the NRC's Division of Biology and Agriculture (B&A), hoped that the proposed NSF might delegate to the Academy the evaluation of proposals and selection of grantees.[80] Although that did not come to pass, NSF did contract with the NRC to administer NSF's pre- and postdoctoral fellowship programs. NSF's relations with the Academy remained close, for biophysicist Detlev Bronk, president of the Academy from 1950 to 1962 had been a member of the NSB since 1950 and served as board chairman from 1955 to 1964.

BMS program officers thus found themselves in a particularly delicate position when it came to their relations with the biology division of the NRC. Paul A. Weiss, new chairman of the languishing B&A in 1951, had grandiose visions of expansion. But committees and projects, to be effective, depended on outside support. Thus B&A joined the ranks of institutions seeking federal grants and contracts. The question became to what extent would BMS foot the bill?

While supporting some NRC activities, the BMS staff, especially Wilson and Consolazio, were wary of Weiss and Bronk and what they saw as NRC pretensions. In Wilson's opinion, the National Science Board, which he generally faulted for conservatism, was dominated by the "National Academy clique." Wilson recalled Bronk as a good chairman of the NSB, but too ready to use the Academy as the instrument for carrying out NSF objectives. He thought the NRC staff "would have liked very much that all the money be funneled over through the Academy and then the Academy run the program." But NSF's biologists "weren't about to do that, because, as I say, the ONR background gave all of us in Bio a feeling of great independence and great nonlimitation."[81] As late as 1983, Consolazio recalled as one of his overlooked achievements that "I worked hard at keeping the NAS out of NSF following the ONR tradition."[82]

BMS funded several NAS projects but was always careful to maintain control over their choice and length. In its first year of operation, BMS supported the NRC's Pacific Science Board (PSB), which sponsored research— much of it biological and anthropological—in the Pacific region, especially in the area known as Micronesia which had recently been under Japanese control. The navy had helped to found the PSB in 1946 and provided early funding for university-based projects through ONR. NSF continued to sponsor PSB through the 1950s. BMS also funded the American Table (research space for American investigators) at the Naples Zoological Station in Italy through a grant administered by a B&A committee.[83]

More central to BMS concerns was its support of the newly created B&A Committee on Photobiology, which was intended to study "biological aspects of radiation at and near the visible spectrum," including ultraviolet and infrared radiation and photosynthesis. Consolazio practically initiated the NRC proposal, for he was eager in early 1952 to demonstrate NSF leadership in photosynthesis. He made sure he had input into the selection of the committee and of the chairman of its first conference on photosynthesis in 1952. Under the NSF contract, the B&A committee supported two invitational conferences on photosynthesis and one on bioluminescence.[84]

The chief source of BMS-NRC conflict in the 1950s arose over a proposal that BMS program directors chose *not* to fund, namely the NRC Biology Council. The story is a complex one, involving strong personalities, several competing institutions, and deeply held convictions. At stake was the old question of who, if anyone, was to speak for biology. In effect, BMS squelched the NRC Biology Council as a perceived threat to NSF prerogatives.

The battle lines were already drawn by November 1951 when Weiss, an embryologist from the University of Chicago, was appointed B&A chairman.[85] A man of vision and a strong adherent of a unified biology, Weiss had ambitious plans for placing B&A at the center of biological policy-making in America. Many biologists, however, found him to be highhanded and authoritarian. In December, Weiss informed Field and Consolazio—then the only NSF staff members in biology—of his plan to reorganize B&A. His scheme called for the creation of six major standing committees to cover all of biology: molecular biology, cellular and microbiology, genetic biology, developmental biology, regulatory biology, and environmental and group biology.

As Field and Consolazio recorded it, Weiss "advocated that these committees should consider such problems as the formulation of a philosophy for science, a policy for science and to be available in an advisory capacity to any of the contributing agencies." They asked Weiss how he could "justify the organization of these committees . . . now that NSF was a going organization? After all, isn't the establishment of a science policy the function of the National Science Foundation?" Weiss replied that the B&A committees could call attention to areas of "over-concentration" or malnourishment. The BMS biologists countered: "Isn't this a function of the NSF also?" Foundation consultants not only evaluated research proposals but also served as "the field men for the NSF, that is, the eyes and ears of NSF, thus bringing to our attention research programs and research people needing assistance." Weiss was said to be none too pleased with the discussion, but he persisted with the scheme.[86]

According to Weiss's memorandum to the Academy in January 1952, each committee of the "Committee Council" was to keep abreast of its own area's current situation, predict trends and recommend further development, and be available for consultation on all matters of biological personnel, education, research, communication, and financial resources. The committees were to publish reports for the guidance of all agencies concerned with the biological sciences including universities, foundations, federal funding agencies, disciplinary societies, and publishers. With support from the Rockefeller Foundation, Weiss had already established in 1951 the Committee on Developmental Biology, his own area of expertise, with himself as chairman. This committee, later offered by Weiss as a model for the others, held meet-

ings, organized conferences, and collected data on personnel, teaching, educational facilities, and research trends.

For Weiss, the Committee Council promised to fill a need for a cohe-
sive force to counter the "progressive specialization and fragmentation" of
biology. He held that the "project system" on which biology was now coming to depend, tended to promote low-risk research in "well-worn chan-
nels" and failed to provide for the broad exploration needed in an immature
science like biology. The NRC, "the agency uniquely suited to serve this
purpose," Weiss maintained, "has not only the opportunity but the obligation to serve as mechanism for the restoration and broad perspective in biology on which the progress of biological sciences depends."[87]

In 1952, Weiss began a campaign to obtain operating support for his
Committee Council from the NSF, ONR, AEC, NIH, and the Rockefeller
Foundation. He estimated initially a need of $120,000 a year for a five-year
trial period.[88] The BMS Divisional Committee, discussing the matter in
April, adopted a motion by John S. Nicholas of Yale which stated that the
committee took "a dim view of the transference of its advisory functions"
or those of BMS advisory panels to the NRC as under the proposed plan.[89]
Frank H. Johnson, program director for developmental, environmental, and
systematic biology, wrote to Weiss on behalf of NSF that "the plan envis-
aged . . . would lead to an extensive duplication of efforts, including major
responsibilities with which the National Science Foundation has been
charged by the 81st Congress, Public Law 507. Moreover, we see neither advantage in, nor justification for, delegating such responsibilities to an organization other than the Foundation."[90]

Warren Weaver of the Rockefeller Foundation upheld NSF's position.
While the proposed scheme left policy making to scientists rather than to
"any nonscientific influence in government or in any other agency," it nevertheless appeared to him to put too much power in one group. "It is at least
in my judgment, questionable whether science should be 'run' by any one
special committee even of scientists," he wrote to Weiss. But more significant to Weaver was the fact that Weiss's advisory committees would compete
with those of the various federal agencies, especially NSF. He maintained
that the NSF Act gave NSF the responsibility "to develop and encourage the
pursuit of a national policy." If it wished, NSF could use the NRC to carry
out this function, but Weaver was unwilling to support the proposal unless
it was implemented in conjunction with NSF and other agencies lest he have
to take sides in "what is in simple fact a sort of jurisdictional dispute."[91]

In his 1953 B&A annual report, Weiss chiefly blamed the NSF for the
slow progress that was being made in implementing his plan. He complained
to Bronk that Johnson's letter was "officious and paternalistic" and that the

"implication of duplication of effort is quite gratuitous." Referring to BMS's adoption of the functional organization in its own programs, he wrote with some sarcasm: "Quite possibly though the NSF might want to appropriate the Committee Council plan as we have designed it, as they already have appropriated the titles of the committees of the proposed Council." But Weiss argued that, even so, the Academy should "go ahead with the scheme independently if for no other reason so as to forestall at least the danger of any monopolistic development."[92]

As his next tactic, Weiss called a meeting in January 1953 of representatives of government agencies. With considerable effort, he obtained minimal funding to activate the Committee Council, now renamed the Biology Council, with himself as chairman. ONR provided the chief source of support, Orr Reynolds, director of its biosciences division, personally processing the proposals without sending them through a panel. An independent thinker and a strong advocate of pluralism in federal support of biology, Reynolds perhaps saw the NRC as a counterbalance to the NSF.[93] In all, Weiss obtained a meager $22,000 from ONR and $10,000 each from the AEC and the air force for the council's operating expenses. A small grant ($1,725) from NSF, followed by a three-year award from the Rockefeller Foundation, enabled Weiss to organize an associated Committee on Educational Policies to deal with precollege through graduate education and adult education in biology.[94]

Finally underway in 1954, the Biology Council held its first meeting on 22–23 January. Sixteen members were named, several of whom—Jackson Foster, Sterling Hendricks, Ernst Mayr—were also familiar advisors of BMS. The council agreed to "study trends of biological research and sociological factors affecting it." Chairmen were named for each of five functional committees, and task forces consisting of members and non-members were formed to deal with specific problems such as public relations, research support, and biological research materials.[95]

The Biology Council soon came into conflict with activities of the American Institute of Biological Sciences, which still functioned as an arm of the Division of Biology and Agriculture. Although AIBS had been created in 1947–48 to speak for biology, Weiss, in effect, was proposing that the NRC itself, through the new council, undertake this function. His annual report for 1954 subtly downgraded AIBS. He found AIBS' objectives "not essentially different" from those of B&A, "of which it is still a dependent offspring." Nor was there "a sharp distinction in the manner of promotion of biology by the Division and its subordinate Institute," although the division "features advisory services on matters of policy, whereas the Institute features operational services—usually the assembling of information that will be useful to all biologists."[96] Weiss objected when NSF Assistant Director H. Burr Steinbach wanted to provide funds to AIBS for consulting on the award of travel grants to international meetings, claiming the money should be given to NRC instead. When Steinbach argued that AIBS was closer to "the grass roots of science" than NRC, Weiss replied that only the NRC could properly handle applications from biological societies that were not members of AIBS.[97]

Frustrated by having no control over the overhead on its grants and contracts since the monies went to the Academy, and concerned that its support from biological societies had begun to erode, AIBS made hasty plans to separate itself from the Academy. John S. Nicholas, an AIBS founder and proposer of the BMS divisional committee's resolution to decline Weiss's proposal, wrote to Weiss that he believed "the calculated risk" of independence was preferable to present conditions. Although AIBS' service functions had thus far predominated, it had never lost sight of its primary conception as a policy-making organization. This broader mission was necessary "to bring unity of action among Biologists."[98]

NSF's Division of Biological and Medical Sciences assisted AIBS' departure from the Academy with a ten-month grant of $17,600 for operational expenses. Moreover, Steinbach postponed the proposed contract for travel consultation until AIBS achieved its independence.[99] Weiss was unhappy with the separation, which he claimed was "jumping the gun," though he admitted it would give B&A more space and "above all, freedom of action." He agreed not to contest it, but warned Frank L. Campbell, executive director of AIBS, that "if there should be any careless spreading of misinformation or distortion of facts (e.g., that the Division has been 'stifling' the development of AIBS)," he would "set the record straight," in public if necessary.[100] Bentley Glass, chairman of the AIBS Governing Board, downplayed conflicts with NRC and presented the move in print as a long-intended event symbolizing a coming of age.[101]

Weiss's Biology Council survived less than four years. While various agencies, including NSF, provided awards for conferences and other specific projects, all NRC efforts at securing long-term operational support failed. In 1955, L. A. Maynard replaced Weiss as chairman of B&A, thus easing tensions with NSF. That fall the NRC made a second and last effort to obtain general support for the Biology Council from the Foundation.[102] When BMS staff brought the proposal to the divisional committee in April 1956, the discussion went beyond the overlap of functions to center on the perception that the Biology Council was the creation of one man. Raising "serious doubts as to the identification with and acceptance of such goals, values, and biases by the field of biology at large," the committee voted unanimously to decline to fund the proposal.[103] In March 1957, NIH's National Advisory Health Council turned down a B&A application for $35,000 a year for five years on the ground that the Biology Council duplicated functions of the various institute councils. Moreover, the Rockefeller Foundation failed to renew funding of the Committee on Educational Policies.[104] The Biology Council held its last meeting in October 1956 and formally suspended operations on 30 June 1957. Its executive secretary, summing up, suggested that the council's failure was ultimately due to "the profound reluctance of any governmental agency (used, as they are, to spending vast amounts of money in jurisdictional disputes with each other) to let go one iota on policy considerations to anyone else, including the National Academy" and their unwillingness to commit funds other than on a project basis.[105]

What can be made of this saga of the Biology Council and its relation to NSF? Weiss frequently pointed out that the council did not duplicate functions of NSF's or other agencies' advisory committees, which did not, in fact, undertake the studies begun or contemplated by the council. NSF's panels and divisional committee, with limited program management funds, could accomplish little in the way of policy. Thus, the Biology Council might have served a beneficial function. But biologists, as fragmented as ever, provided it little moral support, just as they had for AIBS. Though NSF played a key role in assuring the demise of Weiss's scheme, all the other federal agencies also feared a central authority, especially one invested with the aura of the Academy. In the end, no one had the resources to take a broad view of biology.[106]

In Search of a Constituency

Although biology at NSF expanded rapidly, especially after Sputnik, BMS found itself at some disadvantage in the pluralistic system of agencies supporting science. NSF had a certain freedom to fund biology that no other agency enjoyed, but it was unable to articulate a unique purpose or specific goals that might be promoted effectively to Congress and the public. Because BMS supported general-purpose, investigator-initiated, basic research in biology, its spokesmen were hesitant to emphasize any particular areas, however promising, for their social benefit. Without specific biological goals, and unwilling to compete directly and publicly with other agencies, BMS could not make a distinctive case for support of biology. The most cogent argument for raising NSF's overall budget, the Cold War and the shortage of "scientific manpower," was perceived as more applicable to the physical sciences and engineering than to biology.

Unlike NIH, USDA, or AEC, NSF lacked a vocal public constituency for funding biology. NIH could rely on disease-related lobbies such as the American Cancer Society and the American Heart Association as well as the personal experiences of members of Congress and taxpayers with devastating diseases. Farmers and local boosters formed a voice for maintaining USDA supported agricultural experimental stations at universities (although they favored practical crop studies rather than basic research). And citizens who were vitally concerned with the hazards of fallout or the possibility that radioactive substances might supply cures for cancer promoted biology at AEC. Few, however, cared passionately about the biological projects that NSF supported, especially given the dry, impersonal manner in which they were presented to the public.[107]

Biologists themselves failed to act as a viable constituency for NSF. In 1951, before NSF was fully functional, the American Astronomical Society had formed a special committee to promote NSF support of astronomy. In the 1950s, through organization and united action, astronomers were able to obtain funding from NSF for a national astronomical observatory at Kitt Peak and a national radio astronomy observatory at Green Bank, West Virginia, as well as a grants program for individual research.[108] Neither large groups of individual biologists nor the various biological societies interacted with NSF in this manner. Biologists seemed especially resistant to any centralized planning. Even the societies that carried out NSF-supported self-surveys did not focus on ways in which NSF might best support their disciplines. Biochemists, members of one of the fastest growing disciplines, objected to any form of self-survey. Individual biologists, especially those in the newer areas of experimental biology, saw NSF as one more federal funding agency while most biological societies were more intent on obtaining grants for educational projects than lobbying for larger budgets for NSF.

Far more fragmented than physics, chemistry, or astronomy, biology had no single group or small number of groups to speak for it. Those in the best position to do so, the AIBS and the NRC Division of Biology and Agriculture, were hampered by lack of funds and, in the case of the Academy, were deeply distrusted by BMS staff. No agency was willing to provide AIBS or B&A with long-term operating funds to become a centralized and competing source of advice. To some extent BMS's divisional committee and individual panels served to represent biological communities and, with their encouragement, BMS began support of field stations, expensive instruments such as electron microscopes, summer support of research by medical students, genetic stock centers, and curatorial support of museums. But these advisors met too infrequently and lacked the resources to champion NSF interests effectively.

While it briefly appeared that with BOB and administration support a larger percentage of funds for basic research might accrue to NSF, other agencies, congressional committees, and scientists themselves successfully prevented this from happening. NSF was reluctant and unable to use its evaluative function to restrict the scope of other agencies' basic research even though it was clear that much of it was not closely mission-related. At the same time, NSF refused to relinquish any area of science regardless of how well it was supported elsewhere. In biology, these policies resulted in considerable overlap among agencies. Despite some grumbling, BMS program officers, like the biologists they served, were by and large satisfied with the pluralist federal system which provided multiple potential sources of funding and assured that no one peer-review committee or program could exercise monopolistic control.

Despite its broad mission to fund all of basic biology, NSF remained only one of many federal supporters of biology in colleges and universities. The following chapter surveys NSF's relations with these other agencies, especially NIH, and takes a critical look at the available data on federal funding of biologists.

CHAPTER 5

Competing within a Pluralist Federal Funding System, 1952–1963

Waterman, speaking in 1951 at a meeting of the American Institute of Biological Sciences, had declared that although several other agencies supported biology, and some supported basic biology related to their missions, NSF's unique contribution among federal agencies supporting biology "will be its absolute freedom from any limitations with respect to practical application."[1] However, as the staff of BMS began to monitor federal support of the life sciences, it quickly became evident to them that NSF's distinctiveness was not so easy to define.

In 1952, Lou Levin viewed with alarm the extent to which other agencies were funding projects in biology that were only remotely connected to their missions. He estimated that the National Institutes of Health was spending about 25 percent of its extramural funds on research "of very general interest to all biological endeavor," while the biosciences division of the Office of Naval Research directed about 80 percent of its funds to projects that "cannot be considered to be of immediate and direct importance to the Navy, insofar as possible application within a reasonable period is concerned." The following year he reported that "almost every agency is actively and eagerly sponsoring basic research in areas such as enzymology, endocrinology, microbial metabolism, protein structure, photosynthesis, etc." though the linking of such research to the missions of ONR, the Atomic Energy Commission, or NIH was "often very tenuous and only pertinent when an extremely long range viewpoint is taken."[2]

If other agencies were supporting considerable basic research in biology, what then was different about NSF? Assistant Director of BMS Fernandus Payne, in 1953, decided that "a distinction, if it exists, rests in the minds of the research workers. A worker with a grant from the Cancer Institute, even though freedom is given, feels somewhat guilty if he neglects completely all possible practical end results. He feels completely free with a grant from NSF and for this reason prefers a grant from NSF." It might be "straining a point," but "if we believe in basic research, there must be no restraint on the part of the worker. Complete freedom is essential."[3]

In the pluralist and abundant funding system of the 1950s, American biologists could chose among a variety of federal patrons for their basic research, among them NIH, the three branches of the military services, ONR, the Atomic Energy Commission, the United States Department of Agriculture, and the National Aeronautics and Space Administration (NASA). Far from cooperating to plan biological projects, agencies operated independently and in suspicion of one another. Policies were decided at the upper echelons and communicated only after the fact.

This chapter examines how NSF situated itself within an uncoordinated and competitive system of federal funding agencies. How, especially, did NSF relate to NIH which, by 1960, had become predominant in general biology? To what extent did NSF's Division of Biological and Medical Sciences staff perceive NIH as a threat to their own activities? To what extent did they attempt to prevent NIH expansion? To what extent did they reconsider NSF's scope of activity in the light of NIH hegemony?

Finally, what portion of basic research in biology did NSF actually support in the 1950s? Though NSF collected volumes of relevant data, the question is surprisingly difficult to answer because each agency used its own definitions of "basic" and "applied" research and "biological," "medical," and "agricultural" sciences. Politics rather than any objective criteria determined how such inherently slippery terms were employed.

The Military Agencies and NASA

Much has been written about the enormous growth and dominance of military funding of the physical sciences, especially physics proper, in the 1950s and 1960s. In addition to all the applied research they supported, the Department of Defense and AEC paid for over 90 percent of basic research in support of academic physics. Historians have debated whether physicists were exploiting the military, persuading the services and AEC to pay for large amounts of basic research they wanted to undertake on the grounds that it would serve national security, or whether physicists were coopted and their research programs redirected by military money and military interests.[4]

Although defense-related agencies took over the lion's share of the nation's basic research in the physical sciences, their role in funding academic biological research was relatively minor. ONR's biosciences division, under the determined leadership of Orr Reynolds, continued in the 1950s to fund a wide range of biological projects, including imaginative basic research that overlapped the research supported by BMS.[5] But as ONR's budgets did not grow appreciably in the pre-Sputnik era, NSF's program in biology quickly surpassed it in size. By FY 1963, ONR was spending $16 million on the life sciences, of which $11 million was claimed to be basic research, compared to NSF's $42 million (see table 5.2 below). Relations between the two agencies remained frequent and cordial. Throughout the decade, ONR continued to supply personnel to NSF, including Randal Robertson and Wayne Gruner in the physical sciences and George Sprugel and Harve Carlson in the biological sciences.

In the early 1950s, the army and air force established offices similar to ONR for funding academic research. Of the military services, the army spent the most money on intramural and extramural research in the life sciences (see tables 5.1 and 5.2 below). Most of its unclassified individual research support in universities went to clinical studies or more basic projects that might contribute to military medicine, as for example physiologist Edward F. Adolph's studies on hypothermia or microbiologist Michael Heidelberger's "Immunochemistry of Pneumococcal Specificities."[6]

The Air Force Office of Scientific Research (AFOSR), created in 1951, aspired to be an agency for funding undirected research, although at times, for political purposes, it camouflaged its contracts to make them appear more relevant to applied missions. Its Life Science Program used the services of a committee contracted through Johns Hopkins University to assist in reviewing proposals. Through the 1950s, AFOSR struggled with jurisdictional disputes and conflict about its placement and role within the air force. In 1961, the agency was reorganized into five directorates, one of them Biological and Medical Sciences, but despite a similar divisional title, the air force seems to have posed no serious competition to BMS.[7] In contrast to ONR, there was little overlap in support between the army and air force and NSF and little contact between these agencies and BMS.

When the Soviet Union launched its first satellite in 1957, John Wilson hoped BMS might take an important role in funding research in space biology. In his annual report for 1958, he predicted, "In the area of program content, probably the most dramatic development will be the role of biological research in relation to space exploration."[8] The following year, he still expected that despite the activity of the Defense Department and the newly created NASA, the new interest in basic research in space biology would be reflected in the 1960 programs of BMS. By 1960, however, Wilson admitted that BMS had received few inquiries and even fewer proposals in that area. Those judged of quality were transferred to NASA. Wilson instituted discussion of possible mutual projects with NASA, but no collaboration ensued.[9] In 1962, Orr Reynolds, who had moved from ONR to DOD in 1957, went to NASA to take charge of its burgeoning biosciences division. Reynolds, who had always taken a broad approach to biological problems, funded with a free hand basic research on organisms ranging from plants to primates, thereby getting into jurisdictional disputes with his administrative counterpart in NASA's human research–biotechnology division.[10] By 1963, NASA claimed to be spending over $15 million on basic research in the biological sciences, exclusive of medicine (see table 5.2).

The USDA and Support of Agriculture and Forestry

More problematical for NSF than either the military agencies or NASA was the U.S. Department of Agriculture's substantial support of research in the life sciences. USDA funded biologists in state agricultural experiment stations, usually associated with land-grant colleges, through block grants, but it had no competitive grants program to support agricultural research at other universities. Agriculture, as the second major area of application of the biological sciences after medicine, might have offered BMS staff an opportunity to justify increased funding for basic biology. If so, BMS did little to pursue it.

Early on, H. Burr Steinbach, assistant director for BMS, saw poor liaison with USDA as a serious problem and a missed opportunity. He urged NSF to reach out to both agriculture and clinical medicine and strongly recommended that special liaison personnel be hired. Steinbach told Waterman that although NSF did not contemplate a major program in either agriculture or medicine, it seemed "obvious" that "a healthy development of national science policy must take into account relationships between such areas as botany, zoology, physiology, etc., and the applied areas of crop management, animal breeding, disease control and public health."[11] Steinbach's astute advice was not heeded. Though BMS funded some research in horticulture, agronomy, soil science, and plant pathology in agricultural experiment stations and other university settings, it did little in either its programming or testimony to Congress to call special attention to agricultural problems, nor did it establish close working relations with USDA.

BMS's one attempt in the 1950s to study USDA support of biology confirmed program officers' preconception that basic research in agriculture was neglected and deficient, but it led to no significant action. In 1956, as a result of divisional committee concern, BMS undertook a survey of the status of basic research in agricultural experiment stations. Vernon Bryson, program director for genetic and developmental biology, and Rogers McVaugh, his counterpart in systematic biology, after visiting fifteen institutions, took a harsh view of the quality of research at experiment stations. They concluded that basic research was faring poorly because of the intense climate of agricultural politics. "The extent to which pressure groups dominate the scientific work of agricultural experiment stations is almost inconceivable to the uninitiated," they wrote in their "administratively confidential" report. "The farmer is more concerned with parity than with photosynthesis."[12]

After a lengthy discussion, the divisional committee reached "general consensus" that it "might be ill-advised to move further than this report, since to do so would seriously involve the Foundation with matters falling within the scope of the Department of Agriculture." The BMS staff, after some discussion with USDA, reached the same conclusion, though BMS claimed that it would continue to consider proposals for basic research in agriculture at institutions where such funds were difficult to obtain.[13] But BMS was highly selective in what it supported. Howard Teas, program director for metabolic biology, reported to the divisional committee in 1963 that there was "an apparent defeatist attitude" among researchers in agricultural experiment stations "since few of their proposals to the Foundation are funded."[14]

BMS responded in a similar manner to the needs of forestry. The U.S. Forest Service, a part of USDA, funded research in forestry to a modest extent but it did not supply competitive grants. Schools of forestry in universities could not command the federal grant support obtained by other areas of academic sciences. The Society of American Foresters, the chief professional organization, served the needs of practitioners as well as academic researchers; its journal had limited space for research results. To improve the situation, the society, with funding from the Rockefeller Foundation, sponsored a study published in 1955 as *Forestry and Related Research in North America*. Claiming that U.S. forestry research was seriously inadequate, it directed its appeal for greater support of basic research specifically to NSF: "[Although] the National Science Foundation has as one of its principal objectives the stimulation and support of basic research in all fields, little of its effort and even less of its funds have been directed towards basic research in forestry and related fields." To address this "extreme poverty" of forestry, the report "invite[d] the attention" of NSF as well as other foundations and funding groups. It also recommended that the National Academy of Sciences and NSF sponsor "a national conference on forestry and related research."[15]

BMS staff did respond in modest fashion in 1955 by contributing toward the establishment of a new outlet for forestry research, the Society of American Foresters' journal *Forest Science*. But BMS neither sponsored the proposed conference nor made any other overture to solicit a constituency among forestry scientists.[16]

In 1961, when Waterman encouraged BMS to select areas of funding emphasis that might be argued before the Bureau of the Budget and Congress, the staff briefly considered forestry research. Duke botanist Paul Kramer, then rotating program director for regulatory biology, surveyed basic research in forestry and found that it had "developed much more slowly than research in agriculture," but its amount and quality were likely to improve as scientists were better trained and administrators came to appreciate the need for adequate facilities. Since basic research in forestry was supported by the Forest Service, primarily in its own laboratories, by Hatch Act funds in state experiment stations, and by the states, Kramer saw NSF's role as "support of research of staff members of forestry schools." The environmental, genetic and regulatory programs, he attested, were already supporting such proposals, whose number and quality were likely to increase.[17] That is, BMS would fund those forest researchers whose projects the panels considered to be highly meritorious basic research in biology—in Kramer's words, "without consideration of the possible practical applications of what is learned" (although he realized that basic and applied research could not be neatly separated). Viewing forestry, like agriculture, with some disdain as a predominantly applied science influenced by commercial interests, BMS staff in the 1950s and 1960s tried to keep abreast of the status of research but were reluctant to take special measures to aid the field.[18]

The AEC and Support of Genetics and Ecology

Apart from NIH, NSF's most successful rival in biology was the Atomic Energy Commission. Through its Division of Biology and Medical Sciences, AEC funded biological research in the AEC-sponsored national laboratories (especially Brookhaven, Oak Ridge, Hanford, and the Berkeley Radiation Laboratory) as well as the work of individual investigators at universities. AEC contracts were awarded through programs managed by strong program directors who, like their NSF counterparts, were responsible for the final selection of projects.[19] According to AEC data, the total "cost of operations" for "Biology and Medicine Research" at AEC rose gradually from $24.5 million in FY 1952 to $36.0 million in FY 1958 and more rapidly thereafter to $53.9 million in FY 1961.[20] The somewhat differing figures supplied by NSF's *Federal Funds for Science*, which presumably included only the actual support of research, indicate that of the AEC's $37.8 million spent on life-science research in 1959 (about 22% of its total research budget), 57 percent or $21.0 million went to the biological sciences, 43 percent or $16.1 million to the medical sciences, and 2 percent to the agricultural sciences.[21]

NSF staff felt AEC competition most keenly in the area of genetics. To promote worker safety in nuclear installations and to counteract public fears of the dangers of fallout and radioactive waste, AEC supported considerable research on the genetic effects of radiation. The Eisenhower administration's "Atoms for Peace" initiative further encouraged support of geneticists. Some projects were of practical value in attempting to determine human hazard; others simply utilized AEC-distributed radioisotopes for labeling biological material.

One of the AEC's largest biological programs was that carried out through the National Academy of Sciences' Atomic Bomb Casualty Commission (ABCC), established in 1947, to investigate the long-term effects on the Japanese population of the nuclear explosions at Hiroshima and Nagasaki. The ABCC genetics project, led in the 1950s by James V. Neel and William J. Schull of the University of Michigan, produced academically valuable work in human genetics, but at the same time, its effort to minimize the dangers of environmental radiation aroused considerable public controversy.[22]

At issue in this and other AEC-supported research was the question of whether—as Nobel prize–winning geneticist Hermann J. Muller believed— radiation from nuclear testing would increase peoples' "genetic load" of harmful mutations and lead to degeneration of the human species. AEC's attempt in the mid 1950s to prevent Muller—a vocal opponent of the ABCC geneticists and of nuclear testing—from airing his views publicly at a United Nations Conference on the Peaceful Uses of Atomic Energy generated much unfavorable publicity for the agency.[23]

Aside from supporting genetics in national laboratories and through the ABCC, the AEC also significantly supported individual geneticists in universities. In FY 1954, according to BMS data, AEC provided 37.9 percent of federal extramural support in "genetic biology."[24] Although by 1958, NSF was funding a larger number of awards in this field, AEC was sponsoring some of the best known geneticists in America, among them, George Beadle at Caltech, David M. Bonner at Yale, Muller and Tracy Sonneborn at Indiana University, Bentley Glass at Johns Hopkins, L. C. Dunn and Theodosius Dobzhansky at Columbia, and Max Demerec at Cold Spring Harbor.[25] It supported these investigators handsomely, paying the full costs of research, though continuing to encourage university participation.

Agency competition threatened to limit the scope of BMS's genetics program. George Lefevre Jr., rotating director for genetic biology in 1957, was frustrated at the lack of balance in NSF's program, which he attributed in part to the molecular and developmental biology programs' handling of the fundamental areas of biochemical and developmental genetics and in part to competition from AEC and NIH. He reported, "The greatest present weakness of the Genetic Biology Program is its position of being the depository for proposals left after the Public Health Service and the Atomic Energy Commission, with their far greater funds, have skimmed off many important (and expensive) projects." Echoing Waterman's claim to NSF comprehensiveness, Lefevre declared: "Simply because of the availability of mission-related funds, the National Science Foundation should not let work in human genetics, developmental genetics, radiation genetics, and the like go by default to other agencies." NSF must support "independent, non-mission-oriented investigations in each of these areas" so that researchers "will be aware that support for any sort of sound, basic genetic research can be sought from the Genetic Biology Program."[26]

AEC also played key roles in photosynthesis research and in ecology. Through support of the Berkeley Radiation (now Lawrence-Berkeley) Laboratory, AEC funded research on the thermochemical reactions of photosynthesis, which, in 1962, won a Nobel prize for Melvin Calvin, who directed a large number of chemists and biologists known as the biodynamics group.[27]

In the 1950s, AEC practically created the field of radiation ecology, precursor to the ecosystem or systems ecology of the 1960s. Public concern about the dangers of radioactive waste and fallout justified AEC support of ecological studies to track radioactive substances in the environment. Beginning in 1951, AEC funded University of Georgia ecologist Eugene Odum's studies at the site of the AEC's Savannah River nuclear facility. From modest beginnings, Odum built up a large contract program, which helped him establish the university as a premier center for ecological research. In 1954, Odum and his brother, Howard T. Odum, obtained support from the AEC for an ecological study of the coral reef of Eniwetok Atoll, site of atmospheric nuclear testing. This pioneer research, the first to measure the metabolism of an entire ecosystem, served as an exemplar of a new and exciting kind of functional ecology in the 1950s.[28] The AEC funded a second major center for ecology under the leadership of Stanley Auerbach at Oak Ridge National Laboratory. Hired as a member of the Health Physics Division at Oak Ridge in 1954, Auerbach developed a large program of basic research in radiation ecology using the Oak Ridge Reservation as a field site. By the 1960s, his group had incorporated computer simulation of ecosystems and laid the framework for what became known as systems ecology.

With the hiring of ecologist John N. Wolfe as a program director in 1955, AEC greatly expanded its support of ecological research. By 1958,

Wolfe headed the new Division of Environmental Sciences, which funded some sixty projects. In addition to its work in radiation ecology, the radioisotopes that AEC distributed widely to biologists proved to be powerful new tools for ecologists. By experimental radioactive labeling of substances in the environment, ecologists could trace energy flow and matter in food chains and in ecosystems. Thus the growth of ecosystem ecology in the 1950s and 1960s received a major impetus from AEC and the "atomic age."[29]

Although it became a powerful patron of biology, AEC, like the military agencies and NASA, devoted the bulk of its resources to the physical sciences and military technology. The NSF hierarchy, too, was far more concerned with the impact of AEC on NSF support of instrumentation in the physical sciences than it was on the agency's support of biology. Waterman, who always promoted NSF support of science across the board, thus worked hard to break AEC's monopoly of funding accelerators and nuclear reactors. Through delicate negotiations with the Budget Bureau and with the AEC, NSF succeeded in securing funds for supporting smaller university-based accelerators and reactors. Although NSF could not compete with AEC in this "big science" arena, at least it could become one of the players.[30]

The NIH and Its Ascendancy in Support of Biology

By the end of the 1950s, the National Institutes of Health had become the largest source of federal support for biologists and NSF's chief competitor. In the years that followed the Long Report, NIH was more successful than ever in expanding its budget and scope of action. NIH Director James Shannon, the disease lobbies, and the congressional appropriations committees, chaired by John Fogarty in the House and Lister Hill in the Senate, formed a loose coalition, backed by widespread popular support, to increase the funding of medical research. Finding new ways of curing diseases rivaled the Cold War as a justification for federal support of university scientists.[31] In 1956 Congress allowed NIH to award grants for constructing or renovating health facilities. The NIH budget for FY 1957 nearly doubled from $98.4 million to $183.0 million, most of the increase going into the extramural programs. Research grants grew from $38.3 million to $89.7 million, and training and fellowships awards from $17.3 million to $33.4 million.[32] John Wilson worried in his annual report for 1956 about the impact of the "sharply increased" NIH programs, predicting some overlapping particularly in the areas of molecular and regulatory biology. But he then backed away from the potential conflict, suggesting that BMS would not feel "the full influence" of NIH's expanding appropriations "inasmuch as a large portion of the increased funds for the Institutes will be expended for large-scale programmatic research on cancer chemotherapy and on the use of tranquilizing drugs in problems of mental health."[33]

Increasingly in the post-Sputnik era, NIH officials, led by Shannon, made a strong pitch for NIH support of biological research basic to medicine, a position further strengthened by the recommendations of the Bayne-Jones Report in 1958. In August 1957, Health, Education and Welfare Secretary Marion Folsom appointed a committee to advise him on future needs in medical research and education. Both Waterman and Long conferred with its chairman, Stanhope Bayne-Jones, former dean of the Yale University School of Medicine.[34] The Bayne-Jones Report, published in 1958, projected expanding sums for federal support of medical research to 1970. Unlike the Long Report (see chapter 4), it suggested no radical changes for NIH. In particular the Bayne-Jones Committee, consisting largely of medical school and industrial administrators, recommended that both grant programs and direct operations remain in NIH. It claimed that NIH's organization by disease-related institutes was not inevitably a defect, as the Long Report had implied, because the system was administered to support both basic and applied research.[35]

Whereas the Long Report avoided the terms "basic" or "fundamental" research, the Bayne-Jones Report unhesitatingly called for increased emphasis on basic science in both research and education. It assumed without question that such research would be funded through NIH; the report scarcely mentioned NSF support of basic biological and medical sciences. The committee recommended that NIH "encourage research basic to medicine by making funds available for the rigorous training of advanced students for research careers in fields basic to medicine, and by providing research grants under terms and conditions that will encourage fundamental studies."[36]

In the late 1950s, spurred by the Bayne-Jones Report and post-Sputnik congressional enthusiasm for basic research, NIH broadened its fellowship program from specific applied areas to a wide range of biological fields more or less related to medicine. Despite Waterman's earlier efforts to keep NIH and AEC out of general-purpose predoctoral fellowships, NIH's predoctoral fellowship program in biology soon surpassed that of NSF. In 1957, NIH awarded about 300 predoctoral and 300 postdoctoral fellowships. By 1960, it was spending $14 million on 3,950 fellowships compared to NSF's $12.7 million for 3,650 fellowships in all fields of science.[37]

But even more significant for certain areas of biology were the NIH training grants. In 1958, these were still largely limited to clinical areas such as cardiovascular disease, cancer, ophthalmology, pathology, or dentistry.[38] After the Bayne-Jones Report, they too were broadened to cover such fields as physiology, biochemistry, immunology, and genetics. Training grants to departments or faculty groups provided not only stipends to graduate students but also general-purpose funds that might be used for such training-associated amenities as a departmental seminar series. Thus, they gave universities much more flexible support than NSF fellowship awards. Already, in 1957, NIH was threatening to enter into undergraduate and even secondary-school training grants, areas in which the dismayed Lou Levin thought "they really do not have any direct business."[39] The Bayne-Jones Report appeared at the same time that members of the BMS Divisional Committee were struggling with the NSF hierarchy over the issue of BMS training grants. Despite Wilson's and the committee's desire to preempt NIH, the education division and the general counsel prevented them from offering training grants in biology (see chapter 3).

H. Burr Steinbach, now chairman of the BMS Divisional Committee, saw clearly the implications of the Bayne-Jones report for NSF. In November 1958, he recorded a meeting with Waterman and the chairmen of the other NSF divisions, in which he "noted the rather special position" of BMS "in that it existed in an area overlapping to a large extent with that of an extremely wealthy and powerful agency, the National Institutes of Health." By administratively combining its research and training functions, NIH avoided "the many complexities that are inherent in the system adopted by the NSF, where a rather sharp distinction is maintained" between these two activities. Besides having "a great deal of money to spend," NIH was "developing a great advantage" from the new report, which was "rapidly approaching adoption as policy by the various Councils. As a result, NIH programs can be operated in a free-wheeling and adventurous sort of fashion without too much worry about transgressing in other people's areas."[40] Steinbach and other divisional committee members urged Waterman to undertake a similar blue-ribbon study to "help solve the problem of educating the public as to what the Foundation should be doing 10 years from now" and to obtain increased resources. But Waterman was too cautious to pursue the idea, fearing that Congress would not take well to "a 10-year programming effort on the part of the Foundation."[41]

By the late 1950s, NIH had eliminated some of the comparative administrative disadvantages of its early awards. At the beginning of the decade, NIH could only make one-year awards while NSF could give awards for up to five years at a time. Although NIH maneuvered around this provision by making a moral commitment for additional years of support, all unexpended sums still had to be returned at the end of each fiscal year. NIH investigators, compared to NSF grantees, also had difficulty transferring funds between budget categories. These administrative barriers were alleviated in 1956 by order of the Surgeon General.[42] In addition, the NIH over-head, originally set at 8 percent, less than the 15 percent allowed by NSF, was increased to the NSF level.

Indicative of the new public significance given to basic research at NIH was the creation of the National Institute of General Medical Sciences (NIGMS). NIH had always set aside funds for the support of noncategori-cal research. Projects not related to any particular class of diseases—in, say, basic biochemistry, physiology, or genetics—were administered through the Division of Research Grants rather than one of the institutes. In 1958 a Di-vision of General Medical Sciences was established. Then, in 1962, an act of Congress authorized creation of a separate institute for research and training in sciences basic to medicine. The NIGMS budget in its first year of opera-tion, FY 1964, was $104.5 million, roughly two-and-a-half times the bud-get of BMS.[43] Some NSF staff members interpreted the creation of NIGMS as NIH's deliberate response to NSF.[44]

Thus, by the end of the Waterman era, NIH funding covered most of the experimental life sciences, even plant research. In 1965, another study committee, appointed by the White House to investigate NIH activities, stated that while a superficial reading of its legislation might suggest an ori-entation to specific diseases, NIH instead "devotes its principal effort to a broad program of investigation of life processes." The committee, chaired by Dean E. Wooldridge, a Caltech physicist and engineer, provided the stan-dard justification for NIH's approach: "Life science is so complex, and what is known about fundamental biological processes is so little, that the 'head-on' attack is today frequently the slowest and most expensive path to the cure and prevention of disease." Being organized by disease-oriented institutes, the Wooldridge committee claimed, had not prevented NIH from carrying out its "real mission." With many institutes, "there is room for the assign-ment to one or the other of substantially all of the special disciplines that comprise the life sciences."[45]

A critic of the Wooldridge report, Joseph D. Cooper, writing a detailed analysis in *Science,* charged that NIH was becoming more of a "science agency" than a "health agency." He faulted the report for not addressing "the distinction between the functions of NIH and the National Science Foun-dation," noting that "the original concept" for NSF was that it fund "free, non-mission oriented research, including the broad area of biomedical sci-ence. The overall thrust of the Wooldridge Committee would seem to be in the direction of expanding what should be an NSF capability, as stated in the NSF charter, through the structure of NIH."[46] Though some scientists might agree in principle, it was clear that in practical terms the tremendous growth of many areas of the biological sciences in the 1950s and 1960s vitally depended on NIH's superior ability to link biological research to the politically popular imperative of conquering diseases. Few wished to quarrel with success.

Federal Support of Basic Biology: The Politics of Numerical Data

What share of basic research in biology did NSF command? Although NSF tried to collect comparative data on the funding of biology by the various federal agencies, the figures say more about proprietary and defensive agency politics than they provide any objective measure of funds going into "basic biology." Because the definitions of both "basic research" and "biological sciences" were ambiguous, the various agencies were able to define them differently at different times according to ideological and political need.

"Basic research" is a problematical term. Is it defined by the subject matter itself, by the investigator's motive in undertaking the research, or by the agency's motive in supporting it? NSF representatives insisted on a tripartite distinction between basic research, applied research, and development. Basic research, according to NSF, was "that type of research which is directed toward the increase of knowledge in science." Applied research was that "directed toward the practical application of science" while development was "the systematic use of scientific knowledge directed toward the production of useful materials, devices, systems, or processes other than design and production engineering."[47]

Other agencies were less enamored of these distinctions. Agency officials, such as those at NIH, argued that no line could be drawn between basic and applied research. Research that an investigator perceived as basic could at the same time serve an applied function for the agency. Mission agencies resisted NSF attempts to have them separate the research they funded into the categories of "basic" and "applied." In the early 1950s, since it was administration policy that NSF should become increasingly responsible for general-purpose basic research, agencies were understandably reluctant to report too much basic research. After Sputnik, when a premium was placed on basic research to counter the predictions of future Russian domination in science, it became acceptable for mission agencies to classify large parts of their research as basic. Because each agency was allowed to distribute its own research expenditures among the categories formulated by NSF, these issues of public relations were reflected in the data in NSF's annual statistical summary, *Federal Funds for Science* (later *Federal Funds for Research, Development and Other Scientific Activities*).

In FY 1953, the first year for which full data was available, a total of

$182.9 million was spent on federal intramural and extramural research in the life sciences. NSF's minuscule share was $1.2 million, or 0.6 percent (see table 5.1a). But these figures included large amounts of applied agricultural and clinical research. What about basic research? NSF always declared, as a point of pride, that 100 percent of the research it supported was basic. The Department of Health, Education and Welfare labeled 32.6 percent of its research in the life sciences in FY 1953 basic. Other agencies affirmed much smaller portions: 14.3 percent for USDA, 7.0 for the navy, 14.1 for the AEC, and as little as 0.6 and 0.8 percent for the army and air force. Thus, according to the data, NSF's share of "basic research" in the life sciences for fiscal 1953 was 4.4 percent (see table 5.3).

By 1963, in the heyday of congressional support for basic research, agencies declared a much larger portion of their research in the life sciences to be basic. By this time, according to the data, the total for intramural and extramural research in the life sciences had increased over fivefold to $936.7 million. NSF's share was $41.7 million, or up to 4.4 percent (see table 5.2a). But now only the army and air force claimed less than 20 percent of their research in the life sciences to be basic. NIH declared 37.6 percent. Other agencies went well above 50 percent: 68.6 for the navy, 73.0 for NASA, and 87.2 for the AEC. NSF's share of "basic research" was 11.0 percent of a total of $378.1 million spent on the life sciences as compared to NIH's 50.9 percent and AEC's 13.0 (see table 5.2a).

Even more open to ambiguity and political manipulation was each agency's breakdown of research in the life sciences into biological, medical, and agricultural sciences. It is clear from the published data that NSF and NIH used very different practical definitions of "biological" and "medical." According to the NSF definitions, medical sciences were "those sciences which, apart from the clinical aspects of professional medicine, are concerned primarily with the utilization of scientific principles in understanding diseases and in maintaining and improving health." Agricultural sciences were "those sciences directed primarily toward understanding and improving agricultural productivity." Biological sciences were the leftovers—"all sciences other than the medical and agricultural sciences which deal with life processes." The NSF compilers of the first report for FY 1952 admitted with circumspection that "there is a tendency in certain cases for the basic mission of a reporting agency to influence the classification of its funds. Where the mission agency can be closely identified with a particular scientific field, there is an understandable inclination for the agency to consider all of its scientific activities as falling in that field." Thus, they suggested, the biological sciences were probably "understated."[48]

In distributing its grants among these categories, NSF typically attrib-

Table 5.1a   Federal Obligations for Scientific Research in the Life Sciences
by Agency, FY 1953 (in Thousands of Dollars)

	Biological Sciences	Medical Sciences	Agricultural Sciences	Total
USDA	6,155	—	33,255	39,410
Army	20,070	12,875	—	32,945
Navy	5,500	5,905	—	11,405
Air Force	9,860	5,915	—	15,775
HEW	913	44,118	—	45,031
Interior	4,954	—	207	5,161
AEC	14,238	9,841	—	24,079
NSF	1,159	—	—	1,159
Smithsonian	104	—	—	104
TVA	270	—	2,102	2,372
VA	—	5,091	—	5,091
Other	—	147	218	365
Total	63,223	83,892	35,782	182,897

Source: National Science Foundation, *Federal Funds for Science,* III, FY 1953, p. 32.

Table 5.1b   Federal Obligations for Basic Research in the Life Sciences
by Agency, FY 1953 (in Thousands of Dollars)

	Biological Sciences	Medical Sciences	Agricultural Sciences	Total
USDA	582	—	5,074	5,656
Army	—	210	—	210
Navy	795	—	—	795
Air Force	—	125	—	125
HEW	510	14,153	—	14,663
Interior	369	—	37	406
AEC	2,414	981	—	3,395
NSF	1,159	—	—	1,159
Smithsonian	104	—	—	104
Total	5,933	15,469	5,111	26,513

Source: National Science Foundation, *Federal Funds for Science,* III, FY 1953, p. 35.

uted the bulk of its life science grants to the biological sciences, a small portion to medical sciences, and little or none to agricultural sciences.[49] NIH,
by contrast, declared only a minuscule amount of its large number of grants
to be in the "biological sciences." For FY 1953, HEW classed as biological

Table 5.2a   Federal Obligations for Scientific Research in the Life Sciences
by Agency, FY 1963 (in Thousands of Dollars)

	Biological Sciences	Medical Sciences	Agricultural Sciences	Total
USDA	12,141	13,015	69,970	95,126
Army	25,950	31,973	60	57,983
Navy	8,565	7,685	—	16,250
Air Force	7,668	4,514	—	12,182
Other DOD	3,743	2,270	727	6,740
NIH	15,402	495,946	—	511,348
Other HEW	2,042	54,918	—	56,960
Interior	22,572	—	362	22,934
AEC	35,267	19,459	1,711	56,437
NASA	17,497	3,537	—	21,034
NSF	33,966	7,662	40	41,668
Smithsonian	1,361	—	—	1,361
VA	—	26,620	—	26,620
Other	159	6,408	3,493	10,060
Total	186,333	674,007	76,363	936,703

Source: National Science Foundation, *Federal Funds for Science,* XIII, FY 1963, pp. 140–41.

Table 5.2b   Federal Obligations for Basic Research in the Life Sciences
by Agency, FY 1963 (in Thousands of Dollars)

	Biological Sciences	Medical Sciences	Agricultural Sciences	Total
USDA	6,098	8,106	22,516	36,720
Army	3,273	4,540	—	7,813
Navy	4,442	6,711	—	11,153
Air Force	1,735	120	—	1,855
NIH	15,402	177,118	—	192,520
Other HEW	620	12,944	—	13,564
Interior	4,597	—	—	4,597
AEC	30,740	16,968	1,507	49,215
NASA	15,361	—	—	15,361
NSF	33,966	7,662	40	41,668
Smithsonian	1,361	—	—	1,361
VA	—	2,315	—	2,315
Total	117,595	236,484	24,063	378,142

Source: National Science Foundation, *Federal Funds for Science,* XIII, FY 1963, p. 150.

Shaping Biology

Table 5.3   Percentages of Research in the Life Sciences Designated by Agencies to Be "Biological Sciences" and/or "Basic Research," FY 1953 and FY 1963 (in Thousands of Dollars)

Agency	Total Life Sciences	% Biological Sciences	% Basic Research	% Basic Biology	% of Biological Sciences That Is Basic
			FY 1953		
USDA	39,410	15.6	14.3	1.4	9.5
Army	32,945	60.9	0.6	—	—
Navy	11,405	48.2	7.0	7.0	14.4
Air Force	15,775	62.5	0.8	0.8	—
HEW	45,031	2.0	32.6	1.1	55.9
Interior	5,161	96.0	7.9	7.1	7.4
AEC	24,079	59.1	14.1	8.9	17.0
NSF	1,159	100	100	100	100
			FY 1963		
USDA	95,126	12.8	38.6	6.4	50.2
Army	57,986	44.8	13.5	5.6	12.6
Navy	16,250	52.7	68.7	27.3	51.9
Air Force	12,182	62.9	15.2	14.2	22.6
NIH	511,348	3.0	37.6	3.0	100
Other HEW	56,960	3.6	23.8	1.1	30.4
Interior	22,934	98.4	24.4	24.4	20.4
AEC	56,437	62.5	87.2	54.5	87.2
NASA	21,034	83.2	73.0	73.0	87.8
NSF	41,648	81.6	100	81.6	100

Source: Calculated from figures in National Science Foundation, *Federal Funds for Science,* III, FY 1953, and XIII, FY 1963.

Note: The first column gives the total amount obligated by each agency for intramural and extramural support in the life sciences. The second column (% Biological Sciences) represents the percentage of the total obligated (col. 1) that was declared to be "biological sciences." The third column (% Basic Research) gives the percentage of the total obligated that was declared "basic." The fourth column gives the percentage of the total obligated declared to be both basic and biological. The last column gives the percentage of the amount declared biological (col. 2) that is also declared basic.

sciences only 2.0 percent of its total of $45.0 million for the life sciences and only $510,000, or 1.1 percent, as basic biological sciences. Ten years later, the corresponding figure was still only 3.0 percent (see tables 5.1, 5.2, 5.3). Thus, from the published data, it would appear that from 1952 on through the 1960s, NSF supported several times more research in the biological sciences than NIH! Even in 1953, NSF reported supporting 19.5 percent of all research in the basic biological sciences (tables 5.1 and 5.3). This figure went as high as 45.2 percent in 1959, which represented over four times the NIH share.[50] The only explanation for these most anomalous figures is that each agency employed different criteria, based on political considerations, for classifying its research funding. NSF could have been accused of duplication if it claimed to support much medical science. Similarly, NIH would be vulnerable to criticism if it admitted funding a lot of biology that was not directly related to medicine. Because NIH greatly underreported its contribution to biology, NSF's was correspondingly magnified.

By the end of the Waterman era in 1963, when federal support of basic research in the life sciences totaled $378.1 million, NSF claimed to account for $34.0 million (28.9%) of a total of $117.6 million for basic research in the biological sciences, compared to $15.4 million for NIH, $9.5 million for the military agencies, $6.1 million for USDA, and $30.7 million for AEC (table 5.2b).[51] AEC's large figure is partially explained by the fact that it classified 87 percent of its research in the biological sciences as basic. The military agencies (53.1%), especially the army (33.0%), and the Department of the Interior (26.2%) dominated applied research in the biological sciences. All of the other agencies attributed most of their work in the biological sciences to basic research. For AEC, NASA, and NIH, the percentages were respectively 87.2, 87.8, and 100. Throughout the Waterman era, NIH accounted for about 75 percent of basic research in the medical sciences, and USDA for about 95 percent of basic research in the agricultural sciences.

What can be made of such figures? They demonstrate above all that categories are open to varying interpretations. While NSF had an important but not dominant role in funding basic research in biology, the exact measure of that role is elusive. A further disadvantage of this series of data from the perspective of BMS is that it included both intramural and extramural research. NIH and especially AEC spent significant portions of their life science funds on intramural programs. BMS program directors wanted to know what each agency's contribution was in the arena that concerned NSF, that is, federal support of basic research in colleges and universities.

To deal with this problem, BMS began in 1953 to collect its own data. Its annual *Federal Grants and Contracts for Unclassified Research in the Life Sciences* covered the years 1952 through 1958, ceasing when the post-Sputnik spurt made the task of compiling the data prohibitive.[52] Consolazio was in charge of the earliest effort to gather and arrange the data. BMS staff obtained lists of individual extramural grants from agencies and distributed them into BMS-determined categories. These consisted of the BMS functional categories (plus "structural biology") plus categories for medicine (pathology, diagnosis, therapy, etc.), applied research in agriculture (plant, animal, and soil management), technology (techniques), and training awards. (Psychology was handled under a separate compilation.) BMS normalized all grants and contracts to one-year awards including overhead and listed grants by category, state, institution, and investigator. Thus, the raw data provided such useful information as which agencies were supporting which investigators, and how much money by category was going to each institution.

According to the BMS data series, the total amount, excluding fellowships, spent by federal agencies on the life sciences in universities and nonprofit institutions rose from $46.6 million in calendar 1952 to $195.6 million in fiscal 1958 (see table 5.4). The functional categories, which may be assumed to be largely basic biology, grew from $18.3 million in 1952 to $65.2 million in fiscal 1958. Through the 1950s molecular and regulatory biology received the lion's share of funding, roughly three-quarters of all awards, in these categories. Systematic and developmental biology each made proportional gains, while environmental biology grew more slowly. The sudden spurts of funding for training grants between 1954 and 1956 and for facilities between 1956 and 1958 were mostly attributable to NIH.

Unfortunately, this data series did not provide separate figures for each grant, making it impossible to derive total amounts by category for each agency. The one available breakdown by agency, for FY 1954, clearly shows NIH dominance of federal funding of biology, as defined by the functional categories (see table 5.5). HEW accounted for 58.1 percent of molecular biology, 60.4 percent of regulatory biology, 69.0 percent of structural biology, and 81.2 percent of developmental biology. In genetic biology, AEC supplied almost as much money as HEW; together they accounted for 80 percent of the total for this area. The military agencies and AEC (65.3%) led environmental biology. NSF, still a very small agency, provided overall only 4.8 percent of funds in the functional categories. But even in 1954, it led—with 61.3 percent—all other agencies in support of systematic biology.

The BMS compilers set different, but equally arbitrary, boundaries between biological and medical, and between pure and applied, from *Federal Funds for Science.* "Regulatory biology," for example, included several projects that could have been classified as medical, some dealing with specific diseases. And BMS, which claimed to support only basic nonclinical research, listed a few of its own grants under "pathology." While *Federal Funds for Science* overestimates NSF support of basic biology, the BMS data probably underestimates it. In fact, no clear line can be drawn between biological and medical or pure and applied.

The BMS data suggest that many, if not most, leading biologists emphasizing laboratory rather than field research were funded by more than

Table 5.4 Federal Extramural Grants and Contracts in the Life Sciences by Category, 1952–FY 1958 (in Thousands of Dollars)

Category	Calendar 1952	FY 1954	FY 1956	FY 1958
Molecular biology	6,044	7,485	10,186	20,945
Regulatory biology	7,850	12,461	13,732	27,048
Structural biology	593	1,028	1,720	3,148
Genetic biology	1,145	1,577	1,966	3,687
Developmental biology	1,258	1,694	2,319	5,369
Environmental biology	1,181	1,602	2,268	3,372
Systematic biology	285	446	787	1,618
Subtotal	18,356	26,292	32,978	65,187
Pathology	7,092	12,243	14,324	25,900
Diagnosis	678	861	1,575	3,748
Therapy	5,323	6,540	6,435	13,115
Community health	3,024	4,083	4,145	6,447
Plant management	4,021	4,114	7,272	8,680
Animal management	2,771	2,990	5,527	6,834
Soil management	854	1,023	1,634	2,298
Technology	1,347	3,114	5,284	7,312
Methodology	1,308	432	1,202	923
Equipment design	470	656	1,211	1,384
Training	66	74	7,076	15,436
Scientific information	364	874	1,040	2,398
Facilities	954	1,439	1,933	35,979
Total	46,628	64,731	91,638	195,640

Source: National Science Foundation, *Federal Grants and Contracts for Unclassified Research in the Life Sciences,* 1952, FY 1954, FY 1956, FY 1958.

one grant and often by more than one agency. In FY 1958, F. O. Schmitt of MIT had five separate NIH grants plus another from ONR. Irwin C. Gunsalus of the University of Illinois was funded by four separate agencies: NSF, ONR, AEC, and NIH.[53] Even BMS Divisional Committee members relied more heavily on NIH than NSF for funding (see table 5.6).

What then can be said of NSF's role in funding biology in colleges and universities? Through the early 1950s, NSF funded a minor share of extramural biological research, even if its list of grantees contained some of the most illustrious names in their respective fields. By the end of the decade, it

Table 5.5 Federal Extramural Grants and Contracts in the Life Sciences by Agency, FY 1954

Agency	Molecular	Regulatory	Structural	Genetic	Developmental	Environmental	Systematic	Total
A. IN THOUSANDS OF DOLLARS								
USDA	92.0	305.3	4.2	93.2	42.8	34.3	13.1	584.9
AEC	1,416.2	824.2	40.2	597.3	98.6	154.4	—	3,130.9
Army	545.7	1,439.1	63.2	50.9	25.6	469.8	5.7	2,600.0
Navy	639.7	1,000.7	50.0	70.8	79.2	291.6	29.7	2,161.7
Air Force	26.9	1,071.3	137.1	—	—	130.6	31.4	1,397.3
HEW	4,345.1	7,521.7	709.3	647.8	1,376.3	375.2	92.6	15,068.0
NSF	380.7	290.2	23.6	117.4	71.5	98.0	273.1	1,254.5
Other	38.7	8.1	—	—	—	47.7	—	94.5
Total	7,485.0	12,460.6	1,027.6	1,577.4	1,694.0	1,601.6	445.6	26,291.8
B. PERCENTAGES FOR EACH FUNCTIONAL CATEGORY[a]								
USDA	1.2	2.5	0.4	5.9	2.5	2.1	2.9	2.2
AEC	18.9	6.6	3.9	37.9	5.8	9.6	—	11.9
Army	7.3	11.5	6.2	3.2	1.5	29.3	1.3	9.9
Navy	8.5	8.0	4.9	4.5	4.7	18.2	6.7	8.2
Air Force	0.4	8.6	13.3	—	—	8.2	7.0	5.3
HEW	58.1	60.4	69.0	41.1	81.2	23.4	20.1	57.3
NSF	5.1	2.3	2.3	7.4	4.2	6.1	61.3	4.8
Other	0.6	0.1	—	—	—	3.0	—	0.4

Source: William V. Consolazio and Margaret C. Green, "Federal Support of Research in the Life Sciences," Science, 124 (1956): 522–26.
[a]Each column rounds to 100%.

Table 5.6   Sources of Federal Funding in FY 1958 for Members of the BMS Divisional Committee from 1954 to 1966

Name	Field of Biological Sciences	Institutional Affiliation	Federal Patrons
Edgar Anderson	Botany/Genetics	Missouri Botanical Garden	NSF
Marston Bates	Biology/Zoology	U. of Michigan	—
Frank Brink	Biophysics	Rockefeller Inst.	—
Lincoln Constance	Botany	UC Berkeley	NSF
Bernard D. Davis	Biochemistry/Microbiology	Harvard Medical School	NIH, NSF
William F. Hamilton	Physiology	Medical College of Georgia	NIH
C.N.H. Long	Physiology/Biochemistry	Yale	NIH, NSF
William D. McElroy	Biology/Biochemistry	Johns Hopkins	AEC(2), ONR, NSF
C. Phillip Miller	Medicine	U. of Chicago Medical School	AEC, NIH
Clarence P. Oliver	Genetics	U. of Texas	NIH
Frank W. Putnam	Biochemistry	U. of Florida Medical School	NIH(2)
Esmond E. Snell	Biochemistry	UC Berkeley	NIH(2)
H. Burr Steinbach	Zoology/Physiology	U. of Chicago	NSF
Kenneth V. Thimann	Plant Physiology	Harvard	NIH(2)
George Wald	Biochemistry	Harvard	NIH(2), ONR

Sources: *American Men and Women of Science*, various eds.; BMS Annual Reports; *Federal Grants and Contracts for Unclassified Research in the Life Sciences, Fiscal Year 1958* (Washington, D.C.: NSF, 1961).

Note: Psychologist Frank Geldard was omitted because psychology was not included in the listing of grants in the life sciences.

had surpassed the military agencies and even the AEC, but in most of its functional categories of biology, it ran a distant second to NIH.[54]

NSF and the NIH Hegemony

Despite the ambiguity of the data NSF collected, by 1960 it was clear to almost everyone that the National Institutes of Health dominated funding of biology not only in medical schools but in life science departments of main university campuses as well. How then did NSF respond to NIH hegemony in the biological sciences? BMS staff did not attempt to negotiate with NIH or with BOB to demarcate spheres of interest. Nor could BMS hope to obtain budgets comparable to NIH's. Rather BMS's strategy was to compete by not allowing NIH to monopolize any type of grant to biologists outside of medical schools. Thus, if NIH proposed training, facilities, or institutional grants for departments of biology, BMS lobbied the NSF administration vigorously to obtain the right to offer similar awards. Although unsuccessful with training grants, BMS, under the stimulus of NIH competition, pressured the NSF hierarchy into soliciting funds for various forms of institutional grants (see chapter 6).

In 1957, when NIH funding went up sharply, Wilson urged that in order to remain competitive with NIH the size of NSF grants must rise in areas of overlap to match NIH grants. Noting "the ease with which large grants may be obtained" from NIH, Wilson denied "desiring to compete" with that agency but insisted that, for "a healthy state of affairs," there should be "more than one source of funds for medical and biological scientists." His point was that "if NSF granting practices and possibilities are such that the Foundation is not approached by good investigators because of our limited grant size, the net effect is the same as if the Foundation did not exist."[55] Indeed, when the BMS budget doubled in 1959, Wilson, with divisional committee support, deliberately used the additional funds to enlarge the grant size for "high quality projects" rather than to increase the number of projects funded.[56]

NIH expansion meant that BMS received increasing numbers of duplicate proposals, especially in molecular, regulatory, and metabolic biology and, to a lesser extent, psychobiology; by 1958, 10 percent of BMS proposals were also sent to NIH.[57] NSF staff encouraged this duplication even if the investigator, given a choice of patrons, usually selected NIH because its grants tended to be larger. NSF continued to fund many leading scientists who deliberately divided their projects between the two agencies (see chapter 8).

Whether responding to competition with AEC in the physical sciences or with NIH in the biological sciences, NSF officials subscribed to a fixed set of assumptions. First, NSF was the *only* agency to fund basic research for its own sake. Despite evidence to the contrary, NSF officials affirmed publicly that other agencies funded basic research *only* in the areas of their missions. Second, NSF must fund all areas of basic science regardless of other agencies' program emphases. Third, the availability of more than one federal patron benefited science by protecting the autonomy of the individual scientist. It would be dangerous for the progress of science if one agency were to acquire a monopoly over the funding of any area of science.

The phenomenal success of NIH challenged these assumptions at the end of NSF's first decade, but Waterman and the BMS staff held firm to them. In late 1959, Waterman recorded in a diary note that Graham DuShane, editor of *Science,* had inquired confidentially about "a rumor which is going around that NSF would get out of the business of making grants in basic biology." Apparently a remark made at an NIH Council meeting had "spread like wild fire." Its substance was "that NIH has plenty of money for basic biology so why should NSF be in the field at all." Waterman admitted that "occasionally the question arises," and "our reply is that if NSF is going to support basic research in all the sciences, then no area can be omitted."[58] When the divisional committee discussed NSF's relationship to NIH in 1959, Waterman told the members (who fully agreed) "that it was his belief that the Foundation must maintain support programs in every field of the basic sciences." He maintained that "government policy in general agrees with this philosophy, with general basic research responsibilities remaining firmly in the hands of the NSF."[59]

The following year Waterman felt the need to reiterate publicly this justification for continued NSF presence in biology. In the introduction to *Science—The Endless Frontier,* reprinted for NSF's tenth anniversary, he wrote:

> The National Institutes of Health stresses research aimed at the care and cure of diseases, including basic research related to its mission, as defined by Executive Order 10521. The National Science Foundation, on the other hand, supports basic research in this area primarily for the purpose of advancing our knowledge and understanding of biological and medical fields. With more than one source of funds available from the Federal Government, scientists enjoy the broader base of support that is consistent with the American tradition.[60]

It was only in the 1970s that NSF staff began seriously to reconsider these assumptions.

CHAPTER 6

Funding Individuals and Institutions in the 1960s

Opportunities and Constraints

NSF's second decade of funding biology brought to the fore a new generation of leaders, an expanded scope of support, and a new set of opportunities and constraints. The 1960s was a decade of continued expansion of BMS budgets and an even more rapid expansion of biologists' expectations. It was a decade of "big biology"—international programs, million-dollar research ships, expensive controlled-environment facilities, new multistory laboratory buildings, and attempts to raise selected second-level universities into "centers of excellence." It was also a time when older fields of biology clashed with the newer specialties in academia, as departmental structures in the biological sciences were overhauled, and within BMS, as funding priorities among biological fields shifted. While molecular biologists won Nobel prizes for cracking the genetic code, ecologists took the limelight by the end of the decade by promising to alleviate the widely perceived environmental crisis. Then at the end of the 1960s, in the midst of the escalating war in Vietnam, the monumental growth of postwar federal science funding came to an end, creating chaos on campuses and within BMS.

Sputnik brought about widespread acceptance of the ideology of basic research, but it also prompted a reexamination of science policy that led to a greatly expanded institutional structure of federal science advice and oversight. In 1957 the Eisenhower administration created within the White House the position of Special Assistant for Science and Technology (the president's science advisor), a post first held by James R. Killian of MIT, and the President's Science Advisory Committee (PSAC), composed of academic scientists. In December 1958 on PSAC recommendation, Eisenhower set up the Federal Council on Science and Technology, made up of heads of science agencies and charged with improving coordination of federal research and development. Finally, in 1962, the Kennedy administration established the White House Office of Science and Technology (OST), transferring to it NSF's responsibility for national science policy, which the Foundation had, for good reason, shirked.

Congress, finding itself in policy-making competition with the executive branch, formed its own new institutional structures. The House Committee on Science and Astronautics, created in July 1958 and chaired in the 1960s by George P. Miller, Democrat of California, provided a focal point for assessing the entire federal system of research support. Among its duties was oversight of NSF.[1] Concerned with such issues as controlling expenditures and coordinating research, Congress named in 1963 an investigative Subcommittee on Science, Research and Development, chaired by Emilio Q. Daddario, Democrat of Connecticut. It began a specific review of NSF in 1965, which ultimately resulted in amendments to the NSF charter.[2]

The vastly greater scale of science after Sputnik created strains in the university-government relationship that were widely debated through the decade.[3] While prewar and early postwar university administrators had welcomed grants from private foundations and the federal government as supplements to university support of research, by 1960, they had come to expect and depend upon federal funding. They claimed, with justification, that they could not afford to subsidize the costs of greatly expanded federal research on campuses and called for full reimbursement of sponsored research through higher overhead allowances and the payment of faculty salaries for the time spent on research.

Another focus of criticism was NSF's emphasis on supporting individual project research. Some administrators argued that federal money was undermining university authority to the point that grantees had greater loyalty to Washington than to their institutions. The project system was said to have caused serious imbalances on campuses by its emphasis on research at the expense of teaching. Although some biologists satirized the "grantswinging cycle"—a takeoff on familiar metabolic cycles—in which cycles of proposals, projects, and publications generated promotions,[4] most individual scientists liked the project system. Department chairs and administrators, however, called for more flexible funds that could be used for hiring new faculty, purchasing departmental and laboratory equipment, training graduate and postdoctoral students, or bringing in speakers. They especially pointed to the inadequacy of current university teaching and research facilities for handling the increased scale of research and the expected mushrooming of university enrollments as baby boomers reached college age.

The growth of federal funding, public scrutiny, and academic demands in the "Golden Age" of post-Sputnik patronage all led BMS to experience a sharp rise in internal bureaucracy and consequent restraints in grant-making. Pressure grew for more accountability, sometimes creating havoc as illustrated by the "AIBS affair," discussed in this chapter. BMS staff were increasingly occupied with preparing for hearings and providing information to Congress. Program directors resisted calls for rigid fiscal accountability, for payment of faculty salaries, and for geographical distribution, and promoted stratagems to preserve their ideal of funding the best researchers on the basis of mutual trust. Nevertheless, the freewheeling, 1950s style of BMS grant-making gave way to more orderly (bureaucratic) and open procedures, and to funding a broader range of people and institutions.

If BMS staff resisted bureaucracy, they embraced the opportunities afforded by the "institutional support problem" to expand patronage of biology beyond the individual project grant. They were at the forefront of initiating institutional programs in NSF, both to remain competitive with NIH and to maintain NSF's image as the general-purpose patron for basic research. For a decade, a variety of institutional programs including the construction of graduate-level laboratories, institutional base grants, and science development programs, flourished. Biologists were provided with new and up-to-date laboratory buildings, most of which are still in active use. But the growth in the number of biologists encouraged by the project grants, fellowships, and institutional awards of the "Golden Age" would create serious problems for NSF in future years.

A New of Generation of Managers of Science

At the beginning of the 1960s, the "triumvirate" of Consolazio, Wilson, and Levin left BMS, opening the way for a new generation of leadership. Levin, after serving a year as BMS program director for special facilities, became, in 1960, the first head of the Office of Institutional Programs. Shortly afterward, he left NSF to become a dean at Brandeis, but by 1964, he had returned, first as head of the Office of Program Development and Analysis and then back in charge of institutional programs as Associate Director (Institutional Programs). In 1961 Wilson departed to become special assistant to George Beadle, then president of the University of Chicago, only to be lured back two years later to become NSF's deputy director. Finally, Consolazio, disappointed at not succeeding Wilson as assistant director, left BMS in 1961 to join the NSF Science Resources Planning Office. The passage of the triumvirate from BMS was followed in mid-1963 by the departure of Waterman, who had stayed on beyond the normal retirement age of seventy until a successor could be chosen.

Leland J. Haworth, director of the Foundation from 1963 to 1969, was, like Waterman, a physicist. Before coming to NSF, he served as director of the Brookhaven National Laboratory and was a member of the Atomic Energy Commission. Wilson, as deputy director, ran NSF on a day-to-day basis, especially as Haworth was frequently ill and out of the office. In general, Haworth maintained NSF's hallmark dedication to basic science, although he was more open than his predecessor to the possibility of supporting applied research.

The second generation of science managers who led NSF's Division of Biological and Medical Sciences in the 1960s differed in background and outlook from the founders but, as before, the personalities and experiences of individual program directors significantly shaped the policies and programs of the division. Harve J. Carlson (1911–90), the new assistant director for BMS, was the last of the series of BMS personnel brought over from the Office of Naval Research. A microbiologist with a 1943 doctorate in public health from the University of Michigan, Carlson had been associated since 1951 with ONR where he served first at the San Francisco office as a liaison with West Coast grantees, then as scientific liaison officer to the London office, and finally as head of ONR's microbiology branch in Washington. Hired to direct BMS Facilities and Special Programs in 1959, he first replaced Levin as deputy assistant director before being promoted to assistant director in 1961.[5]

Carlson's greatest strength, as he himself recognized, was his ability to get along with a variety of people. Program directors liked his fairness, willingness to delegate authority, and encouragement of new ideas. Until the end of the decade, he managed to bridge the growing gap between the laboratory and field programs in BMS. Not as brilliant as his predecessor, he also differed from Wilson in his greater sympathy toward geographical distribution and wholehearted support of institutional development programs. Carlson's chief weakness was his limited knowledge of current biology. He relied heavily on his staff for input, ideas, and report writing, but was recalled as a very good "front man."[6]

The division's three leading members in the 1960s, aside from Carlson, came to NSF from academia or museums. BMS continued to attract a stream of talented rotators, some of whom decided that administration was more exciting than running a research program. David D. Keck, whose pioneer botanical transplant experiments helped establish the concept of species as natural populations, was head curator and acting director of the New York

Botanical Garden before coming on leave to NSF in 1958 to head the Systematic Biology Program. Carlson's "big picture" style of management was effectively balanced by his choice of the detail-oriented Keck to manage daily business as deputy assistant director from 1961 until his retirement in late 1967.[7]

Herman Lewis came from Michigan State as a rotator for a year and stayed for twenty. Earlier, at MIT, he had listened to presidential science advisor Jerome Wiesner's tales of science in Washington, and when Dean R. Parker, the previous rotator in genetics, suggested he come to NSF, he had come to believe it was a scientist's duty to take a turn at public administration. Lewis headed the Genetic Biology Program from 1962 to 1973. An administrator of strong social conscience, he advocated in the 1960s that NSF begin a program on the ethical implications of science. In the 1970s, he took a major role in the scientific debate over the potential risks of genetic engineering.[8]

David B. Tyler, a physiologist and biochemist who was program director for regulatory biology, had been a member of the Carnegie Institution of Washington's Department of Embryology and director of the pharmacology department of the School of Medicine and Dentistry of the University of Puerto Rico before joining the Foundation in 1961. He took over from Levin the role of chief biomedical liaison for the division. An outspoken maverick, Tyler impressed some colleagues by his brilliance while alienating others by his opinionatedness. An elitist in outlook, Tyler felt strongly that individual project grants should be complemented by flexible grants to departments, which he regarded as the fundamental unit of the university.[9]

Three other long-term program directors came from government. Walter H. Hodge and Jack T. Spencer were previously staff members of the U.S. Department of Agriculture. Hodge, an economic botanist, was initially hired in 1961 as a special assistant in tropical biology and then stayed on until 1973 to direct the Systematic Biology Program. Spencer, a botanist and agronomist, was in charge of Facilities and Special Programs from 1961 to 1968. The Psychobiology Program was headed from 1958 on by Henry Odbert, formerly a psychology professor at Dartmouth, who served as branch chief of the personnel laboratory at Wright Air Development Center, Lackland Air Force Base, Texas, before coming to NSF.[10]

BMS became more bureaucratized in 1964 when, primarily to attract talented people through better salaries, it was reorganized into four sections, each with two programs, one of which the section head continued to manage. In the Molecular Biology Section, run by a series of rotators, the original molecular biology gradually fissioned into separate biochemistry and biophysics programs. Herman Lewis headed the Cellular Biology Section, composed of the developmental and genetic biology programs. Physiological Processes, which included regulatory and metabolic biology, was under the charge of David B. Tyler. The fourth section, Environmental and Systematic Biology, was successively headed by a particularly stellar group of rotators: Robert K. Godfrey, Edward S. Deevey Jr., Bostwick H. Ketchum, and John F. Reed. Outside the sectional organization were Psychobiology and Facilities and Special Programs. No women served as program directors until Lewis selected Ursula K. Abbott of the University of California, Davis, to be a rotator in developmental biology in 1968. However, most of the programs were in large part run by female assistants who, after 1962, were titled assistant and associate program directors.[11]

Bureaucracy also increased with the gradual interposition of an additional level of administrators between the research divisions and the director of the Foundation. In 1958, Paul Klopsteg, formerly assistant director for MPE, became Associate Director (Research). This office, which allocated budgets for the research divisions, acquired more power under the incumbency of Randal M. Robertson, a physicist brought to NSF from ONR in 1961. Like the Office of the Director at NSF, the Office of the Associate Director (Research) was dominated by physical scientists. In 1964, Carlson's title was, in effect, downgraded from Assistant Director for Biological and Medical Sciences to simply Division Director.

Though the new generation of BMS leaders found some of the 1950s freedom to experiment with new types of grants constrained, there was ample scope in the 1960s for the ingenuity of program managers, who continued to propose new ventures for debate within the staff and divisional committee. BMS supplemented individual project grants with "special facility" awards that built new marine laboratories, museum buildings, field stations, controlled environment facilities, and ships. The division participated in two major multi-million-dollar international programs: the International Indian Ocean Expedition and the International Biological Program (see chapters 7–8). Despite a growing divide between its field-based programs and its laboratory-oriented programs, BMS still maintained a sense of internal unity, informality, and congeniality.[12]

Rising Expectations, Rising Constraints

Looking back, scientists and historians have viewed the 1960s as a golden age of science funding. At the time, participants did not see it that way. Some wistfully looked back to an earlier period when expectations and resources were in better balance.[13] On one hand, BMS staff engaged in grandiose "blue sky" projections of future expansion.[14] On the other, they felt considerable anxiety over their perceived failure to keep up with rising demands. Simply put, more biologists wanted more money. The more money that became available, the greater the expectations for further increase. NSF obligations continued to go up in the 1960s, from $132.9 million in 1959 to $320.8 million in 1963, to a peak of $500.3 million in 1968 (see table 3.1). BMS figures rose, too, although at a much slower rate than the overall budget. From $20.5 million in 1959, BMS awards grew to $39.1 million in 1963 and to a peak of $54.0 million in 1967 (see table 3.2).

Though program budgets rose in the 1960s, albeit fitfully, the costs of research appeared to be rising faster. As biology became increasingly "molecularized," the costs of purchasing and maintaining equipment increased. Applications for support multiplied as the first generation of graduate students and postdoctoral fellows, supported by fellowships or by grants to their mentors, set up their own laboratories. The "demand," or total value of requests, of which the legitimacy was rarely questioned, rose from $52.0 million in 1959 to $126.1 million in 1963, to $226.2 million in 1968. At the same time, increased demand for salaries for principal investigators and higher overhead costs reduced the funds directly available for research. From 1964 on, program directors felt squeezed.

In the late 1950s, BMS made an effort to give leading biologists the amounts they requested, or nearly so. At that time, NSF and NIH awards were likely to have been comparable in size. Through the 1960s, over 50 percent of applicants received grants, but as demand outstripped the supply of funds, award lengths were first reduced, then annual rates of support. By mid-decade the average grant was for two years; only a handful of three-year awards were made. (The divisional committee, voicing concern in 1964, "pointed out that certain investigators are now turning to NIH because they can make commitments for longer periods than NSF.") By 1968, NSF had reduced the average length of its grants to 1.75 years.[15] Coherent area grants of the magnitude of Britton Chance's $700,000 were no longer tenable. Renewal awards, even to the best investigators, could not keep pace with the rising costs of research they described in their proposals.

Added to the monetary constraints were the constraints of a growing bureaucracy. In the 1960s NSF lost some of its—perhaps self-indulgent—self-image as a nonbureaucratic agency. Staff frequently complained about the apparent shift from the mutual-trust concept of the "grant-in-aid" as a gift to universities to the notion of a carefully monitored "purchase" of research from university vendors. Two perennially debated issues that contributed to the trend toward purchased research were payment of full and negotiated overhead and payment of faculty salaries during the academic year.

Despite pressure from the Bureau of the Budget for a uniform government policy, NSF had tried in the mid-1950s to retain a fixed overhead rate to avoid complicated record keeping. Overhead went from a flat 15 percent to 20–25 percent. Finally, as of 1 January 1966, NSF acceded to a policy of negotiating the rate with each institution receiving grants. The negotiated rate differed for each institution and theoretically covered the full costs of overhead. BMS grants, unlike those from NIH, covered both direct and indirect costs, all funds coming from the program budgets. At NIH, study sections and councils approved direct costs only; indirect costs were later added and paid from a different account. At NSF, then, an increase in indirect costs translated immediately into less money available for actual research. And negotiated overhead inevitably meant more insistence on accountability.[16]

Equally thorny was the issue of faculty salaries. In the aftermath of Sputnik, PSAC reports and the Committee on Sponsored Research of the American Council on Education pressed for payment of salary for the time that faculty engaged in sponsored research.[17] Mission agencies, especially the Army, Air Force, and NIH, were willing to pay up to 100 percent of faculty salary.[18]

NIH's salary policy was based predominantly on the need to build up research at the nation's state and private medical schools, which, in the early 1950s, were in financial difficulty. Most had small faculties consisting of part-time practitioners. In order to expand medical research capability, NIH willingly reimbursed the salaries of principal investigators for the time spent on research. Full-time medical school faculty per medical school grew at an astounding rate from about 50 in 1951 to 130 in 1960 to 257 in 1970. The total number of full-time medical faculty in the United States grew from fewer than 3,600 in 1950 to over 26,500 in 1970, much faster than the number of medical or graduate students. Medical schools became heavily dependent on research overhead and salary support. In 1968–69, almost half the medical school faculty received some salary support from the federal government. NIH largesse spilled over from medical schools into the main campuses of universities.[19]

Waterman recognized that salary support led to such undesirable consequences as university pressure on faculty to cover their salaries through grants, or Budget Bureau and congressional pressure on NSF to obtain strict records of faculty time spent on research. Yet he yielded to the demand of college administrators, and in 1960 he altered NSF policy to allow academic-year faculty salaries to be charged as direct costs against grants for the proportionate time spent on research.[20] While the Division of Mathematical, Physical, and Engineering Sciences favored salary support, BMS panel members and program directors resisted the new NSF policy. As they saw it, this practice not only transferred what ought to be a university function to the federal government, but it also reduced the amount of funds available for research.[21] Unable to do much about overhead, BMS program directors could indirectly counter NSF policy concerning faculty salaries.

Associate Director (Research) Randal Robertson warned program directors that they had no authority to alter NSF policy on salary reimbursement and that their panels should be "clearly told that this matter is outside their jurisdiction."[22] In response to staff and panel efforts to subvert the directive by reducing faculty salaries in grants, he proposed in 1963 a "radical change," the so-called Robertson Plan, to separate faculty salaries from individual research grants. Endlessly debated through at least 1968, the plan called for pooling part of the research appropriation and distributing these funds as annual block awards for faculty support based on the institution's statement of percentage of faculty time spent on NSF research.[23]

David Tyler, who continued to uphold the traditional notion that research was a normal university function and that research time should be included in the faculty member's regular salary, devised his own elitist solution to the problem. He would provide competitive block awards for salary reimbursement based on a "Graduate Index," or weighted point count of science degrees conferred. Backed by the regulatory biology panel, Tyler insisted that "the research grant is not, and never can be, the proper instrument for a support program for academic salaries."[24]

Despite the appeal of the Robertson Plan, it was never put into operation because of its cost, and faculty salaries remained at the mercy of BMS program directors. Though they paid lip service to the NSF policy, in practice, as part of negotiating budgets downward, they often encouraged grantees, whom they could not afford to fund fully, to bargain with universities to pay their salaries.[25] As a result, the percentage of research funds supporting salaries of principal investigators increased only modestly from 1959 to 1968 (see table 6.1). Wilson, who, as a former university administrator, especially favored salary support as a means of funding higher education, saw BMS resistance as sabotaging stated NSF policy.[26]

The AIBS Affair: The Problem of Accountability

In the early 1960s, Congress became increasingly concerned about abuses of funds by grantees and lack of accountability on the part of granting agencies. From 1959 to 1962, the Fountain subcommittee of the House Government Operations Committee conducted an investigation of NIH that pointed to loose management practices in funding extramural research.[27] Fearing a similar congressional inquiry, the Foundation instituted

Table 6.1   Percentage Distribution of BMS Research Grant Funds by Type
of Obligation, FY 1959 and FY 1968

	1959	1968		
	BMS Total	BMS Total	Molecular	Systematic
Salaries and wages	60.41	49.28	44.67	54.93
Principal investigator[a]	8.62	9.73	5.26	16.39
Research associates	15.28	8.85	12.19	4.79
Research assistants	18.29	7.67	4.60	12.82
Technicians	18.22	9.39	8.87	7.40
Other	—	13.64	13.75	13.53
Permanent equipment	10.86	10.33	15.45	4.68
Expendables	8.65	10.95	14.81	4.91
Travel	4.66	3.37	1.77	9.89
Publication costs	—	1.26	1.46	2.22
Other costs	2.85	5.34	3.71	2.25
Total direct costs	87.43	80.53	81.87	78.88
Total indirect costs	12.57	19.47	18.13	21.12
Average grant size	$21,404	$40,009[b]	$47,156	$25,204

Source: BMS Annual Report FY 1968, Appendix, Tables III and IV-C, NSF Historian's Files,
National Archives and Records Administration.
[a]Includes co-principal investigators and faculty associates.
[b]For an average grant length of 1.75 years.

policies leading to tighter fiscal and administrative control of grants. The
Grants Office, which reviewed all awards for conformance with NSF policies, expected more detail on prospective use of funds and restricted grantees
somewhat in their ability to alter their budgets once an award was made.
Elaborate conflict-of-interest forms were instituted for consultants. In the
1950s, whenever the idea of an auditing capability for the Foundation had
been broached, BMS staff had vehemently objected; in 1961, for the first
time, the Foundation hired a comptroller and began to audit grantee institutions. The American Institute of Biological Sciences became one of the
first casualties.

The "AIBS affair," which came to a head in 1962–63, was aired in the
national press as well as in scientific journals. It came to the attention of
members of Congress who called upon Waterman for an explanation. Since
its independence from the National Research Council in 1955, AIBS had become highly dependent on grant and contract money, especially from NSF.
Among its major projects were the Biological Sciences Curriculum Study

(BSCS), a massive, NSF education division–funded, effort to create new high school biology texts, and a large and expensive film series funded by the Ford Foundation and AEC. When NSF grant funds were delayed because of the vagaries of congressional action, it became difficult for AIBS to keep its paychecks and projects going. Aware that the organization lacked a stable financial base, the AIBS staff approached BMS in May 1962 for a five-year, million-dollar grant for operating expenses, a controversial proposal, which was debated at some length at the October divisional committee meeting.[28]

About the same time that AIBS made its proposal, NSF's zealous new comptroller, Aaron Rosenthal, selected two AIBS contracts for audit.[29] He quickly found evidence of serious fiscal irregularities, the most significant of which was the intermingling of NSF grant funds in a general account used to finance the film series. In addition, AIBS had claimed excessive overhead (though there was some confusion over the rate), used grant funds for nonallowable travel and entertainment expenses, supported other projects with royalties from NSF-funded publications, and failed to repay interest on invested grant funds. The crisis erupted in November when Rosenthal, with National Science Board backing, demanded that AIBS repay more than $330,000. When Rosenthal found the AIBS initial response unsatisfactory, the Foundation forced a reluctant AIBS to cut off disbursements and took steps to transfer BSCS to the University of Colorado, where the project was housed. For a time it seemed that AIBS' leaders would put up a fight, but the new president, James Ebert, a member of the BMS Divisional Committee, vowed to "clean house" and save AIBS from dissolution. At an emergency meeting in January 1963, Ebert pressured the AIBS governing board into agreeing to a plan calling for gradual repayment of the money, a reorganization of AIBS management, and a restructuring of AIBS into an individual-membership organization in order to raise funds from biologists.

The AIBS affair was indicative of increasing misunderstandings in the relationships of scientists, federal patrons, and Congress. Many biologists believed the NSF actions were needlessly harassing and niggardly when the AIBS funds were being spent for science. Previous informal dealings with program officers had led the AIBS staff to believe they were doing nothing wrong. Indeed, standards had shifted from the more laissez faire practices of the past, catching AIBS off guard. Science writer Daniel S. Greenberg, who chronicled the AIBS affair in *Science,* warned scientists that they had better get used to the new order: "The blank check days are going fast, and it is experiences of the AIBS sort that convince Congress that the sooner they go, the better."[30]

In 1963, in one of his last memos to Waterman before the latter's retirement, Consolazio bemoaned the recent changes that had come to NSF:

For the past year I have watched the drift away from more liberal and permis-
sive behavior patterns of the earlier NSF to more rigid and formal practices. . . .
Some tightening up is inevitable, for the times and the abuses necessitated a
change. But I don't see the need for denying those fundamental concepts that
helped create NSF. How else does one account for the shifts that are occurring
away from the mutual trust concept embodied in the original grant to the limitations imposed by the accounting-auditing mode characteristic of contracts
. . . . I feel strongly, as I know you do, that the Foundation is not just another
research-supporting Federal agency. It is different. It is the only Federal agency
conceived as a *Foundation;* therefore, it should not be bound by the pressures or
practices of other governmental groups.[31]

Consolazio's successor in molecular biology, John Mehl, saw the grant-
in-aid being eroded on all sides. Administrators' demands for support of the
full costs of research and "the introduction of the auditor's point of view," he
claimed, threatened to convert the research grant to a contract. Mehl recommended establishment of a few regulations to curb some of the most obvious
abuses such as overly high salaries for research assistants, excessive charges for
travel, and purchase of improper items such as office furniture and air condi-
tioners, but insisted that responsible administration of grants must be left to
the investigators, rather than to auditors, lawyers, or even program directors.[32] BMS program directors through the decade typically upheld traditional values of mutual trust and flexibility in the face of bureaucratic assaults.

"Undue Concentration": The Problem of Geographical Distribution

John Wilson saw the Lyndon B. Johnson era marking a "first break
point" in the character of the Foundation. During Johnson's presidency, he
later recalled, "there came to be an emphasis on a distributive concept of
support versus a competitive concept of support."[33] In what cases was it
proper to use criteria other than the merit of the proposal and the proposer's
track record to distribute research awards? Wilson and Consolazio had taken
the elitist approach that the best science, wherever located, was most de-
serving of support. Johnson and the Congress felt differently and on several
occasions faulted the agency for ignoring the NSF charter's charge to avoid
"undue concentration" of funds. Indicative of growing congressional concern for geographical criteria was NSF's decision in 1964 to restructure its
annual published list of awards. Up to this time, awards for each program
were listed alphabetically by institution. From fiscal 1964 on, awards for each
program appeared alphabetically by state. Thus it became much easier to cal-
culate program spending by state.

No one seriously doubted that BMS concentrated funds in a relatively small number of states. When deputy director David Keck and Howard Teas, program director for metabolic biology, compiled cumulative BMS data on geographical distribution from 1952 to 1963, they found an overwhelming disparity between the "grant-rich" and "grant-poor" states (see table 6.2). During that time, BMS had given over $20 million in grants each to California, Massachusetts, and New York, but under $100,000 each to Idaho, Wyoming, and Nevada. Teas and Keck explained that even though the staff was biased toward funding investigators in "undersupported areas of the country," the problem remained because "the 'poorer' (in NSF-BMS research grant dollars) institutions tend to submit proposals of minor merit no matter how anxious BMS program directors and their panels are to help." They also sent in fewer proposals.[34]

The BMS staff was divided in its response to this situation. Carlson was more sympathetic to geographical considerations than Wilson had been. He and several staff members took seriously the problem of improving not just the good institutions but also the weak ones. Staff discussed such special distribution problems as biology in black colleges, the Southeast, and Appalachia (but took no special remedial action). The solution to the geographical distribution dilemma proposed by Teas and Keck, was a compromise between the needs of weak institutions and the division's commitment to awarding funds primarily on merit.

In 1964, Teas and Keck outlined a new Biological Sciences Assistance Program (BSAP). They rejected out of hand basing awards only on geography since putting money into "third-rate institutions or investigators" was "without appreciable benefit either to the institution or to science." Instead, they offered a plan for "underprivileged" institutions modeled on NSF's Science Development Program (see below). Those receiving less than $100,000 in BMS grants during 1963 could make preliminary arrangements to hire a group of good biologists and then submit a development plan. BMS would provide half the new faculty's salary for four years and would fund permanent equipment, research expenses, overhead, and discretionary funds for the department. The BMS salary contribution was specifically intended to support a half-time commitment to research since too large a teaching load was a typical problem in grant-poor schools. Teas and Keck thought there would be plenty of junior members of large research teams who would benefit from more recognition in a smaller institution. Four years' support (covering three new staff members) was estimated to cost $540,000 per institution.[35]

BMS staff and the divisional committee discussed the BSAP at length, but nothing came of it for the plan found few supporters. None of the op-

Table 6.2   Distribution of BMS Research Grants by State, FY 1952–FY 1963
(States Listed in Order of Dollars Received)

California	23,651,850	Hawaii	1,376,100
Massachusetts	22,115,665	Louisiana	1,366,600
New York	20,993,280	Arizona	1,246,500
Pennsylvania	9,759,830	Utah	1,225,400
Illinois	9,708,550	Rhode Island	1,066,000
Connecticut	5,572,250	Virginia	1,016,300
Michigan	5,470,850	Oklahoma	1,009,000
Indiana	5,453,600	Kentucky	965,800
Wisconsin	5,415,700	Maine	906,900
North Carolina	4,905,150	Nebraska	716,500
Maryland	4,583,375	Montana	414,600
Missouri	3,941,700	Vermont	399,200
Oregon	3,856,000	West Virginia	377,800
Texas	3,703,700	Alabama	371,400
D.C.	3,507,268	New Mexico	328,600
Florida	3,503,280	Delaware	298,100
New Jersey	3,500,350	North Dakota	295,800
Washington	3,346,400	Arkansas	287,000
Ohio	2,981,275	Alaska	277,500
Minnesota	2,826,905	Mississippi	266,300
Kansas	2,198,000	South Carolina	190,200
New Hampshire	2,046,200	South Dakota	154,400
Iowa	2,029,350	Idaho	93,100
Tennessee	1,532,200	Wyoming	82,400
Colorado	1,466,900	Nevada	21,800
Georgia	1,392,350		

Source: Harve J. Carlson to Divisional Committee for Biological and Medical Sciences, 14 October 1964, Table 1, "BMS Divisional Committee Meeting, 28th - 10/2–3/64," Box 20, 70A-2191, RG 307, National Archives and Records Administration.

erative NSF programs for "science development" were geared to grant-poor schools. Some BMS programs, molecular and genetic biology especially, gave geography only tertiary consideration. Their program staffs assumed that if the goal of awarding grants was to get the most good science for the taxpayers' money, then that science was most likely to be found in the laboratories of the leading universities, which in turn attracted the brightest graduate students and postdocs. Only when proposals were otherwise equal in merit should geography be considered.[36] The BMS Divisional Commit-

tee also took a dim view of the BSAP and any other efforts to broaden geographical distribution. The members' consensus, as reported to the National Science Board, was that evaluating proposals by "criteria other than
merit" provided "cause for great concern." "Great caution must be exercised
not to weaken the total academic research structure in any consideration of
redistribution of scientists or creation of new centers of excellence."[37] And
the plan found little sympathy with the NSF hierarchy. Wilson wrote to
Carlson with a tone of warning, "I would hope the issue of geography does
not become overemphasized."[38]

The Era of Institutional Support: New Laboratories
for Biological Research

John Wilson had long hoped the NSF would take on the task of supporting not just research but higher education in general. The agency came
closest to Wilson's goal in the affluent 1960s through its support of a variety
of institutional programs.[39] These were intended to increase the number of
scientists, offset imbalances created by the Foundation's earlier emphasis on
individual project grants, and respond to criticisms of inequitable geographic
distribution. NSF's Office (later Division) of Institutional Programs managed projects to renovate or construct research laboratories, flexible formula
block grants to institutions, the Science Development Program to increase
the number of "centers of excellence," and the Departmental Science Development Program. Related programs in the education division provided
instructional equipment and development awards to undergraduate schools.
It took BMS urging, NIH competition, several influential PSAC reports,
and the expansionist mood of the post-Sputnik era to convince a reluctant
Waterman and the National Science Board to endorse institutional support.
For all that, its heyday was relatively brief. All of these programs disappeared
in the early 1970s.

The institutional program with the greatest and most lasting impact on
biological research was the Graduate Science Facilities Program. From 1960
until 1970, when Congress disbanded the program, NSF spent some $188
million for the construction and renovation of on-campus graduate laboratories in the sciences and engineering. It made nearly a thousand awards in
amounts varying from under $1,000 to over $2 million. Of these, about $44
million (24% of the total) went to the life sciences and another $20 million
to the behavioral sciences. For the space of a decade, universities had the
opportunity to replace old, crowded, and poorly equipped structures with
spacious and up-to-date facilities.

From the early 1950s, Wilson, Levin, and Consolazio had all looked forward to supporting the construction of modern laboratories for biological research and graduate training. A major stimulus for their campaign to convince the NSF leadership to offer institutional support grants was competition from NIH.[40] Between 1956 and 1963, under Shannon's energetic leadership, NIH was able to exploit the popular appeal of health research to convince Congress to establish programs to fund laboratory construction, NIH professorships, and flexible institutional formula grants. Sold to legislators on the basis of the needs of medical and dental schools and hospitals, and at first limited to such institutions, NIH programs soon expanded to main campuses, threatening NSF's image of itself as the general-purpose provider of scientific research.

The Health Research Facilities Act of 1956, which authorized NIH to award matching construction grants, galvanized the BMS staff and advisors into pressuring the reluctant Foundation to create a similar program.[41] In October 1956, the BMS Divisional Committee discussed at length the Foundation's current policy toward facilities in relation to the proposed NIH program. Waterman, still disinclined to enter into institutional support, said that for the near future, at least, it would have to be justified "principally in terms of support for maintenance, operation, construction, or purchase of specific items." He felt that NSF was not yet in a position to "support programs involving general areas having almost limitless needs." In response, the divisional committee formulated a resolution in favor of support for "expensive specialized facilities in biological and medical sciences" including controlled environment facilities, maintenance of systematic collections, and field stations. But they also stated that "expanding government support of research" had to include expanding "general facilities," that is, building and renovating research laboratories. And if nongovernmental support were insufficient, NSF "must inevitably seek to supplement these facilities, and hence should seek appropriations for this purpose." These resolutions, discussed by the NSB, led to the drafting of a staff report to the board on physical facilities and major equipment in February 1957.[42]

As the divisional committee gained experience evaluating the first groups of special facility awards, it increasingly insisted that NSF also support general laboratory construction. Having seen proposals from marine and field stations, specialized research institutes, museums, and biological stock collections, the committee formally resolved in January 1958 that it was "of the first importance to press a program to support facilities for graduate student training and research," graduate schools being "the major national research facility in biology."[43]

By 1958, NIH was operating a $30-million-a-year program. In addition to facilities for medical schools, it was also contributing to biology and zoology buildings for universities and even for a few colleges such as Bryn Mawr and Reed. NSF staff estimated that of NIH's $69.3 million in grants approved since passage of the 1956 act, some $29.6 million (43%) fell into areas of Foundation concern, and the disciplines encompassed appeared to "represent about one-third of the natural science spectrum."[44] Alarmed that if NSF did not act quickly, it would forfeit its role as the all-purpose funder of science, staff drew up position papers for the board in March and May 1958. "If the Foundation makes no further moves toward the support of graduate research laboratories," the staff warned, "NIH, which has already achieved considerable momentum in this area, will fill more and more of the vacuum with respect to these scientific disciplines related to health, including biology, psychology, and important segments of chemistry and physics." Colleges and universities should not have to turn to a "specialized agency" for support of "general purpose laboratories."[45]

Although initially apprehensive over the magnitude of the program and the possibility of federal control of education, the National Science Board adopted a resolution in June 1958 that NSF should support the "equipping, renovation, and construction of graduate level research laboratories." It urged a pilot program of $2 million to $3 million in 1959 and a request for $50 million in FY 1960.[46] The Bureau of the Budget permitted an initial program of $2 million for matching grants in 1960, which was expanded to a high of $30.5 million in 1964.

For the first year, the research divisions (BMS and MPE) handled the Graduate Laboratories Program. Since BMS had a budget of only $1 million for the purpose, awards were small and went primarily for renovation projects. In May 1960, NSF created a separate Office of Institutional Programs, headed by psychologist Howard E. Page (another ONR veteran) after Lou Levin left for Brandeis. (Returning in 1964, Levin eventually became Page's superior as Associate Director [Institutional Programs].) Unlike the offices of Director and Associate Director (Research), Institutional Programs was dominated by biologists and social scientists. Through the 1960s, Joshua Leise, a Yale-trained microbiologist from the Army Research Office, headed the Graduate Laboratories Program.

Even more than with individual project grants, the NSF staff took a major role in deciding what institutional facilities to fund. Staff, including occasionally those of BMS, participated as regular members of site-visit teams and wrote independent reports. BMS staff members regularly provided confidential evaluations of written proposals, making full use of their own informal knowledge of the productivity and potential of the various faculty involved. For staff eyes only, such reviews could be quite frank, gossipy, and cynically humorous. For example, on a poultry science facilities proposal, the program director for metabolic biology wrote: "Their primary goal is to help the state's 100 megabuck chicken raisers with their problems and train graduate students to help the state's future chicken raisers solve their problems. Basic research in science would, I feel, always be subservient to the applied interests. DECLINE." Or, questioning whether a botany department's staff would actually benefit from a proposed modernized facility, the same program director wrote: "As Chairman, [X] has very little time for research . . . I believe that he has only an occasional student and is therefore not likely to be particularly productive. Refurbished laboratories would give him dignity." He rated the proposal low.[47]

The largest awards in biology, ranging from $600,000 to over $2 million, went to the University of Illinois (life sciences), Brandeis (biology), Michigan State (biochemistry; plant sciences), Cornell (biological sciences), Dartmouth (biological sciences), University of Alaska (Arctic Institute and biological sciences), Stanford (biological sciences), Iowa State (botany and forest research), University of Texas (zoology), University of Hawaii (plant science), University of Missouri (botany), Brown (biomedical center), Northwestern (biology), University of Washington (zoology), and Yale (biology). Major awards were also made in the behavioral sciences to such institutions as the University of Illinois and Harvard.[48]

Despite NIH's head start, NSF program officers perceived little direct competition with that agency. Because of NIH, NSF made few awards to preclinical departments of medical schools. To Leise, it appeared that NIH support, in turn, emphasized medical and related schools and limited support of basic biological laboratories to areas related to medicine. As in the research areas, program officers maintained close liaison, and agency staff attended each other's advisory committee meetings. Occasionally, NSF and NIH collaborated on funding large projects such as the Kline Biology Tower at Yale.[49]

Recollections differ on the extent to which the program emphasized established merit or "potential." In any case, a wide variety of institutions and fields of biology received support. Though million-dollar awards went to general-purpose facilities for biology, zoology, or botany departments, many smaller awards supported basic research related to specific applied areas of biology. Grants funded fisheries biology, plant virology and nematology, parasitology, forest botany and zoology, entomology, agricultural biochemistry, psycho-pharmacology, dairy and food sciences, agronomy, horticulture, agricultural economics, poultry science, and watershed management. While relatively well-favored schools received the largest awards, grants were also provided to such institutions as University of North Dakota, North Dakota State, Utah State, Louisiana State, West Virginia University, Rensselaer

Polytechnic Institute, Smith College, and Wesleyan University. Ironically, this program supported a much broader range of schools in terms of geography, reputation, and areas of biology than the Science Development programs.[50]

Flexible Formula Awards to Institutions

NSF's "formula" institutional grants program, in operation from fiscal years 1961 to 1974, was also initially motivated by NIH competition, although the need for such a program was widely felt. Biomedical scientists had advocated general-purpose institutional funds ever since the Bush Report of 1945, most recently in the Bayne-Jones Report. These were to be funds independent of particular research that universities would use in a variety of ways to support science. In 1959, NIH attempted to obtain statutory authority to provide flexible institutional awards to schools of medicine, dentistry, and public health. The plan called for the program to expand after the first year to institutions of higher learning and other research organizations. Waterman supported the bill as long as it limited authority to medically related institutions until broader policy was clarified. He argued to the Bureau of the Budget that "if, as a matter of federal policy, institutional research grants are to be made to institutions of higher education generally, such grants should be made for support of all areas of science."[51]

Wilson wrote the staff paper that resulted in the inauguration of NSF's Institutional Grants Program in 1960, directed by historian J. Merton England.[52] In the post-Sputnik era, administrators argued that the great increase in project grants had led to an imbalance in institutional science activities, a loss of administrative control, and a deemphasis of teaching in favor of research. These new institutional awards, awarded according to a formula, intended to offset the negative effects of project grants, could be spent at the institution's discretion on any form of direct support of scientific research and education.[53]

Initially targeted for research universities, formula grants were extended to a much broader range of academic institutions than the early planners had envisioned. In the first year the formula heavily favored those receiving the most research support. Each institution was eligible to receive 5 percent of all research funds provided to it by NSF during a stated period of nine months up to a maximum of $37,500. The rule was later liberalized to 100 percent of the first $10,000 and tapering amounts thereafter. The addition of two NSF educational programs to the formula base enabled many liberal arts colleges to receive funds for the first time. Wilson, who left the Foundation in 1961 and returned two years later, was said to have been peeved at these alterations of the initial concept.[54] At its height the program had a budget of $15 million and provided awards to a maximum of slightly over $150,000.

The NIH program, delayed until FY 1962, was structured somewhat differently. Its formula was based on an institution's health-related sponsored research received from all agencies. Initially the NIH awards were limited to schools of medicine, dentistry, public health, and osteopathy. NSF thus saw an opportunity to rationalize the two plans, and in effect divide the academic territory between the two agencies. NIH would be responsible for schools for health professionals and NSF for schools of arts and sciences. With Budget Bureau encouragement, NSF attempted to negotiate with NIH on "joint responsibility" for institutional grants, but NIH was unwilling to cede any institutions to NSF. Eventually, NIH expanded its grants to all academic institutions. As no other federal agencies offered general institutional awards, by 1969 NSF had made a successful play to include in its formula base the research sponsored by all other federal agencies, including the HEW Office of Education, *except* for the NIH![55]

In 1961 NSF attempted to forestall NIH in yet another program justified on the grounds of strengthening institutions, that of creating agency-supported science professorships. The plan, originating in the BMS Divisional Committee, called for universities to nominate one candidate per year. Upon appraisal, NSF would recognize outstanding faculty members by naming them National Science Professors and providing their salaries to a maximum of $25,000 annually for five years with the intention to renew. The institution might then use funds saved to create more positions for younger faculty. Although Waterman discussed the plan with members of Congress and proposed a pilot program to the Bureau of the Budget, the idea was eventually dropped because of "the uncertainties involved."[56] A major concern was the impropriety of a federal agency interfering in university appointments.[57] NIH plunged ahead and announced a sizeable Research Career Award Program for fiscal 1962.[58]

Science Development Programs

The Science Development Program (SDP) was, in contrast to the previously discussed institutional programs, initiated by NSF, and only later followed by NIH's Health Science Advancement Awards. Begun in 1963, SDP was a response to congressional concern that research and graduate training were concentrated in too few institutions. The 1960 PSAC report, *Scientific Progress, the Universities, and the Federal Government,* familiarly known as the Seaborg Report after its chairman, Glenn T. Seaborg, nuclear chemist and chancellor of Berkeley, captured the science community's imagination with its pronouncement that in order to provide for the continued expansion of science, the nation should double its present fifteen to twenty "centers of excellence" over the next fifteen years. Seaborg wrote: "In science the excellent is not just better than the ordinary; it is almost all that matters." The United States should "energetically sustain and strongly enforce first rate-work where it now exists," and at the same time "increase support for rising centers of science" in order to "double the number of universities doing generally excellent work in basic research and graduate education."[59]

Planning for the SDP stretched out over the last years of Waterman's directorship. A foundation-wide committee gathered data on a range of colleges and universities and chose eleven to visit for extensive discussions with faculty and administrators on needs, opportunities, and possible goals of an NSF program. BMS staff, especially Carlson and Keck, as well as Consolazio, were heavily involved in the planning process. Long debates by staff and board over eligibility requirements and the agency's inability to persuade Congress to provide the requested $33 million delayed the program, which was finally announced in March 1964. The first awards were given in 1965.[60]

At first the lofty intent of Science Development was to create additional "centers of excellence" by providing large enough sums of money to make a lasting difference. The foremost universities, never formally listed, were discouraged from applying, as were weak institutions. In fact, both extremes embarrassed program directors by sending in proposals. Later in the decade NSF moderated its goal to the more realistic one of providing substantial improvement to schools that already showed existing strength. Groups of academic departments or interdisciplinary areas received University Science Development awards, which they could spend on any combination of faculty salaries, construction of facilities, equipment, graduate students, postdoctoral fellows, support staff, and curriculum development. Program officers argued that even institutions that failed to win an award benefited from the process of self-evaluation and formulation of coherent long-range goals. The short-lived Departmental Science Development Program, begun in 1967 and headed by Consolazio, provided smaller awards, averaging $600,000, for single departments.[61]

From 1963 to the termination of these programs in 1972, NSF spent $233 million on 102 institutions, most of it on 31 universities that received average awards of $6 million. Only $38 million went to the biological sciences compared to $121 million for the mathematical, physical, and engineering sciences.[62] Recipients of awards for over $1 million in the biological sciences included Case Western, Florida State, North Carolina State,

Notre Dame, Purdue, Tulane, Vanderbilt, and the universities of Oregon, Rochester, Texas, and Virginia.

By far the largest awards in the biological sciences went to two schools, the universities of Georgia and Iowa. Georgia received $5,995,000 for bacteriology, biochemistry, biopsychology, botany, entomology, microbiology, and zoology and for general use of the Biological Sciences Division. Under the leadership of Eugene P. Odum and Barclay McGhee, respectively, Georgia had already become a center for ecology and parasitology. The NSF grant was intended to further strengthen biology by supporting the hiring of a critical mass of staff in "key areas of modern biology," particularly molecular and developmental biology and genetics, and the funding of graduate students and postdoctoral fellows, nonprofessional staff, and related facilities. The award also contributed to several structures, including additions to the biological sciences building and a building for plant sciences.[63]

The University of Iowa, a major producer of doctorates in biology, received over $6.7 million to benefit five departments in the College of Medicine (anatomy, biochemistry, microbiology, pharmacology, physiology) and three in the College of Liberal Arts (botany, psychology, and zoology). NSF intended the award to encourage cooperative activities between the two colleges, especially in the interdisciplinary areas of endocrinology, genetics, and neurobiology. Most of the money went toward construction (an addition for the zoology building and partial support of a new building for the basic medical sciences) and initial funding for twenty-four new faculty positions and associated technical support and equipment. In neurobiology, for example, the university boasted of hiring the up-and-coming Rodolfo Llinas and his team of investigators.[64]

Although some SDP awards involved ecology, for the most part they tended to benefit molecular and cellular biology. Except for entomology at Georgia, little attention was directed to agriculturally oriented fields. To an extent, institutions used SDP funds as seed money to provide salaries and facilities sufficient to attract high quality faculty who could, in turn, garner NIH support. University administrators, such as the president of the University of Iowa, pointed with pride to the NIH grants, Career Development Awards, and traineeships his departments had acquired since receiving the SDP award.

NSF was only one of a number of private and federal agencies pouring hundreds of millions of dollars into "upgrading" science activities at universities. NSF's program was preceded by the Ford Foundation's "challenge" grants, which served as a partial model, and by NASA's Sustaining University Program. It was emulated later in the decade by the Department of Defense's Project THEMIS and NIH's Health Science Advancement Awards.

Meanwhile, the HEW Office of Education provided awards to those institutions most truly in need of development, such as black colleges. Several agencies funded the same universities, thus making it almost impossible for later administrators to assess their own agency's results.[65]

Some BMS staff and advisors were frustrated with the institutional programs of the 1960s because they neglected to provide fluid funds to the most important institutions "at the local level," namely university departments. To the working biologists of the BMS Advisory Committee, a number of whom were department chairs, the department, not the university, was the primary level of focus. Except for a few training grants in areas of environmental and systematic biology, NSF had nothing comparable to the flexible NIH training awards. In 1965, David Tyler proposed to Haworth a "Departmental Grants Program," which he claimed would facilitate recruitment of new faculty members, curb "opportunism in grantsmanship, job-hopping, and wheeling-dealing," and restore a measure of responsibility for teaching and research to the department.[66] And in 1967–68, the advisory committee advocated once again training grants and "core" or "sustaining" grants to "superior departments or interdepartmental committees," preferably under the jurisdiction of BMS and funded separately from research.[67]

But the NSF administration, especially Wilson, was not receptive and, in any case, new funds were not forthcoming. Wilson, who favored most forms of institutional support, tried to discourage Carlson from pursuing departmental awards. "I think the weakest area in our university system lies in the general domain of department administration and deanship administration," he wrote, confessing that he had "not much faith that improvement in higher education will come about through the initiative of departmental level organization."[68] Flexible funding for departments in biology was largely left to NIH training grants.

The Legacy of the Era of Institutional Awards

The widespread post-Sputnik acceptance in government of the value of basic research produced high points in the 1960s, since unsurpassed, for the proportion of federally funded university research that was non-programmatic (79% in 1964) and for the dependence of academic research on federal support (74% in 1966). It has been argued that federal awards to increase research capacity in the 1960s were indeed effective for the long term in broadening the number of high quality universities. At the end of the decade research was less concentrated in a small number of states and institutions than it had been at the beginning.[69]

In this heyday of institutional support by federal agencies, biology departments were able to acquire multistory buildings with new research laboratories and equipment; they could utilize discretionary institutional funds to bring in visitors and support other departmental activities; with the aid of NSF and NIH fellowships and traineeships, not to mention research grants, they could train increased numbers of graduate students; and with the aid of NIH professorships and NSF and NIH development grants, they could hire additional faculty. One result of all of these awards was a substantial increase in the numbers of faculty and graduate students in biology. The sudden halting of this heady expansion of biology at the end of the decade led to a crisis on campuses and in BMS (see chapters 8 and 9).

Institutional awards were just one form of million-dollar grants to benefit the biological sciences in the 1960s. More closely under the shaping hand of BMS were the more programmatic "special facilities" and multi-institutional programs discussed in the next chapter.

CHAPTER 7

Promoting Big Biology

Biotrons, Boats, and National

Biological Laboratories

"The large-scale character of modern science, new and shining and all-powerful, is so apparent that the happy term 'Big Science' has been coined to describe it," wrote physicist and historian of science Derek De Solla Price in 1963.[1] Price conceptualized bigness in terms of the exponential growth since the seventeenth century of measurable indicators of the scientific enterprise, such as numbers of scientists, journals, and articles. For Alvin Weinberg, director of the Oak Ridge National Laboratory, who had popularized the term since 1961, "big science" referred to expensive large-scale facilities (such as accelerators and reactors) and programs involving large teams of investigators."[2] Recent historians of science have preferred to characterize big science not just by size of instruments or cost but by the hierarchical organization of scientific labor, multidisciplinary teams, and coalition-building necessary to convince sponsors to fund the project.[3]

The big biology projects described in this chapter do not qualify as big science according to the above criteria. However, these projects were much larger scale ventures than individual research grants, they fueled considerable controversy among biologists, and most important, they were conceived in the image of big physics or astronomy. That is, they were consciously framed as biological analogues to big science undertakings in the physical sciences. BMS program directors did not simply shepherd unsolicited large-scale proposals through the system; they interacted with biologists at all stages to shape the proposals according to their own ideal images of big science. In their enthusiasm, however, they were at times misled as to what facilities biologists would use and the extent to which biologists would engage in orchestrated research.

By 1963, NSF was supporting four "national research centers": the National Radio Astronomy Observatory in Greenbank, West Virginia, the Kitt Peak National Observatory in Tucson, Arizona, the Cerro Tololo Inter-American Observatory in Chile, and the National Center for Atmospheric Research in Boulder, Colorado. The focal points of these strategically located, government-owned, big-science installations were large instruments too expensive for a single university to own. They were funded annually by a separate budget line, managed by independent nonprofit corporations representing consortia of universities, and available to all U.S. scientists with priority of use based on project merit.[4] In addition, NSF had funded a number of expensive, low-energy accelerators and nuclear reactors and, by 1963, was well on its way to becoming mired in the Mohole fiasco—the disastrous big-science project, eventually terminated by Congress, to drill a hole through the earth's crust to the mantle.[5] NSF outlays for large-scale facilities went overwhelmingly to the physical sciences, destroying any semblance of parity between the Division of Biological and Medical Sciences and the Division of Mathematical, Physical and Engineering Sciences.

From almost the beginning, BMS staff aspired to the creation of a national biological facility—a new institution apart from existing universities, separately funded by Congress, and overseen by a "national advisory committee" of scientists. But unlike the astronomers and meteorologists, who achieved not only a monetary commitment from NSF but also internal consensus on the need for their respective national centers, biologists failed to unite behind such multi-million-dollar undertakings.[6] The only proposal for such a national biological facility that was seriously considered was a tropical marine science center in Puerto Rico to be operated by Associated Universities, Inc.

While a true national laboratory in biology proved unattainable, BMS staff made a number of partially successful attempts to provide a modified form of "national biological facility." This was usually an expensive facility, constructed for a single university, which was made available to a larger community of biologists. One form was the large controlled environment facility, deliberately called a phytotron or a biotron to conjure up the image of the cyclotron. Another was the large ocean-going vessel for biological oceanography.

A second model of NSF-sponsored "big science" that the BMS staff sought to emulate was the time-limited international program as exemplified by the highly successful International Geophysical Year (IGY) in 1957–58. For biology, the two most significant international programs of the 1960s were the International Indian Ocean Expedition (IIOE) and the International Biological Program (IBP). Biologists, more so than other scientists, were too heterogeneous not to be deeply divided over such costly undertakings. Some BMS advisors denounced them as "large-scale crash programs" inimical to the tradition of "free-enterprise" in grants-giving.[7]

This chapter looks at several attempts by BMS staff to promote large-scale cooperative ventures: (1) phytotrons and biotrons; (2) biological oceanography projects including large research ships, marine laboratories, and the IIOE; and (3) tropical biology undertakings, especially the failed attempt to found a national tropical marine science center. For comparison, the provision of smaller-scale facilities at inland field stations is also examined. The following chapter will treat the decade's most contentious example of "big biology," the International Biological Program.

Funding "Special Facilities"

BMS made awards for biotrons, boats, and field stations through its Special Facilities Program, set up in 1959 and headed first by BMS deputy director Lou Levin, then from 1961 through its dissolution in 1968 by Jack T. Spencer. John Wilson defined a "specialized facility" in 1959 as one "which is unique in some sense, either in its program or its location, and one which is not found in the usual university or college departments covering the life sciences." "The concept," he explained, "grew out of the fact that it has not been within Foundation policy to provide funds for the renovation, construction, and equipping of *departmental* research laboratories."[8] After 1957 such facilities were funded by a line item in the budget as a small biological counterpart to the national laboratories, nuclear reactors, and computers budgeted by the physical sciences. Initially evaluated by the divisional committee, special facilities were later reviewed by a separate advisory panel, and finally evaluated on an individual basis with the aid of biologists drawn from a large pool of advisors. All, or nearly all, facilities proposals were site-visited by a combination of staff and advisors.

When a program to fund construction of graduate-level laboratories began in 1960, the distinction between a specialized facility and a normal departmental facility became more problematical and also more important, because special facilities awards did not require fifty-fifty matching grants from the institution as graduate laboratory awards did.[9] In many cases, BMS paid all, or almost all, of the costs. After a year, the graduate laboratory program was transferred to the new Office of Institutional Programs, but specialized facilities awards remained in BMS.

Although BMS staff hoped funding for Special Facilities would grow, it reached a height of about $5.6 million a year in FY 1962 and then gradually declined to $2 million in 1968. By far the largest portion (about half) of the nearly $39 million in Special Facilities funds spent from 1957 to 1968 went to marine stations and ships for marine biology (see table 7.1). Only a relatively small amount went to molecular biology and related areas, in part because the experimental laboratory sciences were usually carried out in on-campus departmental laboratories. Although universities such as Duke, Stanford, Wisconsin, and the University of California, San Diego, received substantial sums for oceanography, a large portion of specialized facility awards went to non-degree-granting institutions such as the Woods Hole Oceanographic Institute, the American Type Culture Collection, and non-university museums and botanical gardens. Such institutions were ineligible, by definition, for graduate laboratory awards.

Big biology meant funding not only initial construction costs but also continued operating costs. NSF found it expedient to reserve the facilities funds for construction projects and to use NCE (Not Classified Elsewhere) research funds for other types of nonproject awards such as operational support of facilities and training programs at marine stations. (Members of Congress were more likely to question operational costs.) The NCE category was therefore renamed Special Programs. After mid-decade, congressional appropriations for biological specialized facilities were supplemented by a line item for facilities for biological oceanography.

Compared to graduate laboratory awards, special facility awards had a strong programmatic element and are therefore more interesting to the historian. Most were in areas that BMS especially wished to fund: controlled environment facilities, marine biology and oceanography, inland field stations, and systematic biology. In these areas, staff and ad hoc committees made surveys, wrote reports, and established priority needs and criteria for support.

In general, BMS favored enterprises that served a regional or national rather than a local function. Those requiring the cooperation of various biologists or serving more than one discipline or institution stood a higher chance of being supported than a single-university venture. Some large museums and other forms of non-university-affiliated institutions referred to themselves as national facilities, hoping thereby to obtain long-term operational support. Private laboratories like the Marine Biological Laboratory, the Cold Spring Harbor Laboratory, or the Naples Zoological Station, where investigators from all over the globe converged to carry on summer research, did in fact operate like national laboratories. Other institutions, such as the various genetic stock centers supported by BMS, provided research materials to biologists nationwide. Like their predecessors in the

Table 7.1a BMS Special Facilities Expenditures by Year,
FY 1957–FY 1968

	Annual Proposal Load[a]	Amount Funded
1957	not available	776,800
1958	not available	987,050
1959	21,889,700[b]	2,929,800[c]
1960	10,206,000	2,828,500
1961	28,904,450	2,831,300
1962	34,932,000	5,649,050
1963	26,036,100	5,590,000
1964	21,669,900	4,567,900
1965	20,104,300	4,792,700
1966	17,780,212	4,500,000
1967	21,393,588	3,039,920
1968	19,443,422	2,000,000
Total	$222,359,672	$38,729,170

Source: "Fact Book, Facilities and Special Programs, Division of Biological and
Medical Sciences, National Science Foundation," April 1968, Table B3, NSF
Historian's Files, National Archives and Records Administration. Amounts
for phytotrons and biotron have been added.
[a]New accessions plus carryovers from preceding year
[b]Includes biotron and phytotron proposals
[c]Includes $1,500,000 from Director's fund for the Biotron

Table 7.1b BMS Special Facilities Expenditures by Category, FY 1957–FY 1968

	Amount	Percent
Nonmarine facilities		
Specialized Research Laboratories (including equipment)	3,414,900	8.8
Specialized Research Tools	4,414,100	11.4
Inland Field Stations (including equipment and housing)	3,819,850	9.8
Systematic Biology (facilities)	8,018,120	20.6
Planning Grants	14,500	0.03
Marine facilities		
Ship and Boat Construction or Conversion	5,291,400	13.6
Marine Stations (including equipment)	13,835,900	35.6
Planning Grants	72,300	0.2
Totals	$38,880,070	100.0

Source: "Fact Book, Facilities and Special Programs, Division of Biological and Medical Sciences,
National Science Foundation," April 1968, Table B1, NSF Historian's Files, National Archives and
Records Administration.

Rockefeller Foundation, BMS staff saw themselves in a unique position to advance biology by "cooperative" ventures that broke through institutional and disciplinary barriers.[10]

Big Instruments: Phytotrons and the Biotron

The BMS staff had long entertained the idea of creating and supporting a central biological facility that would be the biological equivalent of AEC's national laboratories or NSF's astronomical observatories. Because these facilities were centered on expensive and unique instruments, the staff was especially receptive to a large-scale biological instrument that would be available to all investigators on a national or regional basis. Construction of the Biotron at the University of Wisconsin in 1959, and twin phytotrons at Duke and North Carolina State University a few years later, was an attempt to realize this vision.

In 1949, Caltech's Biology Division, since the 1930s a center of plant biochemistry, inaugurated the Earhart Laboratory, a series of air-conditioned greenhouses in which plants could be studied in relation to their environment. Scientists could control, through an impressive panel of valves, indicators, and regulators, such variables as the length of day and night, temperature, light intensity, gas composition of the air, and such weather conditions as wind, rain, and fog. Caltech biologist James F. Bonner dubbed the facility the "phytotron," from *phytos,* the Greek word for plant, and "tron," for "a big complicated machine" such as the cyclotron.[11] Frits Went, director of the facility, wrote in 1949, "Any similarity between the term phytotron and such terms as betatron, synchrotron, cyclotron, and bevatron is intentional. Caltech's plant physiologists happen to believe that the phytotron is as marvelously complicated as any of the highly touted 'atomsmashing' machines of the physicists." NSF funded Went's own research from 1952 and began to supply operational support to the Earhart Laboratory in fiscal 1956.[12]

As early as fiscal 1953, botanist Frederick C. Stewart approached BMS to build a phytotron at Cornell. Steinbach told him in 1954 that his proposal had given the divisional committee "a focal point for discussion of large-scale installations of the support of the biological sciences."[13] NSF cited the "urgent need" for a phytotron in the eastern United States in partial justification for a line item in the FY 1957 budget for specialized facilities in the biological and medical sciences.[14]

Once they gained a budget line for specialized facilities, the BMS staff initiated two ad hoc committees of scientists to examine the need for controlled environment facilities and criteria to be used in evaluating proposals.

The first, funded through the American Institute of Biological Sciences and chaired by Kenneth Thimann, presented in March 1957 a survey showing that while many small facilities maintained a constant environment for biological experiments, the Caltech phytotron was the only facility that allowed for environmental variations. Among the advantages the committee cited for building another phytotron was that it would bring together researchers from different fields such as genetics, plant physiology, anatomy, horticulture, and forestry. "In this way it can develop a unified approach to some aspects of biology." The report called for further technical study of specifications.[15]

By the time NSF convened the second committee, funded through the Botanical Society of America, several zoologists had expressed interest in a controlled-environment facility. Thus was coined the term *biotron,* a combination of "phytotron" and "zootron." Because BMS was organized according to programs that encompassed both botany and zoology, the staff looked with special favor on the biotron concept. In 1958, the second committee, which became known as the Biotron Committee, held discussions with biologists on campuses in four areas of the country and prepared an article for *Science* inviting further suggestions for the "planning of national biotron facilities." It listed a wide variety of basic research areas that might be investigated with the biotron: "temperature and photoperiod effects, general interaction of environmental factors, rhythmic and cyclic studies, germination, separation of genetic and environmental effects, mechanisms of adaptation, acclimation, evolution, speciation, dormancy, and hibernation."[16]

Once interest was aroused by the campus visits, BMS let it be known that it would entertain proposals (there was no formal program announcement).[17] Some ten institutions responded, all of which had site visits. Wisconsin emerged as the favorite because it wanted a biotron rather than a phytotron, it planned to use the biotron for basic research, and a variety of departments in the life sciences proposed to cooperate.[18] Even though there were no special facility funds remaining to support the project, Levin was able to extract $1.5 million from Waterman's reserve fund. The board approved the award, the largest BMS had ever made, in May 1959.[19]

Both Wilson and Levin hoped that the biotron would serve a national function. The award specified that competent researchers from other institutions be able to use the facility, that the university would provide housing for visiting scientists, and that there be formed a national advisory committee with members from other institutions. At a Wisconsin conference of biologists, held at NSF's urging in December 1959 to discuss facility specifications and research plans, Levin compared the prospective biotron, "the first large scale biological facility we have entered into," to the national astronomical observatories at Greenbank and Kitt Peak. Unlike these, he noted, the biotron was funded by a grant rather than a contract: "This, therefore, is not a national laboratory or even a regional one in the true sense. However, we did indicate, and it was agreed, that it would approach the capacity of a national or regional laboratory at least until there are others available." Wilson wrote in his annual report, "The enterprise is regarded as a full scale trial on behalf of the biologists of the United States which we hope will demonstrate the scientific potentiality of controlled environment research, particularly in the area of animal experimentation."[20]

The $1.5 million was found insufficient to build the complex structure, officially called "The Biotron" with its forty controlled environment rooms. NSF added another $300,000 to the total cost of $4.8 million, and the remainder was contributed by NIH, the Ford Foundation, and the State of Wisconsin. NSF funded annual operating costs.

The Wisconsin Biotron was followed by an even larger BMS investment in controlled environment facilities—the twin phytotrons at Duke and North Carolina State, called in the 1960s the Southeastern Plant Environment Laboratories. After a feasibility study in 1963, NSF provided $2.85 million, over two-thirds of the estimated $4.3 million needed to build the phytotrons, plus annual operating expenses of $300,000 once the facility opened in 1968.[21] The project, under the charge of Paul J. Kramer, professor of botany at Duke, former NSF program director, and long-time BMS advisor, was particularly attractive to BMS because the phytotrons would serve universities in the Southeast, a region underrepresented in terms of geographic distribution. According to the project summary, each of the "gemini" phytotrons was to be "regarded as a regional or national laboratory to be used for research by staff members and for the training of graduate students from institutions from all over the United States and from abroad."[22]

Despite the enthusiasm with which BMS staff funded them, biotrons and phytotrons as the basis for national or regional biological facilities proved to be largely disappointing. The Wisconsin Biotron, especially, billed as "the largest and most sophisticated controlled environment center in the world," did not live up to expectation. To construct a facility in which a large number of environmental factors could be simultaneously controlled was an exceedingly complex undertaking involving collaboration between biologists and engineers. There were considerable delays in construction, experiments were not initiated until 1967, and the formal opening did not take place until 1970.[23] By the time the Biotron was in operation, smaller manufactured controlled-environment facilities had become available for purchase by individual departments. Much of the basic research envisioned for the Biotron could be carried out in these much cheaper and more accessible instruments.

Few researchers from other institutions took advantage of the Biotron because of the inconvenience of going to Wisconsin and the user fees necessary to defray the high cost of maintenance. Like the Caltech phytotron (which received substantial support from the Campbell's Soup Company) and most phytotrons elsewhere, the Biotron was more readily adaptable to applied research than to basic biology. In the actual research carried out, there was little of the interdisciplinary teamwork that the grant proposal promised. Instead individual researchers or small groups simply rented research space to do "little science." The national advisory committee, which was always dominated by Wisconsin members, was eventually disbanded because there was no plethora of applications from which to choose.

After spending $1.2 million on operating costs, NSF dropped funding of the Biotron in 1976.[24] Two years later NSF terminated support of the North Carolina State phytotron because the facility was used predominantly for applied agricultural research (a fate which some BMS staff had predicted from the initial proposal).[25] The Caltech phytotron, used in the late 1960s largely for Arie Haagen-Smit's research on California smog, was torn down in 1972. Although the Duke phytotron continued to receive NSF operational support and to advertise itself as a national facility, it rarely trained students other than those at Duke.[26]

The "Critical Areas" Concept and Its Critics

In 1960, Wilson announced that while BMS had relied in the past on relative proposal pressure to determine its allocation of funds, future spending would also be influenced by "the emergence of the so-called 'critical area' concept." "Within the past year or so," Wilson reported, "there has arisen, partly through valid natural evolution or national need, and partly through astute promotion, expressions of concern over the lack of support for particular areas of science which for one reason or another are 'emerging' in importance."[27]

Wilson's newfound willingness to set programming priorities for BMS resulted from a memorandum by Waterman to the National Science Board in late 1959 in which he had argued the necessity of identifying and developing "emerging areas of science." In the post-Sputnik era, it became clear to NSF that the argument for expanded budgets for basic research could not be made on the basis of proposal pressure alone, but must be supplemented by active leadership in particular well-chosen areas of emphasis. In response to Waterman's memorandum, Wilson identified three areas to which BMS was prepared to devote special attention: biological oceanography, computers in life science research, and tropical botany. By the following year, tropical botany had broadened to tropical biology, and a fourth area, forestry, was added.[28]

Of these, BMS gave the most attention and resources to biological oceanography and tropical biology, both areas heavily dependent on NSF for support. (By contrast, computers in the life sciences was supported by NIH and forestry by the Forest Service and state forestry schools.)[29] Biological oceanography, the subject of Wilson's allusion to "astute promotion," was a prime example of controversial "big biology" in the 1960s. Tropical biology, targeted by BMS two years before Rachel Carson's *Silent Spring,* presaged the division's increasing involvement in environmental issues. In both areas, NSF activities stimulated new national or international endeavors, involving the cooperation of many universities, and new national organizations. Together these two initiatives represented a shifting of resources away from molecular biology and other NIH-supported areas of the life sciences, to those—systematics and ecology—for which NSF served as major patron. The fashioning of these initiatives further exemplify the crucial role played by federal science managers.

The "critical areas" approach found far more favor with NSF's physical scientists and their advisors than its biologists.[30] While BMS staff expressed ambivalence, the division's advisors were downright hostile to singling out certain areas for special treatment. Responding to Waterman's first raising of the "areas of science" issue in 1958, the divisional committee passed a resolution, reiterated the following year, which "noted with regret that there continues to be rampant the idea that a committee can be constituted to recognize 'special problem areas' or 'programs of special urgency' in a way that implies the ability of such committees to see into the uncharted future." Particularly "fraught with great danger" were such National Academy of Sciences groups as the Pacific Science Board and the Polar Research Committee, which redistributed NSF research funds. Progress in biology depended on "individual initiative," by which the committee meant individuals applying for grants in their fields of choice. The members called upon NSF to "safeguard this free-enterprise aspect of scientific research. By its very definition basic research cannot be programmed." Wilson himself told the divisional committee in 1961 that while "the critical area concept in its best form can be helpful in discovering areas which should be further developed," an "inherent danger" was that it could "develop a 'Madison Avenue' type of effort."[31] The divisional committee's apprehension about large-scale projects or programmed research led to increasing friction between BMS staff and its advisors as the decade advanced.

Big Ships: NSF and Biological Oceanography in the 1960s

Biological oceanography represented BMS's largest venture into "big biology" by far. Prior BMS support of individual research projects in this field was expanded in the 1960s to the funding of four major ships, nearly a dozen smaller vessels, and two new marine stations, the building or renovation of some two dozen marine laboratory buildings, and the sponsoring of the biological work of a major international oceanographic expedition. In all, BMS spent about $19 million on marine facilities from 1957 to 1968 (see table 7.1).[32]

BMS adopted biological oceanography as a critical area in response to an awakening federal interest in oceanography, the initial impetus for which came primarily from the physical sciences and the military. At the formal request of ONR's Earth Sciences Division, AEC, and the Fish and Wildlife Service, the National Academy of Sciences created in 1957 a Committee on Oceanography (NASCO) within the National Research Council's Earth Sciences Division. In late 1959, the first section of NASCO's report, *Oceanography 1960 to 1970,* appeared. Predicting dire military and political consequences if its recommendations were ignored, the report called for a doubling of federal support for basic deep-sea research over ten years. Although the Academy had tried several times in the past to promote expanded resources for oceanography, in the era of Sputnik, the cold war, and the International Geophysical Year with its extensive oceanographic research, the report quickly bore fruit. The oceans represented a vast unknown of obvious military and economic significance, and the oceanography program a parallel in the sea to the burgeoning space effort. The report soon had the support of the President's science advisor George Kistiakowsky and the Federal Council on Science and Technology, and by 1960 several bills to promote oceanography were introduced into Congress.[33]

As federal interest in oceanography grew, a number of biologists became alarmed at the apparent bias of the NASCO study toward the physical sciences. As they pointed out, the only biologists involved were oriented to problems of fisheries and radioactive substances in the sea; little consideration was given to basic research problems in the life sciences.[34] Biologists vociferously objected to the report through the American Society of Limnology and Oceanography, through the AIBS Committee on Hydrobiology, and through interaction with Congress and the federal science establishment. Intense discussion in 1959 and 1960 centered on the alleged "schism" between physical and biological oceanographers. BMS program officers, particularly Wilson and George Sprugel, program director for environmental biology, joined in the thick of the battle, channeling protests of biologists, alerting Waterman and the NSB, and directing Foundation funds to benefit "biological oceanography," a recently coined term that came into rapid usage.[35]

Sprugel, reviewing events for the divisional committee in May 1960, outlined what he and BMS had done in the past year to promote the biological side of oceanography. For example, when biological oceanographers complained that they had difficulty obtaining "ship-time" on oceanographic vessels supported by physical sciences and military contracts, Sprugel took the initiative to meet with Paul Fye, Columbus Iselin, and Bostwick Ketchum of the Woods Hole Oceanographic Institute (WHOI), one of the country's two leading centers for oceanography. He suggested that WHOI submit a proposal for ship-time specifically set aside for biologists, which resulted in an award of $250,000. When the MPE division initiated a facilities award to WHOI in 1959 to build a new research vessel, *Atlantis II*, "steps were taken" by BMS staff to assure that the needs of biological oceanographers were taken into account in the design of the ship. Ketchum prepared detailed requirements for the architects. BMS helped to shore up the American Society of Limnology and Oceanography by underwriting its journal and supporting its Committee on Education and Recruitment in Oceanography to prevent the physical oceanographers from forming a separate organization. Finally, BMS helped to convince the Academy to enlarge NASCO by adding two outspoken biologists, Per F. Scholander and Dixy Lee Ray.[36]

The new attention given biological oceanography, while it stood to benefit some biologists, stirred up considerable resentment in others. Just as many biologists had deplored the sudden rise of NASA and the race to land a man on the moon, they also objected to funneling resources into one area of biology at the seeming expense of others. Even Sprugel was skeptical of the exaggerated claims being made on behalf of oceanography. While he believed the field needed more support, he added in his confidential report to the divisional committee: "Whether the United States will be overrun by barbarians in ten years if the provisions of the NASCO report and the Magnuson bill [on marine sciences] are not implemented or whether more damage may be done in the long run through hurried and unwise over-expansion of the field at the expense of other fields is something else again."[37]

BMS's chief response to the ferment of opinion was to appoint in 1960, with divisional committee approval, an Ad Hoc Committee on Biological Oceanography, chaired by Dixy Lee Ray who was hired as a special consultant. Ray, then associate professor of zoology at the University of Washington and later chair of the AEC and governor of Washington, was the first woman to fill a high-ranking post in BMS. Remaining with the Foundation from July 1960 to 1963, she visited marine stations in Europe and America

with committee members or on her own, and in February 1961 brought together directors of marine and field stations under NSF auspices for a discussion of common problems.[38]

The committee adopted from the outset a broad definition of oceanography in general and of biological oceanography in particular. Its report,
completed in August 1961, argued against what it regarded as common mis-
conceptions—that oceanography dealt only with physical aspects of oceans
and only with the deep ocean. Biological oceanography, the committee declared, was nothing less than "the study of life in the sea," and the sea extended from shore to shore. This and later BMS reports insisted that there
was no significant difference between biological oceanography and marine
biology, the latter an area long recognized as part of biology.[39] If most marine biologists did not have experience with research on the deep seas, that
was because few marine laboratories had sea-going vessels. "This circum-
stance, coupled with the practical problems of cruise scheduling and costs,
has contributed to the fact that biologists have traditionally directed greater
effort to research in shallow areas and along the ocean shore than to the open
sea." Providing biologists ship-time on general oceanographic vessels and
constructing ships specifically for biological research could remedy these difficulties. The committee argued for expanded flexible support of marine
stations, including operational or "logistics" support of shore facilities, coherent area grants, and construction and operational support of research vessels.[40]

In the next few years funding for both individual projects and facilities
in marine biology/biological oceanography expanded rapidly. Program director for environmental biology John Rankin calculated that in the period
1958–65 BMS had spent $19.8 million on 729 individual project grants, the
majority awarded through the Systematics and Environmental Biology programs. Support had grown from $776,000 for 52 projects in 1958 to $3.6
million for 118 projects in 1965. In addition, $4.8 million was supplied
through NCE and Special Programs, $13.9 million from Special Facilities,
and $4.9 million from International Indian Ocean Expedition funds (see
table 7.2a).[41] In the mid-1960s BMS received a separate line amount for
oceanographic facilities plus a share of NSF's funds for IIOE.[42] Jack Spencer
estimated that from 1957 to 1968 the BMS Special Facilities Program had
spent $5 million for boats and ships and $14 million for shore facilities (see
table 7.1b). As a result of increased funding, he claimed, "biology has
emerged to play a highly significant role in the marine sciences. At the outset, biologists had almost no funding for ship-time, whereas today on a national scale they rank on a par with the physical scientists."[43]

The most visible form of BMS support for biological oceanography was

Table 7.2a   BMS Support of Marine Biology/Biological
Oceanography[a] by Program, FY 1958–FY 1965

	Number of Projects	Amount
Developmental	94	2,927,000
Environmental	206	6,324,300
Genetic	11	264,200
Metabolic	40	1,412,250
Molecular	48	1,439,500
Psychobiology	19	628,400
Regulatory	98	3,336,000
Systematic	213	3,430,500
NCE/Special Programs	41	4,801,750
Facilities	57	13,900,415
IIOE		4,921,916
Total	827	43,386,231

Source: John S. Rankin, "Role of the National Science Foundation
in Support of Biological Oceanography FY 1958–1965," January
1966, in notebook "Division and Staff Papers, Book II," Box 3, 76-
172, RG 307, National Archives and Records Administration.
[a]Limnology excluded

Table 7.2b   BMS Support of Marine Biology/Biological Oceanography[a]
by Year, FY 1958–FY 1965

	Project Grants		Facilities		NCE/Special Programs		IIOE
	Number of Awards	Amount	Number of Awards	Amount	Number of Awards	Amount	Amount
FY 1958	52	$775,300	3	$571,950	—	—	—
FY 1959	53	1,671,650	4	596,000	—	—	—
FY 1960	80	1,861,050	8	720,850	—	—	—
FY 1961	82	2,048,700	5	1,496,450	4	$126,600	—
FY 1962	98	2,958,450	12	4,209,100	6	272,600	$500,000
FY 1963	122	3,888,600	10	2,687,300	11	1,091,400	1,626,600
FY 1964	94	2,124,500	8	1,268,665	8	1,268,665	1,991,100
FY 1965	118	3,635,900	7	2,350,100	7	2,350,100	804,216

Source: John S. Rankin, "Role of the National Science Foundation in Support of Biological Oceanography
FY 1958–1965," January 1966, in notebook "Division and Staff Papers, Book II," Box 3, 76-172, RG 307,
National Archives and Records Administration.
[a]Limnology excluded

the building of research vessels. Of twelve large ships NSF constructed or converted from 1958 through 1965, four were specifically for biological oceanography: the research vessel (R/V) *Anton Bruun,* owned by NSF and leased to WHOI during the IIOE; the *R/V Te Vega* for Stanford, also used in the IIOE; the *R/V Eastward* for Duke University; and the *R/V Alpha Helix* for the Scripps Institution of Oceanography (University of California, San Diego). A major attraction for BMS funding of ships was that they could serve as another form of national facility. Each ship met the needs of investigators or students from many different institutions, and, moreover, the scientific programs of each were overseen by national advisory committees of leading researchers. BMS also funded eleven smaller vessels for universities and independent marine stations such as the Marine Biological Laboratory and Bermuda Biological Laboratory.[44]

Almost any marine station of note was able to build or renovate laboratory buildings in the 1960s. NSF made thirty-five awards to twenty-six institutions for marine laboratory facilities; twenty-four of these were for construction or expansion of laboratory buildings.[45] One of the largest construction projects was the building of WHOI's Redfield Laboratory through an award of $2 million in FY 1962. At least two new marine stations were built with BMS support—the University of California's Bodega Marine Laboratory, at a cost of $1.1 million (1963), and Catalina Research Laboratory for the University of Southern California for $500,000 (1965). The Catalina Island facility was intended to be used by over forty colleges and universities in the Los Angeles area.[46]

In addition to continued summer research and training at approximately twenty marine laboratories, BMS supported several larger educational ventures. At the Marine Biological Laboratory, which scientists used almost entirely in the summer months, BMS funded, beginning in 1963, a year-round program in marine ecology under Melbourne R. Carriker. BMS also supported innovative training programs on the Stanford and Duke vessels. Duke's 117-foot *R/V Eastward,* launched in June 1964 as "one of the very few ships designed specifically for biological oceanography," was the centerpiece of a cooperative program serving twenty to thirty colleges and universities in the Northeast, South, and Midwest. These institutions nominated students to participate in short training cruises on the *Eastward.* Stanford's *R/V Te Vega,* a 135-foot two-masted schooner converted by NSF funds for participation in the IIOE, returned to the Hopkins Marine Station, where it served as the focus of a training program open to graduate students from any institution. Its several cruises a year usually lasted about ten weeks, or an academic quarter, and typically carried three faculty members, ten selected graduate students, and two technicians. Students, fully supported by BMS, were to participate as true "research colleagues" rather than as assistants in the expedition.[47]

The *R/V Alpha Helix*, conceived by the brilliant and eccentric comparative physiologist Per F. Scholander of Scripps, was oriented primarily to research by senior investigators. Spencer boasted of it as "the world's only floating physiological laboratory."[48] NSF agreed in 1962 to build the ship, the Physiological Research Laboratory on shore, and its associated pool facilities. The 133-foot vessel, constructed at a cost to NSF of $1.5 million, featured a large main laboratory, electrophysical and optical laboratory, "wet" laboratory, freeze-room, photographic darkroom, and well-equipped machine shop. With air-conditioning and a reinforced hull, the *Alpha Helix* could travel anywhere on the globe, enabling investigators to experiment on exotic marine forms in their native habitats. In its first three years of operation, 1966–68, the ship made expeditions to the Great Barrier Reef of Australia, the Amazon River, and the Bering Sea.[49]

The *Alpha Helix* was for a time one of BMS's most successful attempts to create a national research facility for biology. Roger Revelle, director of Scripps, appointed a National Advisory Board (so named at the behest of NSF) composed of eminent physiologists and biochemists from various institutions. Initially chaired by A. Baird Hastings, the board oversaw the scientific program, selecting the general area of operations each year, soliciting and evaluating research proposals of American and foreign scientists, and helping to make policy for use of the vessel. Each expedition was divided into programs of approximately three months, each overseen by a chief scientist. Program participants were flown in to join the vessel. Scholander wrote in a 1978 autobiographical account, "To myself and hundreds of colleagues from various disciplines and countries who have had the privilege of using this facility, the research vessel *Alpha Helix* stands as a proud landmark of *nonpolitical U.S. generosity*, as a gift to international friendship and scientific cooperation."[50]

The International Indian Ocean Expedition, one of several multinational cooperative projects launched on the model of the International Geophysical Year, represented BMS's largest single endeavor in biological oceanography. The IIOE was conceived at the First International Oceanographic Congress held at the United Nations in 1959 and organized by the Special Committee on Oceanic Research (SCOR) of the International Council of Scientific Unions. A major justification for studying "the world's least known ocean" was the identification of food resources to aid in the development of the densely populated surrounding countries.[51] As with the IGY, NSF coordinated federal participation in the presidentially approved U.S. program, which studied the structure of the ocean basin, the chemistry and physics of the waters, including oceanic currents, the interaction between the ocean and the atmosphere, and the plant and animal populations. From 1962 through 1965, twenty-five nations provided forty-four vessels to participate in the IIOE. The U.S. government spent a total of $20.6 million on the program through FY 1965, the largest share, $15.9 million, provided by NSF.[52]

BMS began preparing for the IIOE in 1960 by sending deputy assistant director David Keck to India to visit marine biological centers and consult on appropriate supporting shore facilities. WHOI biologist John H. Ryther and a subcommittee of NASCO coordinated the biological aspects of IIOE. While all seven participating U.S. ships sampled plankton and some also measured primary productivity, the *Anton Bruun* and the *Te Vega* were specifically devoted to the biological program of IIOE.[53]

When it proved politically infeasible to build a new ship in a foreign country for the expedition, President Kennedy offered the Foundation a former presidential yacht, the *U.S.S. Williamsburg*. Converted at a cost of $900,000 in IIOE funds administered through BMS, the huge 240-foot vessel was renamed the *R/V Anton Bruun* after the recently deceased Danish oceanographer who had promoted international cooperation and served as president of the First Oceanographic Congress in 1959. The *Anton Bruun* departed for India in early 1963 and made nine cruises in the Indian Ocean during 1963 and 1964.[54] The *Te Vega,* under Rolf Bolin, made three cruises around island groups and in shallow waters. Specimens were sent for sorting and identification to a laboratory in Colchin, India, and to the Smithsonian Oceanographic Sorting Center, established in 1963 for the purpose and partly supported by NSF. About 150 marine biologists, American and foreign, participated in biological aspects of the U.S. program. From 1962 through 1965, BMS spent $4.9 million on the IIOE. Ship operating costs alone accounted for $2.6 million.[55]

Spencer claimed that the *Anton Bruun,* of all the ships supported by BMS, was "the only vessel to date which has functioned as a truly national and international platform directly under the control of an independent grouping of U.S. scientists."[56] But as a national biological facility, it was short-lived. After the IIOE, the big vessel was used in 1965 for biological exploration as part of the Southeastern Pacific Oceanographic Program. Then, after much debate over its future, the *Anton Bruun* was retired, primarily because it was too old, too large, and too costly to operate. But even had the ship been more economical, NASCO opposed its continued operation on the grounds that "except under very special circumstances," the NSF should not "either operate a research vessel or directly supervise its operation." An NSF ship should not compete with university-managed ships.[57]

The biological research vessels and their research and training programs, once begun, required ever increasing budgets to keep them going. From 1964 to 1968, some 75 percent, or over $13 million out of a total of $18 million expended by Special Programs, went to providing operational support for ships. The large ships, especially, could be enormously expensive. The *Alpha Helix,* for example, which had berth space for ten scientists, required a crew of twelve; its annual operating costs were about $500,000. The *Anton Bruun* cost approximately $1 million a year to run. The training programs of the *Eastward* and the *Te Vega,* which included ship operational costs, amounted to $400,000 each in 1965. By 1968, when the financial crunch came, rising expenditures for ship operation threatened to crowd out all other forms of facilities support.[58]

Smaller-Scale Biology: Inland Field Stations

While marine laboratories expanded greatly in the 1960s as a result of strong political interest in oceanography, inland field stations—despite BMS interest in them—remained relatively poorly funded. These were facilities for teaching and/or research generally located off campus but without access to the open ocean. They were a diverse group geographically, ecologically, and in type of research emphasized. A majority were located on lakes or sizeable streams and thus provided opportunity for limnological research; a number were mountain stations. Most emphasized ecology and systematics but others focused on animal behavior or environmental physiology. Of those inland field stations regularly funded by BMS, some were attached to universities such as the University of Michigan Biological Station, the University of Virginia's Mountain Lake Biological Station, and the Lake Itasca Forestry and Biological Station of the University of Minnesota. Also included in this category was California's White Mountain Research Station, a facility primarily devoted to studies of high altitude physiology, that BMS had funded since 1953 (see chapter 3). Other stations were independent, such as the Highlands Biological Station, Inc., in Highlands, North Carolina, directed by Thelma Howell, which emphasized ecology of the Appalachians and Blue Ridge Mountains, and the Rocky Mountain Biological Laboratory in Crested Butte, Colorado, under the charge of Robert K. Enders of Swarthmore College. Compared to the great oceanographic centers, most inland field stations were shoestring operations. As Dale Arvey pointed out, among university biological activities "their status is not high, and stations are often dismissed as mere nature study camps."[59]

As it had in oceanography and tropical biology, BMS hired a special consultant in 1963 to survey existing inland field station facilities. Dale Arvey, an ornithologist and former chairman of the department of biology at Long Beach State College, surveyed forty-two stations and published his report to the divisional committee in *BioScience*.[60] Following precedent, Arvey sponsored a conference at NSF in May 1964. It was attended by directors of thirty-two inland field stations plus observers from various federal and private funding agencies. The most divisive issues were whether stations should be coaxed into remaining open year-round (as BMS staff preferred) and whether laboratories at stations should be equipped for physiological as well as observational research. As in other cases where NSF initiative brought together a group with similar institutional interests, the meeting resulted in the formation of an ongoing association, the Organization of Biological Field Stations.[61]

In addition to supporting summer research at field stations, as it had in the 1950s, BMS constructed or renovated laboratories, purchased instrumentation, supplied boats for limnology, and even occasionally provided new housing for investigators. The largest such construction projects in the 1960s were the Laboratory of Limnology of the University of Wisconsin ($480,000 in FY 1960), an on-campus station on Lake Mendota, and the Alfred H. Stockard Lakeside Laboratory at the University of Michigan Biological Station on Douglas Lake ($500,000 in FY 1964). As a condition for funding the latter, NSF required that the university provide funds for winterizing living quarters so that the laboratory could be used on a year-round basis. In all, BMS supported at one time or another some twenty-five stations (not all of which were on Arvey's list) at a cost of $3.5 million for facilities and $800,000 for summer programs and operational expenses (see table 7.3).[62]

Surveying a decade of NSF support of special facilities, Spencer felt that progress in improving inland stations had been relatively slow, a situation he attributed in part to "some lack of interest on the part of terrestrial biologists who should be most concerned with the problem." In 1968, it appeared that not only would facilities funds disappear but also that all the research training programs at inland field stations would be phased out. "I doubt if any greater calamity could befall environmental and systematics biology at this time," Spencer wrote. "While there is great concern about funding international biological programs (as is quite proper) at a multimillion dollar level, it is quite overlooked that the basic field training for beginning and mature graduate students is about to go down the drain."[63] It was not until the new Organization of Biological Field Stations was able to exploit the environmental movement that more funds became available (see chapter 9).

Table 7.3   Summary of BMS Support for Inland Field Stations
(Facilities and NCE), FY 1952–FY 1968 (Total per Station
by Order of Magnitude)

Station	Amount
U. of Michigan, Douglas Lake	$700,200
U. of Wisconsin, Hydrobiology Laboratory (plus Trout Lake)	546,100
UC Berkeley, Behavioral Research Station	367,000
Highlands Biological Station, N.C.	373,850
U. of Minnesota, Lake Itasca	352,400
UC Berkeley, White Mountain	326,800
U. of Oklahoma, Lake Texoma	267,450
U. of Virginia, Mountain Lake	259,950
U. of Texas, Brackenridge Field Laboratory	258,000
Indiana U., Crooked Lake	167,700
U. of Montana, Flathead Lake	149,000
Rocky Mountain Biological Laboratory, Colo.	122,700
Smithsonian, Barro Colorado Island	110,000
U. of Colorado, Science Lodge	73,350
Duke U., Behavioral Research Station	56,400
U. of Wisconsin, Milwaukee Station	50,600
UC Riverside, Deep Canyon Research Station	33,200
Utah State U., Bear Lake	25,000
U. of Iowa, Lakeside Biological Station	21,000
San Diego Museum, Vermilion Sea Station	19,200
U. of Missouri, Prairie Research Station	19,000
Emory U., Lullwater Field Laboratory	18,200
UC Berkeley, Sagehen Creek	16,000
Darwin Foundation, Galapagos Islands	6,500
Total	$4,339,600

Source: "Fact Book, Facilities and Special Programs, Division of Biological and
Medical Sciences, National Science Foundation," April 1968, item U3, NSF
Historian's Files, National Archives and Records Administration.

"A Tower of Biology": The Failure to Fund a National Laboratory in Tropical Biology

David Keck, in a 1961 report to the BMS Divisional Committee, relied
on the familiar frontier metaphor to justify BMS's choice of tropical biology
as a "critical area." The New World tropics, like space, the ocean depths, the

earth's crust, and the Antarctic, represented another vast unknown field for exploration. "This region of the earth," he claimed, "remains as one of the biologist's most challenging frontiers."[64] BMS's initiative in tropical biology predated the full blossoming of the environmental movement, but environmental concerns, expressed in terms of the earlier conservation ethic, were indeed part of its justification. Keck called attention to the richness of tropical flora and fauna, the diversity of environments, the productivity of the tropics, the danger that increasing population pressures would lead to destruction of tropical forests, and the inadequacy of knowledge of the tropics as compared to the temperate zones. He saw an opportunity for the Foundation to provide leadership to a coordinated effort to expand research and training in tropical biology. It was an area that was "not only ripe for a much more intensive effort" but "also one of the most overlooked fields from which results of substantial importance along a broad front can be anticipated."[65]

NSF had a significant impact on the field in the 1960s through its support of a series of seminal conferences that led to the creation of two national organizations and a major expansion of graduate student training in tropical biology. But compared to the division's parallel initiative in biological oceanography, BMS was never able, in this period, to focus its funding for tropical biology or point to major achievements in research supported or facilities built. Despite the many internal reports on tropical biology, at the end of a dozen years, BMS staff were still trying to establish a strategy. The problems with BMS patronage of tropical biology were manyfold, including too many rivalries and competing interests among tropical biologists for effective setting of priorities; misplaced reliance by both staff and grantees on "big biology" models; inability of tropical biologists to convince the NSF hierarchy, let alone Congress, that research in tropical biology served a national need; and the compartmentalization of NSF into research divisions, an educational division, and an international office. Large proposals ran the risk of spanning the boundaries of several divisions of NSF and not fitting into any. Thus, what did get accomplished in tropical biology was shaped and constrained by the organizational structure of NSF.

Expressions of interest in expanded support for tropical biology reached NSF from two groups of biologists with different agendas, one interacting with BMS and the other with the education division of NSF. One strand led to the founding of the Association for Tropical Biology (ATB), the main professional society for tropical biologists, and the other to the founding of Organization for Tropical Studies (OTS), which has since become (along with Smithsonian Tropical Research Institute) one of the two main American institutions for research and training in tropical biology.

On the research side were the systematic botanists who desired better access to tropical specimens. In 1959, William Robbins, Emeritus Director of the New York Botanical Garden approached the National Academy of Sciences to sponsor a small invitational conference at the Fairchild Tropical Garden near Miami. Robbins' immediate aim was to draw attention to the Fairchild Garden as a resource for tropical botany. Keck and Wilson were interested in funding the proposal provided it would address instead "the whole question of improving facilities for research in tropical botany."[66] The Conference on Tropical Botany, which Keck helped to organize, took place in May 1960. It resulted in a report published by the National Academy of Sciences that called for permanent facilities linking basic and applied research; a taxonomic inventory, which would also make use of museums and botanical gardens in the U.S.; impetus given to present indigenous centers of research in Latin America; cooperative undertakings with Latin American scientists; and training of students. To coordinate activities, the group recommended that the Division of Biology and Agriculture of the National Research Council establish a Tropical Plant Science Board.[67] It was primarily as a result of this conference, and Keck's involvement in it, that BMS adopted tropical botany as a "critical area" in 1960.[68]

In 1960 it appeared as if BMS might well spend the $40,000 per year requested by NRC for the Tropical Plant Science Board. However as the Academy's proposal emerged in 1961, it emphasized not research support but exchanges with Latin American students and scientists. Keck discouraged its submission, indicating, in effect, that a proposal concerning "non-research type activities," was outside BMS scope.[69] At the Academy, the proposal met with resistance from the executive board of the Division of Biology and Agriculture. Both H. Burr Steinbach and Paul Weiss were devoted to furthering an integrated "biology" in place of the traditional division of botany and zoology and were hence unsympathetic to a board limited to botany.[70] BMS, too, favored biology over botany. Partially as a result of these considerations, BMS's initiative in tropical botany became transformed in 1961 into an initiative in tropical biology.

Keck, in his 1961 report, advocated as a next step a survey of existing research facilities and programs in the American tropics. From November 1961 to April 1962, he and Walter H. Hodge, a botanist who came to NSF in 1961 from Longwood Gardens in Pennsylvania as a special consultant in tropical biology, made site visits to over fifty research centers including universities, museums, gardens, field stations, and institutes in Mexico, Central America, and South America north of the Tropic of Capricorn. From their travels, they compiled brief descriptions of facilities and their availability to North American investigators.[71]

When the project to establish a Tropical Plant Science Board at NRC fell through, attendees at the first conference planned a follow-up conference on neotropical botany, presided over by John Purseglove in Trinidad in 1962. Like the first conference, it, too, was funded by NSF's Systematic Biology Program, now under the direction of Hodge. The attendees expected to form a permanent association for tropical botany, but Keck and Hodge convinced them instead to form the Association for Tropical Biology. BMS gave the new association a small grant to enable it to get underway. Keck and Hodge published their report on field stations in the Association's first *Bulletin*.[72]

The second strand of interest in tropical biology was concerned with providing training in tropical ecology for American students. In 1959, NSF staff was well aware from having seen draft proposals, that the University of Michigan had been planning for several years to build a tropical science center (CenTrop) for teaching in Mexico in conjunction with a Mexican university. Like many expensive plans hatched in the post-Sputnik era, this one depended almost entirely on federal funding, with very little university contribution. When the Mexican cooperation fell through, Michigan turned to Costa Rica and by early 1962, perhaps as a result of consulting with NSF, began to consider a consortium arrangement.[73]

Other educators were also interested in Costa Rica as a base for training American students. Among them were Jay M. Savage of the University of Southern California, who began an NSF-funded course there for college teachers in 1961, and Leslie Holdridge and Robert J. Hunter, founders of the Tropical Science Center, a private consulting organization in Costa Rica, which was later to organize courses for undergraduates for the Associated Colleges of the Midwest. Hunter, Savage and Raphael Rodriguez of the University of Costa Rica proposed a Conference on Problems of Education and Research in Tropical Biology to be held Costa Rica in April 1962. This group was exceedingly fortunate that the program officer of the Special Projects in Science Education Program in NSF's education division was James Bethel, a tropical forester. Bethel funded the conference, attended all the sessions and served as a moderator for a discussion titled "Ways and Means of Meeting Present and Future Requirements for Education and Research in Tropical Biology." The participants at the Costa Rica conference declared themselves a permanent group dedicated to "the development of a coordinated, cooperative program of education and research centered in Costa Rica."[74] In early 1963, a second meeting was held in Miami at which the Organization for Tropical Studies was founded. Seven universities, including the University of Michigan, were charter members. Bethel, who had since moved to the University of Washington, became a member of the first executive committee.[75]

In order attract NSF funding, OTS deliberately founded itself on the model of the university consortia that managed NSF national facilities in the physical sciences. According to Reed C. Rollins of Harvard, the immediate model was the University Corporation for Atmospheric Research that operated NSF's National Center for Atmospheric Research. Member universities pledged $2,000 a year for five years on the overly optimistic assumption that thereafter OTS would be self-supporting through grants. In mid 1963, OTS projected an ambitious five-year plan calling for $2 million in grants to fund the educational program and research facilities.[76] In addition to requesting funds to run training courses from the NSF's education division, OTS applied to BMS for operational funds.

NSF officials were understandably uncomfortable with the organization of OTS. They felt that the member universities had only a minimal commitment to the organization, that OTS was financially unstable, and that it was run by a small number of people who did not represent a community of biologists. Moreover, the proposal to fund operating costs fell through the cracks at NSF. BMS funded operating costs of a handful of research institutions, but it could not fund an institution primarily devoted to training. The education division generally did not provide operating expenses other than the overhead on project grants. Although the education division began to support specific OTS courses in 1964, OTS officials found themselves in a bind. Rollins, President of OTS, wrote to the OTS board of directors in 1965: "It is clear that we need financial support from three major areas of N.S.F.—facilities, education and research. How do we convince them of our stability and requirements simultaneously?"[77] OTS had no choice but to ask its member universities to continue their annual contributions and rely for remaining support on the fluctuating overhead from project grants, mostly from NSF. By 1968, the Special Projects in Advanced Education Program of NSF was spending $373,000 for eight OTS training courses on such topics as tropical forestry, crop plants, reproductive biology of tropical plants, and dendrology. The main course offered every year, "Tropical Biology: An Ecological Approach," gave participants an intensive exposure to a wide variety of tropical environments in Costa Rica.[78]

Beginning in 1968, BMS underwrote OTS's attempt to begin a research program. However, because of the framing of the program in terms of big biology, the results were mixed. The program director for environmental biology, dissatisfied that BMS support of tropical biology was piecemeal, encouraged OTS to submit a proposal. To appeal to NSF and to obtain a large grant, OTS fashioned its research project as a multidisciplinary, hierarchically organized team project to compare a wet and a dry forest ecosystem in Costa Rica. In all, BMS spent several million dollars on this project before it was dropped in 1976. It enabled researchers to make an inventory at the two OTS sites, Palo Verde and La Selva, purchased in 1968. Donald Stone, later executive director of OTS, claimed this study "laid the foundation for La Selva's preeminence as a field station." Yet even OTS spokesmen admitted that as "big biology," the project was a failure. According to Stone, the administrative mechanisms for the project "failed miserably in pulling the subprojects together in any semblance of an integrated ecosystem analysis."[79]

Despite its key role in supporting conferences and training, and its expanded support of research projects in tropical biology (estimated at $900,000 annually from 1967 to 1971),[80] BMS was able to provide only limited funding for tropical facilities in the 1960s. Perhaps BMS's most significant facility award in tropical biology was a grant of $110,000 in 1963 to the Smithsonian Institution's Canal Zone Biological Area on Barro Colorado Island (now the Smithsonian Tropical Research Institute) to lay an electric cable across Gatun Lake to the research facility. For the first time the station had a reliable source of power, providing lighting and air-conditioning for greatly improved living conditions and also making possible physiological and behavioral research requiring electronic equipment.[81]

While several proposals were said to be in preparation for establishing a field or marine station in the tropics,[82] Carlson and Spencer focussed on a plan of Associated Universities, Inc. (AUI), to both build for NSF and to manage a Tropical Marine Science Center. Few institutions so embodied the notion of big science as AUI, the consortium of universities originated by physicist Lloyd Berkner, which managed the Brookhaven National Laboratory for the Atomic Energy Commission and the National Radio Astronomy Observatory for NSF.[83] AUI's desire to expand beyond its two major projects meshed well with Carlson's and Spencer's eagerness to fund a true national biological facility. Carlson and Spencer no doubt believed that a plan emanating from AUI would have appeal to Haworth, a former director of Brookhaven, and the physical scientists who dominated the NSF hierarchy and National Science Board.

The initiative for what turned out to be AUI's abortive venture into biology came primarily from William D. McElroy of Johns Hopkins, who as a member of AUI's board of trustees, proposed the establishment of a tropical marine biology facility in April 1966. In June, AUI sponsored a one-day conference at NSF under McElroy's chairmanship "to explore the need for a Laboratory for research and education in tropical marine biology which would be national in character and operation." The carefully worded conference report could not cover up the considerable dissent of the participants over whether a national center would drain away funds from existing marine laboratories, where the center should be located, and who should manage it.

Encouraged by NSF, AUI applied in late August for a BMS grant to do a feasibility study and, in mid-September, was awarded $78,000.[84]

Guided by a steering committee of biologists headed by John Ryther and a contract architectural-engineering firm, AUI prepared an official report to NSF in 1967, which recommended that a Tropical Marine Science Center for biological research and training be built in southwest Puerto Rico at an estimated initial cost of $10 million, and an annual operating cost of $3 million.[85] The center would emphasize experimental biology similar to that done at the Marine Biological Laboratory in Woods Hole. Such a project, AUI suggested, verged on "big science" and represented a welcome shift in priorities for big facilities from the physical to the biological sciences.[86]

In 1966 and 1967, AUI made overtures to OTS to participate in the proposed educational program. A representative of AUI attended two meetings of the OTS executive committee. But though initially somewhat positive, OTS soon became disenchanted with the proposal. Bethel, president of the executive committee, reported that while the committee was entirely favorable to expanded resources for tropical marine biology, it was "reluctant to subscribe to the AUI proposal . . . There seems to be no real advantage to OTS, and a risk of lending its name and connections and resources to a project of somewhat dubious origins and sound planning."[87]

Probably the end of the "Golden Age" would have doomed this grandiose plan in any event, but the national laboratory also foundered on the rock of dissension. Although tropical biologists were in accord with the need for expanded facilities for research and training, they could not agree upon where they should be located and who should manage them. The Smithsonian interests promoted development of tropical biology facilities in Panama;[88] OTS wanted a national laboratory in Costa Rica; the University of Miami and the Florida congressional delegation felt that their Marine Science Center should serve as the basis for expanded facilities; and the University of Hawaii and several other university interests also felt threatened by the prospect of a large AUI-managed laboratory. Many individual biologists distrusted AUI because of its domination by physical scientists, or feared that the project would eat up the lion's share of NSF's funds for tropical or marine biology.[89]

Even before the AUI report was submitted, T. Keith Glennan, former president of Case Western Reserve University, member of the National Science Board and president of the AUI, could predict that trouble lay ahead. In an address at the AIBS Annual Meeting in 1967 entitled "A Tower of Biology," Glennan likened the planning of major facilities to the Tower of Babel legend. "Early in our beginnings," his fanciful narrative commenced, "when all men wandered together over all the earth, our forbears made a site visit to the Plain of Sinar. There they envisioned the building of a unique facility which would get them to heaven." But sometime during the construction, "the gods became alarmed at the audacity of the project and its staff" and swooped down and destroyed the edifice by "substituting aimless cacophony for purposeful action." Such he feared would be the fate of the national laboratory in tropical marine biology if the biological community could not be prevailed upon to speak in concert. "For if we do not succeed in this," he concluded, "there is certainly no sense in trying to draw up plans for this particular tower: the gods, having learned something about politics, simply won't put up with a confusion of tongues *before* the project begins."[90]

Despite the negative reactions of many biologists, Carlson and Spencer tried valiantly to promote the faltering AUI plan within the Foundation. Carlson argued that neither the University of Hawaii, the University of Miami, nor the Smithsonian could serve as sponsor of a national facility because it was essential for an inter-institutional program to "be conducted on 'neutral' ground." Carlson and Spencer hoped the plan could be decided upon internally at NSF without further evaluation by outside scientists, since AUI had already thoroughly canvassed the latter. As a last-ditch effort, Spencer proposed that NSF appoint an independent commission to establish criteria for development of a center, after which interested groups could submit proposals for its creation and management. Shortly afterward, Spencer left NSF to become executive director of OTS.[91]

In October 1968, AUI submitted a scaled down proposal for "phase one" of the Tropical Marine Science Center, which would still require $5.5 million for the first three years. Adapting to the current congressional enthusiasm for ecosystem ecology, the proposal now emphasized multidisciplinary ecological research rather than laboratory biology. NSF definitively rejected the AUI proposal in early 1969.[92] By then McElroy, who had initiated the ill-fated venture, had become director of NSF. AUI may have blamed a cacophony of voices for the failure of the center, but it is not clear that there ever was an organized group of biologists behind this "tower of biology"; its chief proponents were AUI, Carlson, and Spencer.

Tropical biology continued to be cited as a priority into the 1970s, supported with stronger and more urgent environmental arguments than a decade earlier and a more sophisticated awareness of how ecology might be brought to bear on agricultural and environmental problems. But the funds were not forthcoming for a major expansion that was always on the horizon.[93]

There is an interesting sequel to this story. In the late 1970s, NSF's Biological, Behavioral and Social Sciences Directorate began to build expanded facilities for OTS at La Selva and to provide annual operating costs. La Selva, managed by a consortium of universities, has since become an international center for tropical research and heir to the planning of the 1960s.[94] If the program managers in biology at NSF succeeded in founding a de facto national laboratory for tropical biology, it was certainly not by any direct route.

Biologists' Ambivalence toward Big Biology

In the decade of the 1960s, BMS staff, employed in a federal agency that gave an increasing portion of research funds to expensive projects in the physical sciences, sought means to generate a similar support for big biology. They held an idealized image of a national biological facility, modeled on the operation of national laboratories in the physical sciences, which they attempted to impose on the large-scale projects they funded. But a national biological laboratory, whether floating on the sea or affixed to the ground proved elusive. One may hazard several reasons this was so.

Most obvious is the heterogeneity and fragmentation of biology. Biologists saw facilities that would benefit areas of biology other than their own as an expensive misuse of funds. Molecular biologists in particular opposed NSF's trend to fund big biology projects in areas of environmental biology such as biological oceanography or tropical biology (see chapter 8).

Second, a successful national facility required continued funding, which could only be assured through a line item in the budget. None of the BMS-supported facilities, except for time-limited programs such as IIOE, had special funding. When money got tight, large projects competed directly with each other and with individual-investigator projects. The *Alpha Helix,* devoted to studying exotic marine organisms in shallow waters, faced severe competition for limited ship-operation funds from mainstream projects in deep-sea biological oceanography by the end of the decade. Strapped NSF program officers could no longer fund wholesale the annual research programs submitted by the *Alpha Helix*'s national advisory board. And, in fact, NSF eventually undermined this board by insisting on evaluating individual proposals (see chapter 9).

Perhaps the greatest barrier to national or even regional facilities was that biologists did not yet need or want them. To the frustration of BMS science managers, biologists and institutions desired their own instrumentation. No biological instruments in the 1960s were so expensive or unique that they functioned like an astronomical observatory or large accelerator. Biologists were unwilling to travel afar to use phytotrons when cheaper, smaller instruments at home could be made to do. Even in the best of circumstances, a "national facility" attached to a particular institution tended to revert to a local facility. The Biotron was used almost entirely by Wisconsin biologists, and almost half the research programs on the *Alpha Helix* through 1970 were led by Scripps scientists.[95]

Finally, even when biological communities agreed on a need, institutions' rival interests prevented the creation of any truly national facilities. Oceanographers opposed the continued operation of the *Anton Bruun* as a national ship because, even with greater efficiency, it would compete with university-owned ships. Even though it was obvious that every institution could not have its own tropical marine laboratory, rivalries between OTS, the Smithsonian, and various universities prevented NSF from making a case in Congress for a national laboratory.

Ironically, the one example of a truly national biological facility that NSF did fund successfully was completely outside of BMS jurisdiction. The biological laboratory and ship managed by NSF's Antarctic (later Polar) Program had all the characteristics of a national laboratory. The complex logistics and limited accommodations for carrying out research in the unique, remote laboratory at the Naval Air Facility on Ross Island, McMurdo Sound, necessitated coordination. A national committee, operating through the National Research Council, selected projects for NSF support. Funded by a separate line item, Antarctic science did not compete with individual project research.[96] Nevertheless, BMS advisors were dissatisfied and complained about the necessity of channeling all requests through an Academy committee.[97]

In the 1960s, through its Special Facilities and Special Programs, BMS played a significant role in building up biological oceanography, marine biological laboratories, inland field stations, and tropical biology. Organizations that it helped to found and new laboratories that it helped to build are lasting legacies of the period. But without the support of the NSF hierarchy and the consensus of communities of biologists, all attempts at "national biological facilities" were ultimately short-lived. Biology functioned very differently from physics, astronomy, or meteorology.

CHAPTER 8

Allocating Resources to a Divided Science

The "New" and the "Old" in Biology

In his 1962 and 1964 BMS annual reports, Harve Carlson turned in his concluding "Prospectus" to the current intellectual and institutional strife among biologists. Like many others, Carlson framed the "breach" as between those who favored extension versus those who resisted the new ideas and techniques stemming from a "revolution" in molecular biology. He wrote of the "competition" in academic institutions between "proponents of the new and the old, of the molecular approach versus the classical approach, of the lab biologist versus the field biologist. In one school, one side dominates; in another the other side dominates. Good people are forced to leave or retire early in order that sweeping innovations may be made."[1]

For many biologists, the 1960s was a period of "crisis in biology." On campus after campus, traditional departments of botany, zoology, and bacteriology were merged to form departments of biology entailing a complete overhauling of introductory courses and major realignments of power. On one hand, the old disciplinary boundaries seemed to be breaking down. On the other, a new division reflecting not only differences in levels of biological organization but also different approaches to biological explanation, was becoming manifest. Journal articles aired such questions as "Are botany and zoology departments passé?" "Is classical biology, especially taxonomy, dead?" and "Have biologists turned aside from still fruitful methods in order to get on the molecular bandwagon?"[2] Balancing the claims on resources of the overlapping subfields of a divided biology became an increasingly delicate matter for the leadership of NSF's Division of Biological and Medical Sciences.

Long sympathetic to the ideal of a unified biology, BMS staff generally welcomed the regrouping of the various life sciences into departments of biology. Though Carlson admitted the brashness of some molecular biologists, he favored speeding the application of molecular techniques to all areas of biology. "The needs of the universities . . . are often blocked by archaic organization of classical departments out of step with the world of today, hence they fail to develop the newer and more significant lines of work," he reported. Money was often a "stumbling block" to change. "The vested interests of established departments work against the formation of new groups that would cut across existing departmental lines and thus create budgetary competitors." Carlson favored NSF aid to help those biologists with "a sound and imaginative plan" backed by administrative support to restructure the life sciences on campuses.[3]

Conflict among academic biologists mirrored growing dissention within BMS. Even before the budget crisis of 1968, the number of biologists seeking support and the costs of research were growing faster than new funds were being made available. Thus the program directors in molecular and genetic biology and related areas found themselves increasingly at odds with their counterparts in systematic and environmental biology over the appropriate distribution of BMS funds. Carlson, who had considerable leeway in allocating funds among programs, eventually got caught in the crossfire.

This chapter looks at BMS patronage in the 1960s in four areas: molecular biology, systematic biology, the plant sciences, and ecology. Patterns of NSF support from the 1950s to the 1970s differed in each. Through most of the decade, BMS gave priority to molecular biology, seen to be the locus of the most pathbreaking life-science research. Despite the dominance of the National Institutes of Health, NSF continued to contribute to the laboratories of the leading molecular biologists and eagerly claimed credit for advances made and Nobel prizes won. For systematic biology, considered by many a backward area, NSF was almost the only federal patron. As a result, its practitioners came to be the Foundation's most organized and vocal biological constituency. NSF aid made possible an intellectual and methodological renaissance in this field. Because of biology's functional organization at NSF, the plant sciences, except for systematic botany, were marginalized, not to become a focus of attention until the 1970s. Ecology, through its promise of solving complex and pressing environmental problems, was transformed from a small discipline to the late 1960s high-profile biological science. As political demands for funding socially relevant research grew more insistent, BMS devoted a larger portion of its resources to ecology, thus unleashing the ire of molecular biologists and geneticists.

Culminating the shift in BMS priorities was its sponsorship of the multi-million-dollar International Biological Program (IBP), primarily devoted to ecology. Though initially skeptical, Carlson and others of the BMS staff came to see IBP as a challenging opportunity to manage a new form of "big biology"—integrated multidisciplinary research using new techniques of mathematical modeling—to study large ecosystems. By emphasizing the IBP's potential contribution to managing environmental problems, ecologists convinced Congress to provide large-scale funding, but this unprecedented largesse for one field at a time when resources appeared stagnant intensified the growing division among biologists within and outside BMS. When budgetary growth came to a sudden halt in 1968, the escalating conflict came to a head.

Molecular Biology: Sharing Patronage with NIH

The quarter-century following World War II has been called "the age of molecular biology."[4] Francis Crick's and James D. Watson's announcement of the discovery of the double-helix structure of deoxyribonucleic acid (DNA) in 1953 led to a rapid series of advances in molecular genetics. Over the next dozen years, scientists elucidated the replication of DNA, the various forms of ribonucleic acid (RNA), the translation of information from DNA to RNA, the mechanisms of protein synthesis, the existence of and deciphering of a linear, triplet code, and the mechanisms of gene regulation. By the mid-1960s, there was widespread acknowledgment of a new era in biology. NSF's *Annual Report* for 1965 proclaimed that "we are now immersed . . . in a biological revolution, the 'molecular or genetic revolution,' which will undoubtedly have social and cultural impact of far-reaching consequences." NSF was proud to state that "the Foundation is an active participant in this new explosion of biological information through its support of many of the most capable investigators in this area." The following year Herman Lewis could write in the annual report of his Genetics Program, "Biology is completing a revolution," for by then, the genetic code had been cracked through the assignment of amino acids to all base triplets of DNA.[5]

Molecular biology had become—as biologists of less favored areas pointed out—a "glamour field" of biology.[6] Each year NSF boasted about its pathbreaking accomplishments in annual reports, press releases, and congressional hearings. By the 1960s, NSF's earlier truncated reports of discoveries by anonymous grantees had been replaced by much fuller expositions of new "breakthroughs." NSF celebrated such mid-1960s achievements as Robert Holley's determination of the structure of alanine transfer RNA, the first example of the sequencing of bases of a nucleic acid; Sol Spiegelman's synthesis of a self-propagating RNA; Gobind Khorana's development of chemical techniques to assemble DNA, which could be used to decipher the genetic code; and Walter Gilbert's and Mark Ptashne's independently arrived at isolation of a protein that repressed gene activity. BMS staff were especially pleased to publicize the 1968 Nobel prize in Physiology or Medicine, which was awarded to grantees Holley and Khorana, who shared the award with Marshall Nirenberg of the National Heart Institute.[7]

In terms of individual project grants, BMS had long favored molecular biology. By the early 1960s, the once-novel term adopted by NSF in 1952 had become widely used, and often narrowed in meaning to molecular genetics. The Molecular Biology Program, under the charge of a series of academic rotators and long-time associate program director Estelle ("Kepie") Engel, continued to apply a broad definition. According to section head Eugene Hess, molecular biology was not a discipline but rather a level of organization or approach to the study of life. Broadly speaking, the program's projects could be broken down into those with a biochemical or a biophysical orientation. The Molecular Biology Program was by far BMS's largest. Its budget rose from $4.0 million in 1959 to $11.8 million in 1967. At the same time, "demand" increased from $11.5 million to $48.8 million.[8]

Under Herman Lewis, the Genetic Biology Program took over substantial support of BMS projects in molecular genetics. Clearly partisan, Lewis claimed that molecular approaches appeared to be pervading all areas of biology from physiology and developmental biology to ecology and systematics, breaking down disciplinary boundaries and contributing to a "unified biology." According to Lewis, nearly everyone agreed "that in the next five to ten years it will be necessary for most biologists, regardless of their areas of interest, to be able to handle biology at the molecular level." The genetic biology budget increased from $1.8 million in 1959 to $5.2 million in 1967.[9]

Both the molecular and genetic biology programs were oriented to supporting the best scientists. Holley and Khorana were only two of a phalanx of past and future Nobelists the programs funded in the 1960s. Among the others were Max Delbruck, Gerald M. Edelman, François Jacob, Arthur Kornberg, Joshua Lederberg, Rita Levi-Montalcini, Fritz A. Lippman, Salvador Luria, Jacques Monod, George E. Palade, Linus Pauling, Rodney R. Porter, Edward Tatum, George Wald, and James D. Watson. BMS supported many other outstanding molecular geneticists as well.[10] It was central to the image of the division, and to these two programs in particular, that a good portion of the country's leading biologists be NSF grantees. For eminent scientists, BMS awards and renewals were practically guaranteed.

Though NSF played a significant part in the spectacular rise of molecular biology, the chief federal supporter of molecular biology was the much wealthier NIH. Fueled by public interest in finding cures for diseases, Congress continued to appropriate much larger increases for NIH than called for in the president's budget. By 1968 the budget of the National Institute of General Medical Sciences (NIGMS), which funded noncategorical research, had risen to $156 million, and NIH's total budget had reached over $1 billion.[11] As its resources grew, NIH expanded into more and more areas of basic biology, even though their relation to medical practice might be tenuous. In the 1960s, NIH supported research in photosynthesis and plant physiology on a broad basis, and even funded the computer taxonomy of Robert R. Sokal. NIH also dominated graduate training in biology through the large sums it poured into departmental or subject-area training grants.[12]

Despite the obvious inequality of NSF and NIH resources, staff members such as Carlson and Lewis did not perceive NIH competition as a serious problem. Lewis recalled, "Everybody here at least viewed the two agencies as complementary and not really redundant. We didn't feel threatened here at NSF by the big agency in Bethesda."[13] As Carlson explained, NSF program officers were always on the phone with counterparts at NIH; they attended each other's study sections and panels. Good communication with NIH at the study section and program level meant that NSF staff always knew which proposals were duplicates and which ones NIH intended to fund. By 1966, BMS received some six hundred proposals a year that had also been sent to other agencies—about 25 percent of the total proposals received. The overwhelming majority of them were duplicates of proposals sent to NIH.[14]

If both NIH and NSF were willing to fund the same proposal, the investigator could take his or her choice. Most of the time, the researcher chose NIH because NIH generally offered a higher annual sum and a longer period of time and was more likely to pay faculty salaries. If the project was at all medically related, NIH also provided a certain sense of continuity. The renewal award documents included a cumulative numbering of years of support for a particular project, high numbers portending a long-term commitment. Investigators used general titles for their NIH grants that they were reluctant to change lest the count begin again from zero. Though most were eager to get on the NIH bandwagon if they could, some biologists who were satisfied with modest laboratories continued to prefer NSF as sole patron because of the flexibility of its awards—for example, the ability to shift funds between budget categories—and congenial relations with program directors.[15]

The different manner in which NIH proposals were peer reviewed practically insured that established scientists would get a higher portion of what they requested from NIH than from NSF. Study sections considered proposal budgets and sometimes reduced them, but as they were evaluating and ranking proposals sequentially and not in relation to a definite funding total, they had little incentive to cut proposals severely. NIH councils in the 1960s rarely deviated from study sections' rankings and budget recommendations and thus tended to fully fund approved proposals until they reached the "pay line," the point score at which the money ran out.

In contrast, NSF panels normally did not discuss budgets in detail or supply rigid proposal rankings. Instead the panels reached consensus on the relative merit of all the proposals under consideration, often placing the principal investigators' names on a board in rough order of merit and adjusting them at the end of the review. The program directors might well make a case for deviating from a panel's ordering; budgets were almost entirely left up to them. In NSF, there was no "pay line." Program directors had the option of negotiating budgets downward so that more biologists could receive some support. As pressure on available funds increased, their tendency was to keep as many scientists at the bench as possible. Lewis recalled that few, if any, of his grantees got all that they requested.[16]

While communication with NIH prevented the *same* proposal from being funded by both agencies, NSF program officers were not averse to the same investigator's receiving dual support for *different* proposals. Most of the prominent scientists listed above were funded by both agencies (though a sizeable minority were not). Their NIH proposals were typically evaluated by such Division of Research Grants study sections as biophysics, biochemistry, cell biology, genetics, or bacteriology. Although funding was most likely to come from NIGMS, several joint NSF/NIH grantees received their NIH funding from such disease-oriented institutes as Cancer (Khorana), Heart (Palade), Arthritis and Metabolic Diseases (Edelman), or Allergy and Infectious Diseases (Luria).[17]

Many of the prominent scientists that both BMS and NIH supported headed large laboratories with senior collaborators, senior and junior postdoctoral fellows, graduate students, undergraduates, and technicians. BMS usually funded only a fraction of a laboratory's work, for to do more would have required a large coherent-area grant, only a few of which it ever awarded. Sometimes investigators set aside a riskier or more basic piece of research for the NSF portion of support, but often no significant distinction was made in the laboratory between NSF and NIH work. BMS program directors did not mind sharing support of large laboratories with another agency. As Lewis explained, since most of the salary support in NSF grants went not to the principal investigator but to graduate students and postdocs, BMS support of an investigator already funded by NIH simply meant that more good people could be accommodated in a productive laboratory.[18]

However, as competition for grants grew, NSF program officers increasingly took existing laboratory support into account in making their decisions. In 1963, for example, biochemist John Mehl, rotator for molecular biology, raised the issue of whether the Foundation ought to give an investigator, however productive, $30,000 to $50,000 more if he already had $100,000 from other agencies, when there were others with more modest needs that were unmet. With the aid of their panels, program officers estimated the appropriate size of an investigator's laboratory—some were judged more capable of handling a larger staff than others—and funded accordingly. Sometimes they declined an otherwise meritorious proposal because they thought the investigator already had enough money from other sources.[19]

Compared to their counterparts in ecology and systematics, program directors in molecular and genetic biology engaged in little infrastructure planning through most of the 1960s. The planning that did take place in fields like biochemistry or genetics was, for the most part, left to NIH study sections.[20] BMS program staff's chief effort went into individual project grants, instrumentation, support of conferences (a large portion of which were in molecular-oriented fields), and international travel. In addition, the Genetic Biology Program (or Facilities and Special Programs) continued to provide operating costs for a number of genetic stock centers.

One unusual award that did reflect planning for molecular biology was the development for market of accurate space-filling molecular models for research and teaching, based on prototypes designed by Linus Pauling and Robert W. Corey of Caltech. In 1960, biophysicist John R. Platt had argued in *Science* the need for sophisticated atomic models that could represent macromolecules. Those in use for small molecules were too clumsy, expensive, and inaccurate to be suitable. That same year the biophysics study section considered such a project, but NIH could not legally fund model development to the point of sale by laboratory supply houses. When NSF's education division considered the project too expensive, the Molecular Biology Program—at Consolazio and Mehl's urging—took on the task. Under an initial $220,000 contract with the American Society of Biological Chemists in 1962, a committee of biochemists developed and prepared model molds and arranged for manufacture and distribution. The plastic models became known as the Pauling-Koltun-Corey (or P-K-C) models. Walter Koltun, originally a consultant to the NIH biophysics study section and later a program director for molecular biology at NSF, designed the freely rotating connectors. Albert Siegel, a later rotating program director for molecular biology, noted in 1968 that the models were "extremely useful" to researchers for representing structural relationships such as those between substrates and enzymes.[21]

NSF's largest contribution to facilities in molecular biology was the construction of general-purpose laboratory buildings for biology departments through the Graduate Laboratory Program in the Division of Institutional Programs. Because they were deemed "normal" departmental facilities, most academic laboratories for research and training in molecular biology were not eligible for BMS special facilities funds. Two notable exceptions, both attractive to BMS because of their interdisciplinary character, were the Laboratory of Chemical Biodynamics (Melvin Calvin Laboratory) on the Berkeley campus, and the Institute for Molecular Biology at the University of Wisconsin.

Melvin Calvin won the 1961 Nobel prize in chemistry for his elucidation of the path of carbon in photosynthesis. This achievement, based on using carbon-14 as a tracer, grew out of E. O. Lawrence's cyclotron research at Berkeley. Long supported by the Atomic Energy Commission, Calvin's research group formed a part of the Lawrence Berkeley Laboratory, one of AEC's national laboratories. In FY 1961, NSF contributed $627,500 to construct a new laboratory building on the Berkeley campus for Calvin's Bio-Organic Group. The Laboratory of Chemical Biodynamics, dedicated in 1964, was separate from the departmental structure at Berkeley and thus eligible for BMS special facilities funding. Unique in design, the round structure featured open workspaces in order to encourage interdisciplinary collaboration. Research activities included studies of photosynthesis, the chemical origin of life, neurochemistry, radiation chemistry, and pharmacology.[22]

In 1962, when budgets were still expanding rapidly, Harve Carlson wrote, "There has been a general consensus in the Division in the last few years that it would be desirable to give impetus on two or three major campuses to the establishment of special organizations devoted to a multi-disciplinary approach to molecular biology."[23] The first and only such award of this kind was that of $600,000 given in 1962 to Harlyn O. Halvorson of the University of Wisconsin to build an interdepartmental Institute for Molecular Biology. The university had recently begun a graduate program in molecular biology, and the new building provided offices and laboratory space for a number of molecular biologists who were also salaried members of one of the biological departments at Wisconsin.[24]

Also benefiting molecular biology were the various forms of support provided to two venerable independent institutions at which molecular biologists and graduate students from all over the country congregated in the summer months: the Cold Spring Harbor Laboratory on Long Island, New York, and the Marine Biological Laboratory (MBL) in Woods Hole, Massachusetts. The Carnegie Institution of Washington's genetics laboratory at

Cold Spring Harbor had long been the summer gathering place of geneticists such as Luria, Delbruck, and Watson. Future Nobelist Barbara McClintock was a member of the small permanent staff. The independent Cold Spring Harbor Laboratory was formed in 1963 by the merging of the former Carnegie Institution facility and the Long Island Biological Association. NSF continued to provide partial funding for the laboratory's celebrated annual symposium and began also to provide partial operating support.[25]

The Marine Biological Laboratory, unlike many marine biological stations, was primarily oriented to biophysicists, neurobiologists, and physiologists, many of whose research programs were supported by NIH. MBL carried out seminal experimental work using marine invertebrates—especially the Woods Hole squid with its giant axon—as model systems. NSF provided MBL with a variety of forms of aid. Among these was its support of MBL's collecting vessel and general operations. NIH had converted a number of the traditional MBL summer courses into training courses that carried student stipends; NSF supplemented these by funding summer training in invertebrate zoology and marine botany and a year-round training course in systematics and ecology under Melbourne Carriker.[26]

In 1966, BMS gave MBL $2,150,000, one of its largest facilities grants of the decade to replace nineteenth-century wooden structures with a new building to house its training courses. The NSF award along with $50,000 from the Richard K. Mellon Foundation was matched by a $2.5 million grant from the Ford Foundation. Over half the Ford money went to build Swope Center, the dormitory/dining hall complex that was part of the original MBL proposal, while all the NSF funds went into the four-story Loeb Teaching Building. NSF Director Leland J. Haworth and the National Science Board were reluctant to provide so large a grant to a nonuniversity institution and especially to pay two-thirds of the cost of a laboratory that would be fully used only in the summer, while a comparable laboratory building at a university would be occupied throughout the year and matched on a 50–50 basis. Only after Haworth had canvassed all the NSF divisions for other priorities, was the award belatedly made.[27]

Budget pressures in the molecular and genetic biology programs continued to rise as more biologists completed their training and began new laboratories and as instrumentation grew more complex and costly. Molecular biologists became alarmed by the end of the 1960s as they perceived BMS priorities shifting toward field biology at the same time that NIH, also under budget pressure, curtailed its support of basic biology. Ray D. Owen, Caltech geneticist and immunologist, wrote to Harve Carlson in 1968 that the BMS Advisory Committee, which he chaired, could afford tolerance as long as our base was increasing; some of us could nod approval at tropical marine stations, big boats with big budgets, or computerized natural history even though all these things looked like costing a lot of money, without going very deeply into the substance of what was likely to come out of them, as long as we could hope also to see a Spinco [brand of ultracentrifuge] in every corridor . . . . With the reduction in available funds, this easy tolerance began to be strained. Our different fields became competitors, in a most relevant kind of struggle—to some parts of it, at least, such as biophysics and biochemistry, almost a struggle for survival of a large fraction of the current enterprise.[28]

Systematic Biology: The Dominance of NSF Patronage

Although NIH dominated support of molecular biology, as early as 1952 NSF became the primary federal patron of systematic biology. In 1959, Wilson reported that NSF grants had played a significant role in the "resurgence" of that field. The program staff estimated that for 1967 BMS spent $6.6 million on basic research, curatorial support, and facilities, about 75 to 80 percent of the total federal awards in systematics.[29] Because institutional structures and the character of research activity in systematic biology differed greatly from those of molecular biology, BMS provided a different pattern of support, thus insuring not only the survival of systematic biology but contributing also to a renaissance of ideas and methodology.

University mergers of botany and zoology departments to form departments of biological sciences placed systematic biology in a precarious position. On many campuses, large natural history collections were abandoned or given away, the most notorious example being the transfer of major collections at Stanford to the California Academy of Sciences.[30] Positions occupied by traditional botanists and zoologists were given to molecular and cellular biologists. Systematists felt that the merged departments, dominated by laboratory studies, neglected the whole animal, thereby distorting the biology taught to students. Only a few universities—foremost among them Harvard, Michigan, Berkeley, and Kansas—maintained substantial collections and faculty to provide graduate training in systematic biology. Many systematists worked in museums or botanical gardens that did not award degrees, such as the Academy of Natural Sciences of Philadelphia, the American Museum of Natural History (New York), the Field Museum (Chicago), and the Missouri Botanical Garden (St. Louis). Most of these were private institutions dependent on endowment funds and public support. Systematists complained that on one hand, there were not enough researchers available or being trained to serve as needed experts on the world's biota, and on the other that there were too few full-time positions available.[31] Compared to the glamour of molecular genetics, systematic biology seemed to many scientists a backwater. However, despite skepticism among some program directors in the newer areas of biology, BMS was committed to maintaining research and graduate training in systematic biology in a core of institutions.

The Systematic Biology Program grew from $2.6 million in 1959 to $5.6 million in 1967, the last year for which comparable figures were available. At the same time proposal "demand" rose steadily from $5.1 million to $16.4 million. The program funded a wide range of academic and nonprofit institutions, but with a concentration on the major museums, botanical gardens, and academic centers of research. Until Congress forbade NSF patronage of other federal agencies in 1964, it even supported curators' research at the National Museum of Natural History of the Smithsonian Institution (twelve awards for $234,600 in 1963).[32] Typically, awards in systematic biology were much smaller than in other BMS programs; many of them to smaller institutions were for amounts under $10,000. (For a comparison of spending patterns in systematic and molecular biology, see table 6.1.)

BMS Planning for systematic biology reached a high point in the late 1950s. The staff and divisional committee appointed an ad hoc Committee on Systematic Biology under the chairmanship of Ernst Mayr, which presented a report on the needs of systematic biology in 1957.[33] As in marine and tropical biology, NSF contributed to the founding of national systematic biology organizations. In 1957 and 1958, BMS brought together administrators of natural history museums at the Academy of Natural Sciences in Philadelphia and the New York State Museum in Albany. By the following year, the participants had created a formal organization of directors of systematic collections, which, in turn, led to the formation of the more broadly based Association of Systematic Collections in 1972. The Systematic Biology Program also arranged to have directories of specialists in plant and animal taxonomy compiled under the auspices of the Society for Systematic Zoology and the American Society of Plant Taxonomists.[34]

BMS tailored forms of support to the particular needs of systematic biology. Increased research activity in museums created a strain in curatorial resources. It was easier to obtain funds for expeditions than for the more routine work of organizing and caring for collections. Based in part on the 1957 needs report, BMS began, in 1959, to award grants for curatorial support and maintenance of collections. Separate from research awards, these were made under the NCE or Special Programs designation. Jack Spencer, director of Facilities and Special Programs, estimated that from 1964 to 1968 BMS spent $1.1 million on curating and maintenance of collections, or about 6 percent of Special Programs funds during this period.[35]

To help fill the gap in systematics and ecology left by NIH training grants in other areas of biology, the Systematic Biology Program, beginning in 1965, gave a handful of training grants to academic institutions. These were block grants to support research training and thesis field work for graduate students and, at Harvard, also postdoctoral research. The first three "research and training" awards were made to Reed C. Rollins, director of the Gray Herbarium at Harvard, for "the development of an integrated program of research and instruction in evolutionary biology" ($193,000 for three years), to Theodore Hubbell of the University of Michigan for systematic and evolutionary biology ($130,000 for two years), and to William A. Clemens, University of Kansas, for systematic and evolutionary biology ($110,000 for two years).[36] Similar awards were made in environmental biology. When David Tyler, who headed the Regulatory Biology Program, argued (unsuccessfully) for the expansion of his pet idea of "departmental sustaining grants" to high quality biology departments, he emphasized the stimulation provided by training grants in systematic biology: "We are informed in unequivocal terms that through these grants there has been a markedly increased tempo of inquiry, a healthy ferment in graduate student–faculty relationships, and a recruitment to the field of more qualified and dedicated devotees."[37]

The survival of systematic biology, both in universities and within NSF, put a premium on new orientations and new methodologies. Program officers felt the continued need to counteract the image of the systematist as "an unimaginative biological stamp-collector" carrying out taxonomic studies through the decades with an unchanging methodology.[38] The federal granting system, in general, discouraged any area of science from exploiting the same methodology for many years; in a highly competitive environment, panels looked for some degree of innovation.[39] In the 1960s systematic biology became increasingly identified with population biology, behavior studies, and evolutionary biology. A rash of new techniques and theories were adopted, sometimes leading to acrimonious debates among rival groups of systematists.[40]

BMS program directors, who had to contend with somewhat disdainful colleagues in molecular biology (not to mention the physical sciences), encouraged experimentation with new forms of taxonomic analysis. Although they continued to support checklists, inventories, and monographs on plants and animals based on classical morphological methods, they liked to boast of a panoply of modern methods for determining evolutionary relationships, such as employing computers, electron microscopes, behavioral studies, ecological relationships, population biology, serology, electrophoresis, comparison of nucleic acids, and other biochemical techniques. In their annual report for 1968, the systematic biology staff proclaimed: "Taxonomy is now in the hands of specialists, whose specialties now are often not the taxon—not the family of insects or the class of mollusks, for example—so much as the experimental techniques. We identify ourselves as morphologists, cytologists, genecologists, biochemical taxonomists, biometricians, or population biologists. By the cooperative efforts of this spectrum of specialists the future advances will be made." So many "enterprising systematists" were now using the new tools "to answer questions that would have been beyond reach a decade or two ago, that it seems no exaggeration to speak of a 'revolution' now in progress in taxonomy."[41]

Of special attraction to program officers in the 1960s were "numerical taxonomy" and "molecular evolution." Numerical taxonomy was the controversial research program, spearheaded in America by Robert R. Sokal of the University of Kansas, to reform classification by using computers to analyze a large number of taxonomic characters. NSF funded Sokal as well as Charles Heiser, Rita Colwell and others who were using this new methodology. NSF especially liked to publicize research in "molecular evolution," such as that of Charles G. Sibley of Cornell, comparing DNA or other macromolecules in organisms—sometimes in widespread groups of animals, as, for example, mice, monkeys, and man—to determine relative evolutionary relations. Such dramatic examples of molecular methods illuminating systematic biology not only enjoyed popular appeal but also lent strength to BMS's rhetoric of a unified biology.[42]

Through the Special Facilities Program, where, unlike NSF's graduate laboratory program, matching funds were not required, BMS supported the renovation or construction of museum and herbaria buildings. In all, the facilities program provided $8.0 million to systematic biology between 1957 and 1968, about 21 percent of its total awards (see tables 7.1 and 8.1). Many of these awards were quite significant in their long-term effects.

While BMS supported construction at both independent institutions and universities, three large academic awards merit special mention. A grant of $1 million in 1962 enabled the University of Michigan to build a four-story annex to its venerable 1920s museum building on campus. Described in the proposal as "a regional facility for research in biosystematics of animals," the annex was intended to provide laboratories for taxonomic and evolutionary studies utilizing live animals.[43] At the University of Florida, substantial collections had been housed all over campus as well as in the Florida State Museum in downtown Gainesville. In 1966, a joint award of

Table 8.1   BMS Facilities and Special Programs Support
of Systematic Biology Collections by Institution,
FY 1957–FY 1968

American Type Culture Collection	$1,282,100
Museum of Comparative Zoology, Harvard	1,025,000
U. of Michigan, Museum of Zoology	1,000,000
Field Museum (Chicago)	987,700
U. of Florida	800,000
New York Botanical Garden	744,800
American Museum of Natural History (N.Y.)	487,000
Bernice P. Bishop Museum (Honolulu)	471,600
U. of Kansas	362,500
California Academy of Sciences	332,400
Missouri Botanical Garden	310,000
Academy of Natural Sciences (Philadelphia)	289,000
U. of Michigan, Botanical Garden	254,500
U. of Nebraska	225,000
Los Angeles County Museum	212,600
Smithsonian, Sorting Center	196,600
Fairchild Tropical Garden (Fla.)	153,200
Peabody Museum, Yale	105,100
Rancho Santa Ana Botanical Garden	89,600
Carnegie Museum (Pittsburgh)	79,200
Museum of Northern Arizona	68,000
Panhandle Plains Historical Museum (Tx.)	50,200
U. of Hawaii, Lyon Arboretum	33,900
San Diego Natural History Museum	31,400
Michigan State U.	25,100
Santa Barbara Botanical Garden	21,000
Total	$9,372,500

Source: "Fact Book, Facilities and Special Programs, Division of Biological
and Medical Sciences, National Science Foundation," April 1968, item K1,
NSF Historian's Files, RG 307, National Archives and Records Administration.

BMS and the Division of Social Sciences for $1,112,650 enabled the university to bring the collections together in a new museum located near its botany and zoology buildings. Florida was thus able to become the largest center for systematic biology in the southeast. The low-lying structure, suggestive of Florida Indian mounds, with most of its offices and storage space below ground, won considerable architectural notice.[44] A third award, for

$635,000 in FY 1968, partially paid for a five-story wing to Harvard's nine-teenth-century Museum of Comparative Zoology to house a "national fa-cility for biological animal systematics." As in Michigan, the more modern auxiliary areas to systematics were stressed; the new wing provided space for Richard C. Lewontin's research on population genetics and E. O. Wilson's on animal behavior as well as the more traditional collections of fishes and insects.[45] These awards to university museums helped to assure a continuing supply of graduate students in systematic biology.

Included in the 1960s under the rubric of systematic biology was the American Type Culture Collection (ATCC), an independent nonprofit in-stitution which maintained and distributed cultures of microorganisms for biological and medical research. Incorporated in 1925, the ATCC struggled on a shoestring until the postwar era, depending for survival on funds from the Society of American Bacteriologists (now the American Society for Mi-crobiology), the Rockefeller Foundation, other public and private agencies, and the good will of a succession of host institutions. In 1954, BMS gave its first grant to ATCC: $11,000 to establish a collection of bacteriophages. A major NSF award of $865,000 in 1961 enabled the ATCC, then located in a three-story brick house in Washington, D.C., to build modern, more per-manent facilities in Rockville, Maryland. From 1957 through 1968, BMS spent $1.3 million on ATCC facilities and operational support. A microbi-ologist by training, Harve Carlson was especially partial to the ATCC as grantee. After leaving NSF, he served for many years on the ATCC board of trustees.[46]

Compared to other areas of BMS interest, staff and divisional commit-tee planning for systematic biology was not substantial in the early 1960s. Systematic biology did not have the benefit of a paid consultant such as Dixy Lee Ray or Dale Arvey, or of a new ad hoc committee of advisors. Nor was systematic biology put forward as a "critical area." But systematic biologists were sufficiently organized that as BMS funds began to level off in the mid-1960s, they made preparations to gain more leverage through a series of well-publicized reports. The directors of systematic collections, led by Wil-liam C. Steere of the New York Botanical Garden, requested funds from BMS for a study of resources and needs, hoping thereby to gain more recog-nition from NSF and Congress. In 1968, BMS also funded a feasibility study by the National Academy of Sciences of a proposed Institute for Systemat-ics in Support of Biomedical Research. And systematists prevailed on the Committee on Environmental Quality of the Federal Council on Science and Technology to establish a panel on systematics and taxonomy, which produced its own report in 1969. As the decade advanced, systematic biology aligned itself with growing concern over the extinction of natural populations and pollution of the environment. Systematists were able to portray their museum and herbaria collections as irreplaceable "national resources."[47]

In 1968, however, before the reports appeared, the situation looked bleak. David M. Gates, director of the Missouri Botanical Garden, reporting on a preliminary survey of directors of major collections for the BMS Advisory Committee, estimated that museums and herbaria needed $60 million for new buildings and a $10 million increase in annual funds for curatorial research. Yet Jack Spencer faced the prospect of no funds at all for facilities and curatorial support in fiscal 1969. He asserted that non-degree-granting institutions such as museums would be especially hard hit by the loss. Not wanting to be "unduly pessimistic," he wrote, "I would predict that it will be many years before a flexible funding program can be re-established to assist these institutions. This is the more tragic since these same institutions have not participated in the post-Sputnik blush of science support which has been experienced by the university community."[48] Though, after the 1960s, NSF could no longer build museums, the downturn in funding curatorial support proved temporary. The environmental movement, coupled with the series of reports, led to renewed and much expanded BMS curatorial support in museums and herbaria in the 1970s (see chapter 9).

Plant Sciences: A Non-Category at NSF

In the 1970s, the field variously known as plant sciences, plant biology, or botany in its broadest sense, became, and has since remained, an area of priority within biology for funding at NSF. In the 1950s and 1960s, however, the plant sciences were relatively ignored. BMS's organization along functional lines was deliberately intended to break down the barriers between botany and zoology; each program was designed to fund research on both plants and animals, as well as on microbes. The plant sciences thus were to be everywhere in BMS but nowhere in particular. Because the plant sciences were a noncategory at NSF, BMS maintained no records on how much plant research was funded.

In practice, except in systematic biology, by the mid-1960s research on plants formed a minor part of the functional programs. While it would be a daunting if not impossible task to reconstruct plant science funding, some idea of the relative importance of plants can be gleaned from looking at the composition of the BMS senior staff before 1970. Of the assistant (or division) directors, only Lawrence Blinks (1954–55) was a botanist. No other plant scientist headed biology at NSF until Mary Clutter assumed charge of the Directorate of Biological, Behavioral, and Social Sciences in 1989.

However, David Keck, formerly of the New York Botanical Garden, served as deputy division director in the 1960s. All but a handful of BMS program directors were zoologists or microbiologists. Of the succession of rotators who headed molecular biology in the 1960s, only one, Samuel Aronoff (1963–64), worked with plants. For most of the two decades, regulatory biology was managed by biochemical endocrinologist Louis Levin and animal physiologist David Tyler. Paul J. Kramer (1960–61) was the only botanist to direct the program. Metabolic biology, which in the 1970s became the chief funder of plant physiology, was directed by rotators from its creation in 1958 until the arrival of animal physiologist Elijah Romanoff in 1969. Only three directors—Aubrey Naylor, Howard Teas, and John M. Ward, spanning the period 1961–65—were plant scientists. None of the rotators who successively headed developmental biology were plant scientists. Nor did any plant scientists head genetic biology, despite the fact that departments of genetics were often located in agricultural schools. In environmental biology, where plant ecologists might be expected, botanists (John E. Cantlon, 1965–66, and Philip L. Johnson, part of 1968–69), directed the program for only two years. Psychobiology, unlike the other programs, was never expected to fund plant research.

Surprisingly, all but one of the directors of the Systematic Biology Program were botanists: William C. Steere, Rogers McVaugh, A. C. Smith, David Keck, Walter Hodge, Robert K. Godfrey, and Kenton Chambers. Richard F. Johnston (1968–69) of the University of Kansas was the lone zoologist. Part of the reason for this unlikely imbalance could be that in the 1950s and 1960s rotators often recruited their successors.[49]

Although many botanists were not happy with their place in universities, there was little public complaint in the 1960s about botany at NSF. Plant physiologist and former rotator Paul Kramer specifically noted to botanists in 1966 that although BMS had "a very unconventional organization which makes no mention of botany or zoology, yet it has worked satisfactorily through a decade of dealing with botanists and zoologists."[50]

Indeed, systematic botany was relatively well cared for at NSF while the more experimental areas of plant physiology, biochemistry, and genetics had the advantage of several federal patrons, including ONR, AEC, and NIH. Even in the 1950s, NIH funded such prominent plant physiologists as Kenneth Thimann, James Bonner, and Arthur W. Galston, while the AEC supported Kramer. (All were also supported by NSF.) By the mid-1960s, NIH was sponsoring over 250 projects on plants. In 1965, AEC funded the new Michigan State University/AEC Plant Research Laboratory headed by former Caltech botanist Anton Lang. NIH, AEC, ONR, and NSF all supported considerable research on photosynthesis, which might be carried out with either plants or bacteria. NIH, for example, funded Daniel Arnon of Berkeley and Martin Kamen of the University of California, San Diego, while AEC had long provided for Nobelist Melvin Calvin.[51]

Despite the ready availability of grants, botanists—and plant scientists generally—felt on the defensive in the postwar era. Even in the 1950s, they worried about the growing number of introductory biology courses replacing introductory botany and zoology. The new courses, they claimed, more often than not emphasized zoology at the expense of botany.[52] The mergers of academic departments in the 1960s, which resulted in, for example, the loss of the formerly strong botany department at Yale, were a heavy blow for many botanists. Everyone agreed that botany as a separate area of university study was declining, but there was much contention over what to do about it. Some suggested a change of name to "plant science," "plant biology," or "phytology," since "botany" recalled the stereotype of "pressing and drying and naming flowers."[53] Some encouraged botanists to become more involved in teaching "biology" to insure that plants would be significantly included. Others urged botanists to revamp their courses to make them more exciting, up-to-date, or relevant to societal needs.[54]

Some defenders of botany called for an all-out effort to preserve or restore the botany department. One of the most ardent, William L. Stern, wrote in 1969 that "from the botanical standpoint, *biology departments have not been a success anywhere!* It has almost become a dictum among botanists: 'if you want to ruin botany, establish a biology department'!"[55] Other botanists maintained just as vociferously that plant research must identify more fully with biology, especially with the recent cellular and molecular trends. James Bonner of Caltech argued that the life sciences should be divided, not into botany and zoology, but rather along such functional lines as biochemistry, biophysics, genetics, and ecology. He replied to Stern, "I would say that my dictum is, 'If you want to save botany from dissolution, join it up with the mainstream of modern biology.'"[56] Thus within the plant sciences, as within the biological sciences generally, there was a clash between the "new" and the "old" specialties.

A number of botanists called for united action to convince university administrators and the public that botany was an exciting and valuable area of study. Members of the Botanical Society of America made several attempts to form a federation of plant science organizations, all of which floundered because of the difficulty of working out how this federation would relate to the American Institute of Biological Sciences. To a modest extent, NSF supported these efforts to preserve botanical autonomy. For example, in 1957 it funded a Botanical Society of America committee to meet in Washington to discuss the place of botany in undergraduate and high school curricula. The NSF education division supported for many years a summer workshop in botany for college teachers of botany or biology.[57] Most significantly, in 1961 BMS funded a feasibility study for "the centralization of the plant sciences," one of the attempts to form a federation that was reluctantly abandoned when AIBS was reorganized in the wake of the AIBS Affair.[58] Yet much more NSF funding went into the teaching of integrated biology at the undergraduate level. The NSF-funded Council on Undergraduate Education in the Biological Sciences (CUEBS) promoted a core course in biology, which proved a sore point to a number of defenders of botany.[59]

Although BMS provided special facilities for botanists, when given a choice, BMS preferred supporting facilities that encompassed both botany and zoology. For example, in 1959, Paul Kramer's application for a regional phytotron at Duke University was rejected in favor of a biotron for plant and animal research at Wisconsin. The Duke phytotron, funded a few years later, turned out to be the more satisfactory investment for NSF. The BMS predilection for biology over botany can especially be seen in the area of tropical biology (see chapter 7). The 1960 initiative in "tropical botany" had become by 1961 an initiative in "tropical biology."[60] In 1962, NSF funded a conference of tropical botanists in Trinidad, whose organizers intended to form a new organization for the study of neotropical botany. However, by the meeting's end, the group, consisting entirely of botanists, had formed the Association for Tropical Biology. According to attendee William Stern, NSF's representatives to the conference, Keck and Hodge, both botanists themselves, insisted that as a condition of initial NSF support, the organization must be expanded to tropical biology.[61]

It was not surprising, given botanists' power in systematic biology at NSF, that this was the one area in which the Foundation agreed to fund a large-scale project solely for botany. The failure of Flora North America, a monumental undertaking by systematic botanists to describe all of the vascular plants of North America, is discussed in chapter 9.

Although no federation of plant societies emerged, botanists did unite in the 1960s through the National Academy of Sciences to produce two policy reports on the "plant sciences." Both emphasized the valuable contributions to society made by plant scientists and the challenges and opportunities of future basic research. The first report, undertaken in 1962 by the Academy's Section of Botany suggested that botany was a critical area that was being neglected in the current rapid growth of the sciences. "It is imperative," the report maintained, "to have a balanced development of all aspects of science because in reality one part that is neglected will delay the progress of those farther ahead."[62]

The second report, *The Plant Sciences, Now and In the Coming Decade,*
published in 1966, was a product of the Academy's Committee on Science
and Public Policy (COSPUP)—its only 1960s study on the biological sciences. COSPUP's plant sciences panel, chaired by Kenneth Thimann, made
a far less compelling case than had the chemists in their COSPUP report of
the previous year, *Chemistry: Opportunities and Needs,* because they did not
address the relation of the plant sciences to other sciences or to current or
projected federal funding patterns. After the usual appeal to the social and
economic utility of plant research, the report detailed the research trends and
wish lists of the various botanical subfields. Its "total costs to federal granting agencies"—an expansive $1.5 billion for research support, training, and
facilities over the next decade—was simply the sum of the subpanels' requests. The report presumed that since plant scientists had expressed "requirements" for increased federal support, their "needs" deserved to be met
by the taxpayers.[63]

NSF paid little heed to either of these documents. Only after the funding crisis of 1968–72, did BMS staff recognize plants as an area of programmatic concern (see chapter 9).

The Rise of Ecology and the International Biological Program

Ecology, a weak and relatively minor biological science in the 1950s,
grew to prominence with the burgeoning environmental movement in the
1960s. Whereas NSF's role in medicine and agriculture, the traditional applications of biology, was limited by the prior existence of mission agencies
in those areas, the Foundation was well positioned to serve as a key supporter
of basic biological research related to managing and preserving the environment. In the 1960s, NSF's largest investment in ecology was the International Biological Program. Championed by Harve Carlson for its innovative
methodology, promise of applicability to environmental management, and
multidisciplinary, multi-institutional teams of investigators, the IBP also became the most divisive example of big biology at NSF.

From 1954 to 1964 the Environmental Biology Program, the primary
NSF program that supported ecology, was under the charge of George
Sprugel Jr., who had been trained in economic zoology at Iowa State University. Josephine K. Doherty, Sprugel's "professional assistant," played a major role in the functioning of the program.[64] In his early annual reports in
the 1950s, Sprugel claimed that basic research in ecology was emerging from
a "state of lethargy." Many proposals had been of low quality, in part, he
claimed, because ecologists had little experience in seeking funding for basic research; some had to be convinced that they did not have to justify the

utility of their projects in order to receive support. Sprugel and his panel sought to encourage newer areas of ecological research in place of qualitative studies of community ecology, population surveys, or single factor studies of ecological change. In 1956, the panel compiled a list of the most promising areas of fundamental research which included, quantitative analysis of populations, dynamics of ecology, energy balances, physiological responses to environmental stresses, study of specific environments such as the tropics, and animal behavior including orientation, migration, homing, and imprinting.[65]

In a wide-ranging discussion of the program in 1956, the panel defined environmental biology as "ecology in its broadest sense." That encompassed ecology proper as well as biological oceanography and limnology—fields that used ecological methods though practitioners might not identify themselves as ecologists. In the 1950s, NSF supported individually such noted ecologists as the Eugene and Howard T. Odum, G. Evelyn Hutchinson, F. Herbert Bormann, Thomas Park, Alfred E. Emerson, Paul B. Sears, and Lawrence B. Slobodkin.[66] The Environmental Biology Program also funded institutions benefiting ecologists. It provided support for summer research at inland and marine stations, it underwrote the founding of the *American Journal of Limnology and Oceanography,* and it funded a study committee of the Ecological Society of America to examine the status of ecology and how it might be more effectively applied to problems of natural resources.[67] Although the AEC sponsored pioneering projects in radiation and ecosystem ecology, NSF's environmental biology program was from the outset a significant source of federal patronage for basic research in ecology.

In the 1960s, the International Biological Program, in conjunction with the environmental movement, transformed ecology from a relatively small segment of NSF biology to its largest component. From 1968 through 1974, the federal government spent about $43 million on the IBP, $39 million coming from NSF. The bulk of this new funding went to ecology and specifically to a new subfield known as systems ecology. Seventy percent of the U.S. contribution was spent on five large "biome" studies in which the ambitious goal was to design and field test computerized models of large ecosystems (biomes).[68] Before, during, and after the IBP, biologists have battled with one another over whether the IBP was a highly innovative experiment that established systems ecology or a politically motivated diversion of money that could have been much better spent elsewhere.[69]

The initial conception of the IBP differed markedly from the eventual U.S. contribution to the program. Early discussions were primarily motivated by a desire to emulate the highly successful International Geophysical Year (IGY) of 1957–58, which, to the disappointment of a number of biologists, had incorporated little biology. As early as 1956, the environmental biology panel discussed the lack of biological research in the IGY plans, and "a suggestion was made that we really need an International Biological Year in which certain world-wide measurements of biological phenomena can be made simultaneously."[70] In 1959, more serious discussions of an international biological interval or program were held in Europe under the auspices of the International Union of Biological Sciences (IUBS). While proposals in a number of areas of biology were floated, it was early realized that an IBP that covered all of biology would be too amorphous and unwieldy. Field studies providing comparable data lent themselves much more readily to a global effort than did laboratory biology. In 1961, the International Council of Scientific Unions (ICSU) established an international planning committee, whose report, approved by IUBS and ICSU in November 1963, proposed an "International Biological Programme" with the overall theme of "The Biological Basis of Productivity and Human Welfare."[71]

The planners initially envisioned the IBP as a five- to seven-year program providing basic biological data on a global scale that could eventually be applied to alleviate the threat of rapid population growth and its resulting disturbances of natural communities and depletion of natural resources. The planning committee established subcommittees on productivity of terrestrial, fresh-water, and marine communities, human adaptability, and the use and management of biological resources. In each of these areas, comparable measurements were to be made in different environments using agreed-upon methods. It was thought that the IBP would subsume and coordinate a lot of existing research.[72]

Since it was widely assumed that NSF would be asked to take an important role in channeling federal funds for the IBP, Foundation officials were consulted from an early date. In 1961, Waterman responded with his typical caution to an inquiry from Hiden Cox of AIBS: "While a biological project of a broad world nature could no doubt aid the progress of biological research to some degree, one must balance the investment that such a project would require against the same investment made otherwise." Waterman suggested that although the project might improve "international collaboration in matters descriptive, systematic, and ecological," it would "not necessarily advance biology as a whole to any marked degree."[73]

Reactions of American biologists to the international plans as they unfolded from 1961 to 1963 tended to be lukewarm or negative. A number saw the IBP as an unfortunate example of "me-too." Molecular biologists, finding nothing in the IBP of interest to their own research, generally opposed it altogether. Even ecologists thought the program vague, the research old-fashioned, and the international framework likely to produce too much regimentation. Americans felt that they had little input into the planning document. Until NSF was appointed by the president's science advisor to serve as lead agency for the federal contribution to IBP in 1965, Harve Carlson and George Sprugel, shared the overall critical reaction.[74]

Despite their own initial doubts, Carlson and other sympathetic staff members played a key role in getting IBP underway. In 1963, they funded an ad hoc committee of the National Research Council, chaired by Stanley Cain, to gauge the extent and nature of American biologists' interest and to recommend changes to the proposed IBP. BMS staff actively participated in all discussions of this committee. Deputy division director David Keck attended the First General Assembly of the IBP in Paris in July 1964 at which the initial IBP plan was modified in the light of examination by groups from the various participating countries. Following this meeting, the National Academy of Sciences set up a U.S. National Committee on the IBP (USNC/IBP), chaired by Roger Revelle of the University of California, San Diego, and funded primarily by NSF. The Foundation also funded a large part of the operating expenses of the IBP international coordinating committee.[75]

These staff efforts to promote the IBP were increasingly resisted by the BMS Divisional (after 1965, Advisory) Committee, representing the full spectrum of basic biology. In 1963 and early 1964 it called the IBP vague and in need of clarification and claimed it was "still too early in the planning stage to judge how much of value will result from such an international approach to what is essentially a non-urgent, applied problem."[76] In 1965, when it became clear that the IBP would go forward, several members complained that its current scope was too narrow to elicit general enthusiasm: "It was felt that the scientific problems related to productivity and human adaptability, though of utmost importance, involve only a few areas of biological research and thus, the IBP, as now conceived, will not be of interest to scientists working in the many other areas of biology." The committee passed a resolution urging the USNC/IBP to broaden the program and to add "a few additional distinguished biologists working in other areas" to its ranks.

The National Science Board, however, dominated by nonbiologists and unsympathetic to IBP, strongly opposed any attempt to broaden the program; instead, the board favored initial concentration "on two or three of the most urgent problems and preferably those whose effective study requires international collaborative effort."[77]

Carlson and other BMS colleagues were significant factors in the transformation of the initial plans for the American contribution to IBP into innovative big biology. In 1963, as U.S. planning got underway, Keck made it clear to the National Academy that NSF would not give it a block grant for IBP as was done for IGY. NSF would instead insist on its own competitive evaluation of individual IBP projects. This decision was not surprising given that Carlson, like his predecessor Wilson, thoroughly distrusted the Academy as a powerful institution wanting to take over policy-making from NSF. As chairman of the federal government's Interagency Coordinating Committee for the IBP, inaugurated in 1965, Carlson presented IBP in a favorable light to other funding agencies and achieved a measure of cooperation, but to the disappointment of the USNC/IBP, neither NSF or any other agency was willing to fight for a line item or set aside funds for IBP.[78]

By 1966, the year before the U.S. program was to begin operation, the U.S. National Committee found itself in a frustrating position. IBP at this point was to consist of a series of smaller ecological projects within the general framework adopted by the international community. But if NSF would not provide block funds for the Academy or give IBP proposals special standing, then the USNC could do little beyond certifying unsolicited projects (many of them ongoing) as relevant to IBP when certification conferred no acknowledged advantage. On this basis, it was impossible to organize diffuse projects into a coherent program.[79]

Faced with this deteriorating situation, the Academy decided to make radical changes in the IBP concept. In October 1966, the USNC/IBP held a seminal meeting at Williamstown, Massachusetts, to which it invited potential project directors and funding agency representatives, in order to take an intensive look at the proposed program. There it was agreed that a major component of the U.S. program should be USNC/IBP-initiated large-scale programs embodying new, multidisciplinary research.[80] Although several "integrated research programs" were planned during the next two years, the most exciting were the "biome" projects grouped under the heading "analysis of ecosystems." From the Williamstown meeting, systems ecology—the analysis of ecosystems by means of computer modeling, at that time largely a new methodology—became established as the core of the U.S. IBP. This novel approach to ecology had developed out of AEC-supported studies in radiation ecology at such research centers as the Oak Ridge National Laboratory under Stanley Auerbach and the University of Georgia under Eugene Odum.[81] Carlson encouraged the refashioning of IBP as big biology by funding Frederick Smith of the University of Michigan—chosen by the USNC/IBP to head the Analysis of Ecosystems Program—to organize the biome projects to the point that proposals could be submitted to NSF. Studies of six biomes were projected: western coniferous forests, eastern deciduous forests, tropical forests, grasslands, tundra, and desert.[82]

The USNC/IBP took a second bold action by mounting a direct appeal to Congress. Its leaders, Roger Revelle and W. Frank Blair, arranged hearings on a resolution of congressional endorsement in the summer of 1967 before Representative Daddario's Subcommittee on Science, Research and Development. Estimating a total cost of $199 million for the U.S. portion, the USNC/IBP urged Congress to provide line items for IBP in federal agency budgets. Carlson and Keck, who had become enthusiastic supporters of the integrated research projects, testified on behalf of IBP and for inclusion of a line item for NSF. Keck asserted NSF's strong interest in IBP "because of its importance to mankind" and requested $20 to $25 million for BMS "to fund IBP projects which are first rate." At subsequent House hearings in 1968 and 1969 and Senate hearings in 1970, Carlson and other BMS staff continued to testify, emphasizing the benefits to be derived from innovative integrated multidisciplinary attacks on complex ecological problems.[83]

Advocates presented IBP as a necessary means of combating the widely perceived environmental crisis. Whereas early in the 1960s, the chief environmental issue had been management of dwindling natural resources, by the end of the decade, the focus had become pollution—of water, of soils, and of the atmosphere. The influential President's Science Advisory Committee report, *Restoring the Quality of Our Environment* (November 1965), in which Revelle had participated, was just one indication of growing federal involvement.[84] Adopting the then familiar metaphor of nature as a self-regulating cybernetic machine, proponents claimed that the biome programs would reveal how the machine functioned so that, through human intervention, it could be stabilized.[85] Carlson was an active player in the series of complex interactions that led to the passage of a House resolution on IBP in 1969, a companion Senate resolution in 1970, and the acquisition of a line item in the 1970 NSF budget.[86]

In the face of congressional enthusiasm for IBP, the recalcitrant BMS Advisory Committee in September 1967, then chaired by clinician and cell biologist Harry Eagle of Albert Einstein College of Medicine, reiterated its previous stand that IBP proposals should receive no special favor and passed a resolution that IBP projects "are expected to compete on merit grounds within the limits of regular Foundation budgets." Also, requests for funds for IBP beyond the current program budgets should not be approved "until additional funds specified for this purpose are on hand." Finally, the committee insisted that it "be informed of the selection mechanism of projects for this large program." W. Frank Blair, a new advisory committee member and soon to be chairman of the USNC/IBP, cast the lone dissenting vote.[87]

Despite the opposition of his division's academic advisors, and before a budget line item was secured, Carlson set about making BMS's first large IBP award. If Carlson would not give block grants to the Academy, he made it clear he would give them to individual biome project directors for proposals that reflected BMS's concept of big science. NSF developed criteria emphasizing coherence, coordination of research, and synthesis of results, which were said to have created difficulties for drafters of biome proposals. The organization of the first project—Grasslands—as a multi-university managed team effort was worked out by direct negotiation between Carlson and project director George Van Dyne, thereby undercutting the authority of the USNC/IBP.[88] When the National Science Board balked in 1968 at approving the first installment of up to $700,000 for fifteen months of a projected $7.4 million for the Grasslands biome project, Carlson and the staff convinced the board by appealing to its big science features—planning, complexity, interdisciplinary cooperation, multiple-institution participation, and novel team research—in addition to its environmental relevance. Grasslands would involve over fifty senior investigators, including plant, animal, and microbial ecologists, range scientists, hydrologists, meteorologists, and systems analysts. Five academic institutions in Colorado and Wyoming as well as several branches of the USDA were to participate in the project, cooperating in an intensive study of Colorado's short-grass prairie. According to the BMS justification, "For the first time, systems large enough to be really interesting, because man's activities *in* them provide some of the most significant parameters, would be studied in their totality, as input-transformation-output systems."[89]

BMS staff serving laboratory biology and members of the BMS Advisory Committee understandably resented the rapid rise of ecology at the seeming expense of other areas of biology. In FY 1968, $700,000 from regular BMS funds were added to the Environmental Biology Program giving it an 11.4 percent increase while molecular biology and genetic biology suffered 7.7 and 7.0 percent decreases, respectively. Even excluding the IBP activities, environmental biology would have fared better than the others, with only a 0.7 percent decrease. Later in 1968, without discussing the matter with his staff, Carlson created a new BMS program, Ecosystem Analysis, initially under the charge of ecologist Philip L. Johnson, to handle IBP projects.[90] In November 1968 the advisory committee warned that while necessity had called for support of the Grasslands project from regular division funds, "further momentum" of IBP must be contingent on "substantial additional funding."[91] Dissention among biologists escalated even more in the 1970s when annual budgets for individual biome projects rose to over $1 million (see chapter 9).

A Divided Biology Faces the Budgetary Crisis

Herman Lewis, program director for genetic biology, devoted much of his 1968 annual report to "the disproportionate emphasis given to environmental biology," while support of genetic and molecular biology, which served to vitalize progress in all of biology, was reduced. Evidence of favoritism toward ecology was everywhere, he claimed. A new program, Biological Oceanography, had just been added to the Environmental and Systematic Biology Section. Even the BMS Advisory Committee was now stacked with four environmental and systematic biologists, while an ecologist (Frederick E. Smith, director of the IBP Analysis of Ecosystems Program) had replaced a biochemical geneticist on the National Science Board. Lewis readily dismissed the excuse that the environmental crisis justified the large increase in environmental biology. "The big push from the point of view of national priorities is to develop programs to abate or control air, land, and water pollution. How much do the kinds of projects the NSF supports contribute to this?" he inquired rhetorically. No more than genetic biology projects contribute to problems of inheritable disease. The relation was, at best, indirect. In its embrace of environmental biology, had the Foundation "followed a course of opportunism or been unduly influenced by the persuasiveness of special interest enthusiasts?"[92]

As the NSF budget went flat in 1968, signaling, as we now know, the end of real growth for over a decade,[93] every group became concerned with protecting its turf. Infighting within BMS was only a symptom of larger divisions within the Foundation as a whole. Biologists on the advisory committee criticized the education division for wasting money better spent on research for "retread" programs for high school and college teachers. The research budget, the committee complained, was also being eroded by NSF administrators through payments for indirect costs and faculty salaries that should rightly be considered institutional costs and paid for elsewhere (if at all) in the Foundation.[94] Finally, both BMS staff and advisors blamed their lack of funds on the perceived dominance in NSF of the physical sciences.

Part of the problem, NSF biologists, and molecular and genetic biologists in particular, realized, was attributable to NIH support of biology. NSF's emphasis on field biology had been justified in part because NIH was not funding it to the same extent as laboratory biology. In the larger sphere, the favoritism shown the physical sciences and engineering had, in turns, been justified in part because NIH poured large sums into the biological sciences. BMS produced a series of charts to show that NSF had neglected biology over the past decade. Biological research as a percentage of all NSF obligations had fallen from 15.4 percent in 1959 to 10.3 percent in 1968. As a percentage of the research budget, BMS had fallen from 44.5 to 29.8 percent.[95] Reviewing this data in the BMS Advisory Committee's annual report to the NSB in December 1968, chairman Ray D. Owen of Caltech suggested:

> What appears to be happening is a re-patterning of the Foundation's role in the support of science. Previously fostered in very large part by other government agencies, the physical sciences have turned with increasing success to the Foundation to further their basic enterprise. Perhaps largely because, under intense fiscal pressure, it has been felt that the biological and medical sciences have been more adequately provided by mission-oriented government agencies, the Foundation's stake in research in the biological and medical sciences has been permitted a proportionate lag . . . . The Foundation has a unique role, which it cannot responsibly abdicate . . . . Biology can no more than other sciences wisely be left for its nourishment to agencies other than the one whose prime purpose is to further basic research and education in the sciences.[96]

Battles over the rise of funding for ecology were only one of many manifestations in BMS of the chaos surrounding the end of the Golden Age of science funding.

CHAPTER 9

Forging New Directions after the Golden Age, 1968–1972

For Deputy Director John Wilson, the law amending the NSF charter, signed by President Johnson in July 1968, marked a second major breakpoint in the life of the Foundation. It symbolized the end of his hope that NSF would eventually support not just research, training, and facilities in the sciences but also higher education in general. Instead the Foundation would expand in a different direction and incorporate applied research alongside of basic research.

The Daddario–Kennedy legislation, revising NSF's charter and culminating years of investigation of the Foundation's activities by Emilio Daddario's Subcommittee on Science, Research and Development, both authorized and encouraged NSF to support applied research and enacted two other major changes that Wilson strongly opposed. It stipulated presidential appointment not just of the NSF director but also of the deputy director and four assistant directors, one of whom was an assistant director for research. In addition, it subjected the agency for the first time to authorization hearings before Daddario's subcommittee. For Wilson, all of these provisions were detrimental because they left NSF more vulnerable to partisan politics. In an interview in 1980, he recalled, "The Foundation, I thought, was an absolute gem of an agency up to that time, because we had avoided the pitfalls of the political mainstream." Wilson felt that, in order to accommodate Daddario, the Foundation had sold its birthright for "a mess of pottage." The impending new law was Wilson's cue to depart NSF for good. His appointment as vice president of the University of Chicago under President George Beadle was announced in March 1968.[1]

The years 1968–75 were a time of rapid change and turmoil. The post-Sputnik consensus regarding the role of basic science in society had broken down, and the steady expansion of support for research and for science education had come to a halt. The war in Vietnam, the end to automatic deferment of graduate students from military service, and campus unrest, along with social and environmental concerns, engendered a vocal antiscience sentiment and public pressure for scientists to work on "problems of society." With the advent of the Nixon administration in 1969, it became clear that the Foundation would undergo major shifts of policy.[2]

In this era of transformed priorities, support of biology could no longer be defended by the old metaphor of stockpiling knowledge that might be useful one day to agriculture or medicine. Nor could the need for training more biologists be argued. Additional funds had to be justified by their potential contribution to the solution of complex problems of current concern to Congress and the public: environmental degradation, loss of ecological diversity, indiscriminate use of chemical pesticides, drug addiction, cancer, overpopulation, an inadequate world food supply, and dwindling energy resources. After 1970, arguments for new money for biology were couched almost entirely in terms of taking over projects dropped by other agencies or "new initiatives" in areas related to practical concerns. Yet, however applied the titles of these new initiatives might seem, BMS director Harve Carlson and his staff remained committed to funding only basic research.[3] Negotiating the obscure boundary between basic and applied was no easy task and could sometimes lead to battles among staff, reviewers, and project directors.

How advocates of biology at NSF coped with the tumultuous changes marking the end of the Golden Age of science funding is the subject of this chapter. BMS floated a number of new and expensive initiatives during the period 1970–75. Some, despite the expenditure of considerable staff time and resources, went nowhere. Other ventures in big thinking proved at least partially successful in terms of garnering support. Those to be discussed in this chapter include human cell biology, regulation of pest populations, neurobiology, support of systematic collections, plant sciences, and the biome projects of the International Biological Program. Through formulating and reformulating these initiatives, BMS staff attempted to mediate between research that was, on the one hand, basic and scientifically significant, and on the other, could be sold politically for its relevance to public goals. Resulting from the budget crisis and consequent reliance on initiatives to gain new money was a very different pattern of support of biology from that of the 1950s and 1960s.

Biology, the Budget Crisis, and Nixon-Era Science Politics

The end of the Golden Age meant not only an end to growth, but also altered priorities for federal patronage. For twenty years the federal government had supported science on the premise that the nation needed more scientists. Many NSF programs were geared to attracting more college students into majoring in science and attending graduate school. At the same time, the availability of federal support encouraged universities to set up new doctoral programs. By the end of the 1960s unemployment among recent Ph.D.s was widely evident. The Nixon administration, through the Office of Management and Budget (OMB, formerly Bureau of the Budget), cited unemployment as one justification for ending direct support of institutions or any form of NSF grant that would tend to increase the number of scientists.

The implications of this policy shift were enormous. NSF ended its traineeships (institutional grants to support graduate students) and greatly reduced its fellowship programs. The Graduate Laboratory Improvement Program was terminated in 1970, and science development awards were slated for phase-out. The Graduate Education in Science Division curtailed summer and academic year institutes for high school teachers and training courses at such institutions as the Organization for Tropical Studies, and the Office of Scientific Information Service greatly reduced its support of scientific journals. It seemed that the only permissible awards remaining were basic research grants. These alone could be justified as improving the quality of science without increasing the quantity of scientists.[4]

The Daddario-Kennedy Act was followed a year later by the "Mansfield Amendment" to the Defense Appropriation Act which forbade the Defense Department from supporting research not directly relevant to its mission. Other mission agencies, although not directly affected by the amendment, also narrowed the scope of their support, thus giving rise to the so-called dropout problem. As science reporter Daniel Greenberg explained, "In Washington bureaucratese, a dropout is not a student who voluntarily quits school; he is a scientist who involuntarily loses his research grant, not because he is no longer productive, but because the funding agency has redefined its program in such a way that the scientist's work is no longer 'relevant.'"[5] The Department of Defense, Atomic Energy Commission, National Aeronautics and Space Administration, and the National Institutes of Health all dropped projects that were potentially assumable by NSF. Since the abandoned scientists flocked to NSF for funds, the dropout problem placed tremendous pressure on NSF resources. But it also presented a renewed hope—that NSF would finally be able to assume a larger portion of the nation's basic research support.

In the 1960s, the National Academy of Sciences and congressional committees had begun to refer to NSF as a "balance wheel" in the federal government. The metaphor cut two ways. At its best, it gave NSF ultimate responsibility for the health of the basic sciences. If other agencies cut back their support of basic research, the Foundation's budget ought in theory be adjusted to "achieve the optimum federal science budget." At its worst, an interpretation the staff resisted, NSF was to serve merely as a "gap-filler" for other agencies. After the passage of the Mansfield Amendment, NSF pressed valiantly for increased appropriations to fund mission agencies' castoff projects.[6] Although the dropout argument won NSF a larger budget, the increase was much less than desired. To the biologists at NSF, it seemed that Foundation leaders gave priority to easing the crisis in the physical sciences.

A focal point of controversy in this chaotic period was NSF director Leland J. Haworth's successor, William D. McElroy. Professor of biology and director of the McCollum-Pratt Institute at Johns Hopkins, McElroy was the only NSF director before the appointment of Rita Colwell in 1998 to be a biologist. Noted for his research on luciferase, an enzyme responsible for bioluminescence in fireflies, McElroy had been a grantee of the Molecular Biology Program since 1954 and was a former member of the divisional committee. Described by Daniel Greenberg as coming across like "central casting's stock entry for a ward politician," McElroy quickly acquired a reputation as an "operator." This characterization was reinforced by his decision to leave NSF before half his term had expired to become chancellor of the University of California at San Diego. While Waterman and Haworth had tried to sustain an image of the Foundation as apolitical, McElroy actively sought out members of Congress and engaged openly and adeptly in the political fray.[7]

McElroy also angered old-timers by his conception of management, which differed markedly from that of his predecessors. To fill key posts, he brought in professional administrators from NASA. Many of the long-term Foundation staff had contempt for these seeming interlopers who thought they could manage a science agency without knowing anything about science or scientists. To Wilson, McElroy was "a disaster." He "gave the place away to the professional administrators."[8]

McElroy was instrumental in moving NSF into large-scale funding of applied research. To respond to the congressional directive to be socially responsive, Haworth had proposed a new program with the cumbersome and heavy-handed title, Interdisciplinary Research Relevant to Problems of Our Society (IRRPOS). While some older staff recalled the oft-quoted dictum, attributed to Vannevar Bush, that applied research drives out basic, others saw IRRPOS as a legitimate extension of the Foundation's scope. Launched in late 1969, IRRPOS functioned in a manner similar to the basic research programs. That is, university investigators initiated proposals for interdisciplinary research in any area that bridged basic and applied science—for example, energy, pollution, environmental planning, or technology assessment. Proposals, judged on quality of research and potential contribution to social needs, were handled by programs in the research divisions, the funds to be taken from moneys made available for IRRPOS.[9] However, IRRPOS's successor, RANN, initiated by McElroy without full consultation of the National Science Board, was a very different matter.

When Congress appropriated more funds for applied research, McElroy incorporated IRRPOS into a broader program in a separate division titled Research Applied to National Needs (RANN). RANN was administered in a way that many program directors found alien to the Foundation tradition of grants-in-aid. Under the direction of a former NASA administrator, RANN was a managed program with goals, deadlines, and deliverables. Applied research areas were decided upon in advance and appropriate research directly solicited. As science writer Greenberg predicted, old-timer NSF staff were likely to have "an immunological reaction" to RANN. In an informal interview in 1974, Wilson declared that "RANN is a kind of prostitution of the technology kind of thing—prostitution in its worst sense." Despite the scientific community's continued debate over whether NSF had any business engaging in directed applied research, RANN grew rapidly, reaching $80.8 million by fiscal 1975.[10]

Biology at NSF never quite recovered from the events of 1968–72. BMS was never again able to support as high a ratio of applicants as it did in the years of post-Sputnik growth. In 1968, BMS was still funding 51.4 percent of applicants, but only by cutting drastically the length (average 1.75 years) and amount of their grants. Despite these limitations, BMS still maintained a large clientele of leading scientists, many of whom also had awards from other agencies. It did so by funding a whopping 86 percent of renewal applications, as compared to only 31 percent of a greater number of new applications.[11]

A major pressure on BMS resources was the substantial growth in the number of biologists during the 1960s. According to NSF data, the annual production of Ph.D.s in the biological sciences grew rapidly and steadily from 1,244 in 1961 to a high of 3,653 in 1971, a near tripling. In 1969, before the Nixon administration discouraged further growth, BMS staff based their case for more funds on the plight of the young investigator. In documentation presented to Daddario's science subcommittee, they explained that between 1960–61 and 1965–66 the number of biology doctorates had increased at a rate of 15 percent a year, which they calculated would result in a 20 percent increase of potential grantees between 1968 and 1970. Inability to provide support for new researchers, they argued, made for ineffective use of scientific manpower and discouraged young people from becoming biologists.[12]

An additional squeeze on the available resources for biology came from dropouts from other agencies, primarily NIH. After the Mansfield Amendment, NIH discontinued research in photosynthesis (including the work of such veteran NIH investigators as Daniel I. Arnon), plant physiology and pathology, and soil microbiology while ONR dropped research on circadian rhythms. It seemed for a time that NIH would be concentrating increasingly on clinical research. Carlson saw BMS's assumption of a number of scuttled researchers as an example of the "balance wheel" concept in action.[13] BMS was partially successful in gaining more funds in 1972 by using the dropout issue (see table 9.1).

Indices of "success" continued to decline through the 1970s. (The length of award, however, was increased somewhat by shifting from multiple-year to "continuing" grants—supporting each year from current-year funds.) Responding to new constraints and opportunities, BMS created its own dropouts, long-term grantees who were denied further support. Several great but aging biologists were saved from declination only by William Consolazio's personal intervention with McElroy to obtain funds from the director's special reserve. Other leading biologists had their annual level of support reduced. Letters of complaint filled McElroy's files.[14] By 1974, BMS was funding only 40 percent of its proposals and the figure continued to decline.[15]

One casualty of changing times was a proposed award to the Marine Biological Laboratory (MBL). Consolazio, as director of the Division of Institutional Development, had worked with MBL, (in his view "the single most important biological resource in the world") to arrange a science development grant, despite the fact that such awards had never been intended for non-degree-granting institutions. In 1971, MBL applied for $4.7 million over five years to enable it for the first time to operate year around and become a "modified national laboratory." While Consolazio pushed for the award, Assistant Director for Institutional Programs Louis Levin questioned a "national laboratory" without national control. Despite negotiations lasting over two years and a visit by top Foundation officials to Woods Hole, the controversial proposal had to be declined. MBL was told that NSF could presently fund only research proposals.[16]

The phasing out of institutional awards also applied to special facilities.

Table 9.1   BMS Obligations, FY 1967–FY 1975 (in Millions of Dollars)

	1967	1968	1969	1970	1971	1972	1973	1974	1975
Cellular[a]	10.33	10.02	9.28	8.68	8.89	11.90	11.90		
Ecology and Systematic[a]	7.44	8.65	7.96	8.60	8.50	10.28	12.42		
Molecular[a]	10.75	10.34	9.88	9.76	10.73	12.65	12.50		
Physiological Processes[a]	7.85	11.18	10.04	9.53	9.68	10.63	10.92		
Neurobiology[a,b]						4.47	4.46		
Psychobiology[a]	4.96	4.27	4.02	4.30	5.56	4.07	4.02		
Subtotal	41.33	44.46[c]	41.18[d]	40.87	43.36	54.00	56.22	58.93	77.88
Special Facilities[e]	1.67	1.71	0.88	0.92	0.98				
Biological Oceanography[f]	6.85[g]	6.05[g]	6.03[g]	3.66					
IBP[h]	0.50	0.70	1.22	4.00	7.50	9.44	9.20	8.82	
Total	50.35	52.92	49.31	49.45	51.84	63.44	65.42	67.75	77.88
Number of awards[i]	1,099	1,161	1,223	1,107	1,428	1,682	1,614	1,677	1,760

Source: NSF Annual Reports, FY 1969–FY 1975. Each annual report provides comparative data for the past three years. These figures differ depending on which annual report is consulted. Where there is a discrepancy, I have chosen the figures reported at the latest date. Thus, the figures reflect sums spent rather than sums initially budgeted.

[a]Figures not available from *NSF Annual Reports* for 1974 and 1975.

[b]The Neurobiology Program did not begin until 1972.

[c]Data from *NSF Annual Report*, FY 1970. As of *NSF Annual Report*, FY 1969, this figure was 40.83, a decrease from the previous year.

[d]Data from *NSF Annual Reports*, FY 1970 and FY 1971. As of FY 1969, this figure was 38.31.

[e]Funding for special facilities ended in 1971.

[f]The Biological Oceanography Program was not part of BMS after 1970.

[g]The larger figures reflect a share of the funds for ship facilities. The lower funds reflect research support only.

[h]IBP was discontinued after 1974.

[i]Does not include Biological Oceanography.

BMS had retained a line item for special facilities even after the demise of its Special Facilities Program in 1968, but here again, priorities were altered. While a few awards were made in systematic biology, such as $600,000 to the Missouri Botanical Garden in 1970 for herbarium and library facilities, most of BMS's limited funds went to sophisticated instrumentation for biochemical and molecular approaches to biology. BMS supplied equipment for X-ray structural studies, microtubule studies, a computer analysis of the neural bases of behavior, and for the isolation and characterization of protein venoms in marine animals.[17] The $1.7 million available in 1967 dwindled to below $1 million in 1969 and then disappeared altogether with the rest of the Foundation's special facilities in 1972 (see table 9.1).

BMS's limited educational ventures were likewise terminated by shifts in Foundation policies. Summer training at marine and inland field stations, supported since 1952, had to be abandoned. And the NSF hierarchy told BMS in 1970 to phase out its handful of NIH-like training grants in systematic and environmental biology. BMS was able to compensate somewhat by instituting awards for dissertation support.[18] While the NSF hierarchy discouraged any form of award that would increase the number of graduate students, it was still permissible to provide funds to improve the quality of graduate research. Thus the Section of Ecology and Systematic Biology began to offer small awards for graduate students that provided no tuition or stipend but simply paid for research expenses. Such awards were especially needed in these fields where, unlike biochemistry, students did not typically pursue their dissertation research as a by-product of their mentors' grant-supported programs. Instead, they were likely to carry out independent projects requiring off-campus fieldwork. By 1972, the section was providing some fifty dissertation awards.[19]

Finally, due to the realignments of this period, BMS lost oversight of biological oceanography, a field it had fostered from the beginning. When Facilities and Special Programs folded, biological oceanography, including both individual research and facilities grants, became a separate program within BMS. In 1970, however, over Carlson's forceful protest that "a biological oceanographer is primarily a biologist and secondarily an oceanographer," it was switched to the Division of Environmental Sciences. It was joined there to the remainder of oceanography in the hope that a united program would become more visible and be better funded.[20]

This change meant that biological oceanography was severed from marine biology and limnology which remained in BMS. The large vessels that BMS had supported were placed in jeopardy when more narrow oceanographic criteria for funding were applied and when graduate training in biological oceanography was curtailed. The Hopkins Marine Station of Stanford University, for example, had to terminate its biological oceanography program that NSF had supported for a decade.[21] And the *Alpha Helix,* once lauded as a unique floating physiological laboratory, was under fire by oceanographers for ignoring the needs of graduate students and emphasizing research in shallow seas (or worse yet, on land near shore camp). In 1969 the ship was outfitted with a winch and A-frame and forced to alter its program to spend part of the year on deep-sea cruises. After 1972, NSF discontinued block grants for research funding of the ship and reviewed all programs separately.[22]

Although facilities were no longer supported and the individual grant programs of the 1960s stagnated, the period 1969–73 was hardly one of inaction for biology at NSF. Far from it, these years were characterized by bold attempts to wrest additional funds from the NSF hierarchy, the Nixon administration, and Congress. At least some of these new and often controversial initiatives flourished.

Bill McElroy had vowed that as long as he was director, the Foundation would not merely react passively to external changes but would actively set new directions for American science.[23] To demonstrate the Foundation's intent of enlarging its role, he urged the research divisions to establish special "emphases." Carlson and the BMS staff came to realize that to obtain additional funds from both the NSF hierarchy and Congress, it was politically necessary to point to novel and significant areas needing support. After 1970, arguments for new money for biology were couched almost entirely in terms of new initiatives and taking over dropouts. There was no dearth of fresh emphases or of big thinking in the McElroy years. Though budgets were static, the staff planned new ways to spend hypothetical millions. They hoped some of these ideas would catch fire as had the International Biological Program.

In early 1970, BMS staff produced a long-range plan, which, based on recent rapid advances in biology and the claim that the great problems facing society were essentially biological, predicted a forthcoming Decade of Biology. The document, echoing McElroy, declared that BMS could no longer operate in a passive way and would exercise leadership in projecting needs and allocating resources for the next five years. For "undirected" research projects exclusive of biological oceanography/marine biology, the staff proposed a budget of $91.5 million by 1975. They allocated for biological oceanography/marine biology and for related facilities, $26.2 million and $5.6 million respectively. In addition they called for funding undirected team research, directed team research in specified areas, research centers, operational support of resource centers, dissertation research, and major equipment and specialized facilities. With the addition of the IBP, the grand total came to $187.0 million annually by FY 1975. Assuming slow growth of the rest of the federal budget for biology, Carlson claimed that the projected figures would enable the Foundation to double its share of federal support of academic biology from one sixth to one third.[24] The reality was to be much different. BMS's actual budget for FY 1975 was $77.9 million plus another $4.3 million for biological oceanography, which was by then part of the Division of Environmental Sciences (see table 9.1).

In place of the divisional annual report, McElroy instituted in 1971 a formal internal review of BMS, for which the staff prepared a printed, illustrated brochure to highlight grantees' achievements and to argue the case for the division's "emphases." There were eleven of these for the first review before the Foundation's new Program Review Office. Several of the emphases had an overt social-relevance bent: biological regulation of pesticides, fertility and reproduction, drug tolerance and dependence, molecular biology of the human cell, and learning and memory. IBP and Flora North America (see below) were large team projects. The Institute of Ecology was an ambitious project initiated by the Ecological Society of America.[25] Other emphases were directed to the infrastructure of biology: operational support of resource and research centers, including museums, stock centers, tissue and cell banks, and biotrons; centers for the study of macromolecular structure, to provide high-technology instrumentation for molecular biology; and dissertation research.[26] Despite the effort staff invested in furthering them, a number of these Nixon-era projects never came to fruition. Some fizzled, others were embarrassing failures, but some were surprisingly successful.

New Initiatives: Some Failures

Among the failed visions of this period were research on drug addiction and dependence, centers for macromolecular structure, and Flora North America. In the case of the first two initiatives, necessary cooperation with NIH fell through. Moreover, NSF could expect little political support for new money for research considered to be within the province of NIH. The centers for macromolecular structure also failed because there was an insufficient constituency for such regional centers among biologists. Flora North America collapsed through overambition and for lack of a credible connection to an issue of social relevance.

Drug Addiction and Dependence

David Tyler first proposed the idea that NSF should initiate a "crash program" in "drug addiction and dependence" in 1969.[27] While obviously relevant to coping with the Vietnam War–era drug culture, this was a most unlikely topic for a BMS emphasis. It seemed not only too applied but much better suited to NIH. Although the Center for Narcotic and Drug Abuse of the National Institute of Mental Health (NIMH) dominated federal support of research on drug use, the BMS staff argued before the Program Review Office in 1971 that NSF could play a role by funding work "focused on understanding the basic physiological and psychological mechanisms involved in the development of drug tolerance and dependence." For several years, BMS tried to coordinate a program with NIMH. A formal joint committee prepared a document, *Collaboration Between NIMH and NSF—A Proposal,* but as with previous joint efforts with NIH, it came to naught. BMS funded a few projects, but by 1973 the initiative had to be abandoned because, when NIH got added funds for drug research, there was no special niche for BMS. John Mehl, BMS deputy director explained that "a large number of proposals in this area were reviewed with the conclusion that no very promising avenues were failing to be explored at the basic research level for lack of funds."[28]

Centers for Macromolecular Structure

The chief initiative promoted by the molecular biologists on the BMS staff was for instrumentation. Such funds as remained in the line item for specialized facilities before it disappeared in 1971 were redirected to purchasing instruments for molecular and cellular approaches to biology. As thinking big sometimes paid dividends, the program directors in the Molecular Biology Section of BMS hoped to obtain more funds for instrumentation through the creation of centers for the study of macromolecular structure which, like phytotrons and biotrons, they conceived as regional or national facilities.

Led by Eloise Clark, program director for biophysics, the staff proposed three levels of centers—departmental (electron microscopes, recording spectrophotometers, small computers), regional (a high-voltage electron microscope, a 220 nuclear magnetic resonance spectrophotometer, X-ray diffraction equipment), and national. The last, admitted to be several years in the future, "would center around an instrument of unique design or a facility that requires a special site, for example, a neutron diffraction apparatus or a high-resolution scanning electron microscope." BMS requested a budget of $5 million for 1972. By mid-1971, several proposals were on hand, some in the million-dollar range.[29]

From the beginning, BMS officials recognized that coordination with NIH would be necessary. They held meetings with representatives of both

NIH and AEC in 1969 and 1970 to explore the idea of joint funding but, as in the case of drug dependence research, collaboration proved impracticable. The centers continued to be emphasized until 1972, at which time BMS proposed to support "a series of small-to-large centers located at strategic places throughout the country, each of which is properly and adequately equipped with the specialized instruments necessary to perform certain types of research." By 1973, with no special funds forthcoming, the idea was allowed to drop.[30]

In retrospect, not only was collaboration with NIH difficult to achieve, but the idea of centers for macromolecular structure was premature; despite the growth in the complexity and expense of instrumentation during the 1960s, molecular biologists were not yet prepared to champion or use regional facilities.

Flora North America

No doubt the worst funding fiasco to befall BMS in the Nixon years was the Flora North America (FNA) Program. This grandiose project was launched with fanfare in 1972 after years of planning only to crash to the ground six months later, creating considerable embarrassment for NSF. All the work invested in designing the complex computer system and in organizing the editorial apparatus was abandoned, and the project's full-time staff, including eight recently hired Ph.D. and Ph.D.-candidate botanists, suddenly found themselves out of work.[31]

FNA began modestly enough. Soon after the completion of a Soviet *Flora* and the appearance of the first volume of *Flora Europaea* in 1964, members of the American Society of Plant Taxonomists envisioned a similar manual of the vascular plants of North America north of Mexico. By 1967, the name Flora North America was selected, the Smithsonian Institution offered to serve as host, and the American Institute of Biological Sciences agreed to administer the proposed fifteen-year program. Involved with FNA from the beginning, Stanwyn G. Shetler, associate curator of plant taxonomy at the Smithsonian, eventually became its project director, and Peter Raven, associate professor of botany at Stanford and soon to be director of the Missouri Botanical Garden, headed the FNA Program Council.

Although the initial focus was on a series of volumes similar to the *Flora Europaea,* planners soon recognized that to handle the vast amounts of data, computer support would be necessary. As the project evolved, creation of a linked system of databases began to take precedence over the hardcover volumes. By the time of the final grant proposal to NSF in 1971, the printed Flora was envisioned as a by-product of the Taxon Data Bank, a text that could in theory be generated at any time. In retrospect the plan was overly ambitious, but computerization was very appealing to NSF. In this period predating the personal computer, a project linking traditional taxonomy with the latest advances in computer technology was an exciting and radical departure. FNA was billed as "a computer-age scientific enterprise in the realm of systematic biology," "the first electronic encyclopedia of plants in history," and a model for the rest of biology. BMS was willing to support only the botanical aspects of FNA, but the Office of Science Information Service (OSIS) of NSF was eager to fund development of the data-processing system. From 1968 to 1971, BMS and OSIS awarded AIBS planning grants amounting to nearly $400,000.[32]

Horizons continued to expand as the project moved toward its operational phase. The "FNA Project" had become the "FNA Program," which was seen as including many potential "projects." It was to last into the indefinite future, expanded as funds and opportunity permitted. "As a next step," the proposal declared, "one can visualize a 'Biota North America Program' including fauna as well as flora; then the geographic restriction might be dropped as the effort expands worldwide."[33] Understandably reluctant to make an indefinite commitment, NSF strongly encouraged the 1971 transfer of FNA from AIBS to the Smithsonian, an operating agency that could eventually take over the project. Following protracted and difficult negotiation, a contract for $625,000, of which $340,000 was supplied by BMS, was finally submitted for the board's approval in September 1972. In all, NSF proposed to contribute up to $1.4 million over three years, after which the Smithsonian would take full charge.[34]

After six years of planning, the operational phase of Flora North America was officially launched on 1 October 1972. More staff was hired, including associate editors for each of the editorial teams. By early 1973, the project had twenty-two people on the payroll, all but Shetler on grant money. Some seven hundred botanists in America as well as many others overseas were prepared to take part in the editorial and advisory work. Canadian funds had been secured from the National Research Council and Department of Agriculture of Canada. All was set for "the greatest cooperative effort of systematic biology in our century."[35]

The bubble burst quickly. By the terms of its agreement with NSF, the Smithsonian was to phase in funding beginning with $200,000 in FY 1974 and increasing to full support of about $1 million a year by the end of the grant period. When the agreement was discussed in June with OMB examiners for the two agencies, no reservations were expressed. Yet when the Smithsonian tried to add a new line item of $200,000 in its budget for FY 1974, OMB disallowed it. According to the Smithsonian, OMB had directed it not to start any new science programs; prospects seemed no better for the following year when the Smithsonian's installment was to increase to $750,000. Thus, NSF learned from the Smithsonian on 24 January 1973 that the agency felt forced to back out of its commitment.[36] NSF had little choice but to back out as well. As Mehl informed NSF Director H. Guyford Stever (1972–76), "The cognizant Divisions of the Foundation have concluded that there is no prospect of carrying through with the proposed plans, and have asked for plans to abort the project in such a way that the waste of NSF funds will be minimized." Faced with the prospect of no funding at all, the FNA Program Council met on 8 February and voted to suspend FNA indefinitely.[37]

To the botanical community, the cancellation came as a sudden shock, which was followed by considerable lamentation and recrimination. Shetler and Smithsonian officials claimed that their agency had committed itself only on condition that Congress appropriate money "earmarked specifically for FNA"; thus the demise of the project was due to OMB's refusal to authorize seeking new funds. Science reporter John Walsh's article in *Science,* "Flora North America: Project Nipped in the Bud," likewise castigated federal budget officers for having "blighted" the program. Others, doubting that the Smithsonian had been taken by surprise by OMB, faulted it and in particular Secretary S. Dillon Ripley for acting in bad faith and irresponsibly. Though botanists tried in succeeding months to restore funding by direct action of Congress, nothing could be done to resurrect the project.[38]

New Initiatives: Some Qualified Successes

While some post–Golden Age attempts at thinking big failed miserably, others came to fruition, providing large sums of money for particular projects or programs that could be sold to the OMB and Congress on the basis of relevance. Two of these areas, "regulation of pest populations" and the investigation of human cell biology, seemed to many observers foreign to the previous framework of BMS-supported basic research. Public interest in alternative methods of pest control was fueled by the "pesticide crisis," and in human cell biology by the hope of controlling diseases, especially cancer. A third success, research in neurobiology, crystallizing in the late 1960s as a distinct field of biology, was less readily tied to a pressing social issue but could be related to popular interest in behavior, learning, and memory. A fourth area, support of systematic collections, long advocated by museum administrators, was finally made possible by an appeal based on the environmental crisis. Finally, through their relation to the world food supply and energy, the plant sciences, long hidden in functional programs, for the first time became a specific category for NSF funding. Through program director initiative and perseverance, these emphases resulted in the formation of three new BMS programs: Human Cell Biology, Neurobiology, and Biological Research Resources. These, along with plant sciences, and the IBP, represented the high points of an otherwise frustrating period for NSF biology.

Biological Control of Pest Populations

BMS was far more successful in funding initiatives related to agriculture than to medicine. This was in part because the U.S. Department of Agriculture (USDA) did not have a competitive grants program, nor did it encourage sustained fundamental research at state agricultural experiment stations. Experiment-station scientists, in order to carry out more basic research, had sought support from other federal agencies, including NSF and NIH.[39] Thus, NSF found an opening to solicit funds from Congress to support basic research underpinning societal issues related to agriculture.

Since the appearance of Rachel Carson's *Silent Spring* in 1962, the excessive use of broad spectrum pesticides such as DDT was a matter of heated public debate. Carson documented how indiscriminate pesticide application had destroyed non-targeted natural enemies of crop pests and poisoned the environment. The resulting furor over pesticides encouraged renewed interest in "biological control" of crop pests through the classical method of identifying and introducing natural predators as well as through newer means. In 1965, the President's Science Advisory Committee report, *Restoring the Quality of Our Environment,* urged greater federal support for research on alternatives to chemical pesticides. In response to the growing interest, BMS staff highlighted in the *NSF Annual Report FY 1966* its support of research on two more recent forms of potential biological control: insect juvenile hormones, which could be altered to serve as insecticides, and pheromones, chemical attractants secreted by insects, which could be synthesized to control mating behavior. In 1969, NSF's Special Projects in Graduate Education Division inaugurated a cooperative training program in "pest population ecology" which, guided by leaders in the field. was to train graduate and postdoctoral students in pest management, especially biological control, at four universities.[40]

David Tyler and John L. Brooks, program officer for general ecology, suggested in 1970 a divisional emphasis on biological regulation of pest populations, approached through both ecology and biochemistry. According to BMS staff member James H. Brown, explaining the initiative to the NSF Program Review Office in 1971, a superabundance of pest species was due to human altering of the "natural ecosystem" by planting single crops, which offered certain species an "unnaturally" large food supply. He argued the need "to develop other methods which are effective and yet harmless to nonpest species and not detrimental to the environment." Scientists were "just beginning to explore the possibilities for increasing mortality by the use of disease vectors, parasitism, predation, and interference with the reproductive process." BMS's role was "to develop the basic knowledge required to exploit the potential of biological control."[41] BMS claimed to have spent $1.4 million in FY 1971 plus $100,000 through IBP on the regulation of pest populations, primarily insects. In NSF annual reports and press releases, BMS featured grantee research involving hormonal control of insects, insect viruses, and pheromones.[42]

NSF's largest endeavor in pest control research in this period was the Huffaker project, "Principles, Tactics and Strategies of Biological Pest Regulation in Crop Ecosystems." Although funded as a part of IBP from 1972 to 1978, it took on a life of its own. More than three hundred researchers from eighteen universities, USDA, and private industry participated in this big biology multidisciplinary undertaking that involved the coordinated support of NSF, USDA, and the recently founded Environmental Protection Agency (EPA).[43]

Carl B. Huffaker, an ecologically oriented entomologist in the Division of Biological Control at the University of California, Berkeley, sought to develop a systematic study of an alternative pest control strategy based on the ecology of agricultural ecosystems. His initial proposal to NSF in 1969 dealt primarily with the classical biological method of introducing natural predators of five agricultural pests. However, Charles F. Cooper, NSF program director for ecosystems analysis, and the proposal reviewers greatly modified this plan. Cooper encouraged Huffaker to emphasize the theoretical components of the research, to include systems analysis and mathematical modeling of ecosystems, to organize the project around crops rather than pests, and to broaden the approach to what was to become known as "integrated pest management" (IPM). The control strategy would include not only the introduction of parasites but also the use of hormones and pheromones, the artificial introduction of predators, crop breeding for resistance, and the selective use of chemical pesticides. With his first NSF planning grant for the IBP project in 1970, Huffaker organized a series of workshops to draw up a broad-based research proposal, which, in final form, involved the study of the structure and function of six major crop ecosystems—alfalfa, citrus, cotton, pine trees, pome and stone fruits, and soybeans—and their associated insect pests. NSF presented the project in a press release as an analogue of the IBP biome projects which would examine, in place of natural ecosystems, ecosystems created by human activity.[44]

In early 1971, Carlson held a briefing with entomologists, senior NSF staff, and representatives of other agencies on prospects for funding the Huffaker project and other research on hormones and pheromones. On NSF's initiative, USDA, the primary federal patron of research on pest control, signed an official agreement setting up a working relationship whereby the Foundation would support basic research on pest regulation and small-scale field testing, while USDA would carry out extensive field testing.[45] Though BMS was clearly interested in the Huffaker project, financing the large-scale undertaking remained a problem. The USDA, through its Agricultural Research Service (ARS), had its own biological control project emphasizing pest elimination through the "sterile male" technique and, for both bureaucratic and ideological reasons, was reluctant to support IPM. There had been long-standing conflict between the scientists in USDA-ARS and those in the state agricultural experiment stations.[46]

In their funding quest, Huffaker and colleagues such as Ray Smith and Perry Adkisson, following in the footsteps of Roger Revelle and Frank Blair, deliberately sought out political support in Congress and the executive branch. In part through their efforts, the Council on Environmental Quality, established by Nixon in 1970, reported favorably on integrated pest management in 1972. The submission of Huffaker's proposal coincided with congressional debate over new pesticide legislation and Environmental Protection Agency hearings on the banning of DDT. Responding to the favor shown IPM by environmental groups, Nixon made it a key part of his environmental program. A revision of the federal pesticide law in 1972 included a directive to the EPA to support IPM. In February 1972 Nixon, through his assistant for domestic affairs John Erlichmann, specifically instructed NSF to "launch a large-scale integrated pest management research and development program" in cooperation with EPA and USDA, that is, to fund the proposal.[47]

The Huffaker project was initiated in 1972 with a first-year grant of $1 million ($300,000 from NSF and $700,000 from EPA) plus another $911,500 in individual grants from USDA. The NSF contribution to the $11 million–$13 million federal undertaking amounted to over $1 million a year. However politically attractive, the project remained problematical for BMS since it verged on applied agricultural science when BMS staff were committed to funding only basic research. During its course, project managers and NSF staff and reviewers came to disagree over methods and goals. Yet the project was in retrospect one of the most important that BMS sponsored in the 1970s. Significant for its transcendence of disciplinary, geographic, and crop-specialty boundaries, this project was said to be the first major investigation based on the now well-established integrated pest management concept.[48]

Human Cell Biology

On the face of it, Human Cell Biology (HCB) was an unlikely candidate for an NSF program. Why should an agency focused on basic research, with programs organized primarily along functional lines, direct attention specifically to the human cell? Wasn't HCB, as the BMS Advisory Committee asked in early 1972, more appropriate for NIH than for NSF?[49] That such a program did get established was due to an unusual set of circumstances.

HCB originated about 1968 out of discussions of the budget crunch by members of Herman Lewis's genetic biology panel. In many laboratories, it seemed, senior scientists were not taking on new graduate students because of the lack of opportunities for jobs or funded research. Impressed by the manner in which ecologists had promoted IBP before Congress and, in effect, imposed a budget for IBP on NSF, Lewis became convinced that a "big biology" approach could garner more funds for genetics. Thus, he and the panel proposed to generate a politically salable interdisciplinary program capitalizing on the recent progress in molecular genetics. With IBP as a model, they contemplated a national effort focused on the multidisciplinary analysis of a single cell. NIH was expected to be predominant among several agency supporters.[50]

After these initial explorations, Charles Yanofsky of Stanford, a genetics panel member and also president of the Genetics Society of America, obtained a grant in June 1969 to conduct a feasibility study for a National Program on the Exhaustive Study of a Cell. When the proposed program was discussed during a symposium on DNA replication at the Cold Spring Harbor Laboratory, the response was mostly negative; biologists argued that a targeted program was not needed, that it would divert money from other research, and that it would lead to too much control of science from Washington.

Yanofsky and Lewis tried a new approach, assembling a panel of younger scientists to formulate the program. Twelve young but already well known investigators met in late 1969 for two days at the Rockefeller Institute for Medical Research with Lewis, Yanofsky, Norton Zinder, and James Watson. Robert Haselkorn of the University of Chicago emerged as spokesman for the group, which decided to focus the program on the human cell. Lewis was pleased with this choice, for he had long predicted that the genetics of the human cell would soon blossom as a research area. Tissue culture techniques had so advanced that it was now possible to do the kinds of research in human cells that formerly were done only in bacteria or yeast. Because Lewis still envisioned the program as too large for any one federal agency to support, the fact that the human cell seemed more appropriate to NIH than NSF did not seem to be a barrier.[51]

After several additional meetings, funded by Yanofsky's grant, the young scientists drafted in 1970 "A Proposal for a National Program for the Molecular Biology of the Human Cell," which recommended a multidisciplinary, coordinated attack on "the organization and expression of genes in human cells, including the regulation of RNA and protein synthesis, chromosome organization and replication, and somatic cell genetics"; "the structure and function of human cell membranes"; and the development of supportive facilities for cell production and cell banks.[52] Lewis created a steering committee of prominent scientists including Paul Berg, Gerald Edelman, Yanofsky, Watson, and Zinder. In May 1970 Haselkorn sent McElroy a prospectus that described the "explosion of knowledge of the molecular biology of bacteria and their viruses" over the past twenty years that, "when applied to man, will constitute a giant step toward preventing and correcting human diseases and malformations."[53] That is, the proponents of HCB justified the program primarily by its relation to human disease. In particular, they hoped to link basic research on the human cell to the Nixon administration's war on cancer, which culminated in the National Cancer Act of 1971.[54]

The prospectus appealed strongly to McElroy, who was eager to move ahead with or without NIH cooperation. Without consulting the NSF Board first, McElroy put the new program into the FY 1972 budget request under preparation. Lewis recalled that NSF had a budget before it had a program. To his disappointment, Harve Carlson and other NSF staff failed to convince NIH officials to participate in a joint budget presentation of the program for FY 72; Lewis knew from the outset that a program limited to BMS would be much too small. With NSF as the sole sponsor, the Human Cell Biology Program was inaugurated in December 1971 and began awards in FY 1973 at a level of $2 million. Though BMS staff in 1970 had projected a budget of $9.2 million in FY 1972 growing to $50 million in FY 1976, its appropriations remained modest. As of FY 1976, the program's budget had only risen to $2.4 million.[55]

HCB was an unusual program for BMS. Unlike the other research programs of the division, HCB projects were to be collaborative and multidisciplinary, with special attention to funding young investigators. A major feature of HCB was the support of regional stock centers for the growing and distribution of mammalian cells in culture. After a special competition, BMS funded two: one at MIT and one at the University of Alabama. The former, especially, fulfilled its function as a regional center. After a panel ranked proposals for use of the cells, successful investigators sent in seed cells, which were grown in quantity in the facility. The HCB program tried to collaborate with NIH on supporting a cell bank facility. However, approached to share funding, NIH officials favored a much larger enterprise than NSF could afford. For the program's first two years, the two agencies split the cost of the Human Cell Bank in Nutley, New Jersey, but afterwards NIH assumed the entire amount. As Lewis recalled, it was difficult for NSF ever to collaborate with NIH on an equal basis because the two agencies had such unequal resources.[56]

In conjunction with the Human Cell Biology Program, NSF sponsored a pioneer conference in January 1973 at the Asilomar Conference Center in Pacific Grove, California. It was a precursor to the famous Asilomar conference on recombinant DNA two years later. According to Lewis, this meeting, sometimes referred to as Asilomar I, was strictly NSF-initiated. Lewis had been concerned with the risks of human infection at the planned regional cell culture centers, which would be growing cells and viruses for use in cell fusion techniques on a large scale. Paul Berg, a member of the HCB panel, organized the three-day workshop where about one hundred participants assessed the risks and precautions needed for working with tumor viruses and animal cell cultures. The proceedings, published as *Biohazards in Biological Research*, were widely distributed among researchers.[57]

Later that year, the discovery by Charles Boyer and Stanley Cohen of a new technique for cutting and splicing DNA (recombinant DNA) touched off a more extended debate among scientists over biohazards. After a discussion on this topic at the Gordon Conference on Nucleic Acids in the summer of 1973, and its organizers' subsequent publication of a letter in *Science,* Philip Handler, president of the National Academy of Sciences, asked Paul Berg to head a committee on risk assessment. Lewis was a member of this committee, which drew up a statement, published in scientific journals in June 1974, urging a voluntary moratorium on certain types of experiments.[58]

Berg organized a second conference at Asilomar in February 1975, this time with international participants and members of the press, to discuss the potential and hazards of recombinant DNA. While some called this meeting Asilomar II, it became known to the world as *the* Asilomar Conference. The HCB panel consulted on the conference, and Lewis was made a member of the planning committee. NSF contributed $10,000, although the bulk of support came from NIH.[59]

Though the Asilomar organizers intended to restrict discussion to scientists and place boundaries on the scope of the issues to be debated, the publicity generated by the conference opened the recombinant DNA controversy to public involvement. The consensus document that emerged from

Asilomar divided gene-splicing experiments into low, moderate, and high risk and called for the use of safety measures appropriate to each level. A fourth category of experiments were not to be undertaken at all with current containment capabilities. This document served as a basis for the development of federal agency guidelines by Lewis for NSF and by NIH.[60]

BMS's Human Cell Biology Program can be considered only a qualified success, for it did not become the large-scale national program its initiators envisioned. It did, however, involve NSF in an integral manner in the recombinant DNA debate and in broader discussions of ethics in science. Lewis's long-time concern with science and social responsibility led to the creation in February 1973 of a new NSF program, Ethical and Human Value Implications of Science and Technology (EHVIST).[61]

Neurobiology

Neurobiology, defined by the BMS Psychobiology Program in 1970 as "the study of the central nervous system and its relations to both the environment and behavior," emerged as a new synthetic field of biology in the late 1960s, linking genetics, biochemistry, biophysics, neuroanatomy, neurophysiology, and psychology, as well as other areas of biology.[62] In 1971 BMS created the first funding program in the federal government intended to support basic research in a broad spectrum of the field.

Herman Lewis, predicting that "neurobiology will probably occupy the position in the last half of this decade that molecular biology has had during the first half," first proposed a neurobiology program in 1965. He felt NSF could play an important role in getting this new discipline off the ground "through staff and advisory panels who are experts in this field." Other program directors, including David Tyler and Psychobiology Program Director John F. Hall, disagreed. Tyler argued that BMS already covered this research and that establishing new programs in faddish areas (which would take certain types of proposals away from existing programs) might work to a disadvantage by attracting "fund-chasing entrepreneurs" or setting up a "cult of high priests who tend to evaluate proposals not on the basis of whether good experimental procedures are proposed, but whether they meet current fashions and contain the popular catch words."[63] The matter was temporarily dropped.

A few years later, despite the funding slump, conditions had become much more favorable for a new program. In 1969, the field had further coalesced through the creation of the Society for Neuroscience. That same year, with Director McElroy's request for new directions and the shift in interest to more socially relevant research, the proposal was put forth anew. This time psychobiology staff, Henry S. Odbert and James H. Brown, as well as Tyler, agreed on the need for a program. Neurobiology had become a distinct area of biology, they argued, and it was no longer sensible to partition it according to the technique used, whether biochemical, neurophysiological, behavioral, or genetic.[64]

BMS inaugurated its Neurobiology Program, directed by James H. Brown, in 1971, spending in its first year some $4.5 million.[65] Neurobiology acquired proposals from several other BMS programs, especially psychobiology and regulatory biology, and such longtime Foundation grantees as Gunther Stent (who had just shifted to invertebrate neurobiology), Theodore Bullock, C. A. G. Wiersma, Antonie van Harreveld, and Rita Levi-Montalcini. It contributed to Frank O. Schmitt's Neuroscience Research Program at MIT, which organized a set of workshops that synthesized the field by producing bulletins and an especially effective series of handbooks that served as the first textbooks in neuroscience. In addition to funding many leading neurobiologists as individuals, the program also assisted several team projects through long-term coherent-area grants such as one given to William F. Battig and five senior colleagues at the University of Colorado to study cognitive factors in human learning and memory.[66]

Although NIH funded considerable research in neurobiology, the NSF program had a significant impact, according to Program Director Brown, because of the planning undertaken by the panel. NIH, Brown recalled, was too big for effective strategic thinking about the new field; it had several study sections related to neuroscience but no focal point. NSF had the flexibility to call together the leaders of the field in a panel and ask them what most needed to be done. NSF influence in neurobiology in the early 1970s was thus, according to Brown, "way in disproportion to the dollars."[67]

BMS attempted to obtain larger budgets for neurobiology and psychobiology by a divisional emphasis on "learning and memory." Brown argued: "There is probably no area which so completely permeates and influences the diverse aspects of our complex society as does the topic of learning. We are constantly learning, in school, at work, at home, and in our social relationships." Through recent advances in molecular biology, progress was being made in understanding the physiological mechanisms of learning, memory, and forgetting. With coherent-area grants Brown hoped to support "centers" that would serve as a "badly needed bridge between the large and growing body of basic research data on learning and the various educational problems that our society faces." The emphasis was only a modest success. BMS spent about $2.9 million in 1972 on projects relevant to learning and memory, but of this amount, only $500,000 represented "new money."[68]

Aid to Systematic Collections

A substantial success of the early 1970s was the establishment in 1973 of the Biological Research Resources (BRR) Program, one of whose chief functions was the support of systematic biology collections in museums and botanical gardens. At a time when NSF was retreating back to the traditional individual research grant, the new program was able to provide exceedingly broad-based institutional funding. Five-year awards paid for curatorial staff and assistants, equipment for storage of collections and to improve access to them, computer costs, supplies, travel of visiting experts, binding of library journals, and even limited renovation of buildings.

From 1959 to the late 1960s, BMS had provided only a small amount for curatorial support of collections. Even this was in jeopardy after 1968 when, after the termination of Special Facilities and Programs, support reverted to the Systematic Biology Program where it competed directly with research grants. However, systematists, NSF's most organized and vocal biological constituency, took steps to counter the deteriorating situation.

A large part of the impetus for the new BRR Program came from the "Steere Report" which appeared in 1971. *Systematic Biology Collections of the United States: An Essential Resource,* named informally after William C. Steere, chairman of the editorial committee, was initiated by the Conference of Directors of Systematic Collections. Funded by BMS, the report was directed primarily to NSF. The BMS staff, in fact, worked with the committee to add more punch to the summary and recommendations.[69]

The success of the Steere Report lay in systematists' ability to capitalize on the environmental movement:

> Everywhere today there is growing awareness that in our unbalanced relationship with the natural world—signified by rampant starvation, heedless exploitation, appalling pollution, and disappearing species—we are edging ever closer to the tolerance limits of the delicate complex fabric of natural law. . . . The health of the world's ecosystem depends squarely on keeping as much diversity in the natural world as we possibly can. Because knowledge of the kinds of creatures in our world is fundamental to real understanding of their interaction, the great specimen collections are the very cornerstones to studying, comprehending and living within the world ecosystem.[70]

The nation's collections, it was argued, had experienced rapid growth and vastly increased use through NSF-supported research, but personnel and funding for maintenance and responding to service requests had not grown to meet the added burden. These irreplaceable national resources were endangered by a shrinking base of endowments and local support. The report optimistically recommended a federal contribution of $19.8 million a year for ten years for construction and renovation of space, new professional staff, curatorial support including salaries of technicians, and equipment including computers to unite the collections into a "national information resource." This largesse was to be directed primarily for the benefit of the twenty largest institutions that the Conference of Directors of Systematic Collections represented.[71]

The report's most persistent advocates at NSF were Walter Hodge and William E. Sievers, director and associate director for systematics, and David M. Gates, former BMS Advisory Committee member, now (at the initiative of Carlson and Keck) a member of the National Science Board. Director McElroy, primarily a molecular biologist, was not especially sympathetic to systematic biology. His response (drafted by Consolazio, who was also unsupportive of systematics) to Gates's letter of advocacy was entirely discouraging: "We have had to make some rather difficult choices respecting what and what not to support with our basic research funds. The taking on of new responsibilities such as the logistic support of major facilities is just out of the question at this time." McElroy complained that the systematic biologists had failed to do their homework and that the report "conveys the impression of being an open-ended and bottomless pit."[72]

Nevertheless, with the persistence of Gates and the BMS staff, a program to benefit systematic collections was soon established. The plight of the nation's systematic collections was argued sympathetically before the BMS Advisory Committee, NSF Program Review Office, National Science Board, and finally OMB and Congress. Walter Hodge told the Program Review Office in 1972 that the institutions maintaining the "small number of truly major reference collections for systematic research" could "no longer meet entirely the burgeoning costs of collection, maintenance, or of servicing loans." Since NSF was the major federal funder of research in systematic biology, it was "appropriate for the Foundation to support financially a fair share of the national service activity of the major museums." For the benefit of NSF's nonbiologists, he avowed that "such reference collections serve biology in somewhat the same fashion that a Bureau of Standards serves the physical sciences."[73] Gates was an especially effective advocate before the physical scientists on the NSB because he himself was a biophysicist trained as a physicist who nonetheless made the case for whole animal biology. He recalled that at the board discussion, when astrophysicist Robert Dicke scoffed at "a bunch of sparrow skins," he argued convincingly that the older collections contained invaluable comparative environmental data from a less polluted era.[74]

From the beginning Harve Carlson and BMS planned to combine support of systematic collections with other forms of "resource centers" in a single program.[75] At the same time as they advocated the needs of systematists, the staff argued for funds for living organism stock centers (which included the former genetic stock centers plus the American Type Culture Collection), biotrons and phytotrons, and field stations.

Other than systematists, the inland field station biologists were probably the only group of life scientists in this period to direct a campaign for support specifically at NSF. Under a grant to AIBS, M. Dale Arvey, former BMS special consultant now chairing the biology department at the University of the Pacific, headed a new project to survey inland and marine stations, the number of which had considerably increased through the 1960s. The survey report, *The Role of Field Stations in Biological Research and Education* (1971), included data on existing stations, their projected needs, and the text of a symposium on biological stations as "a national resource for teaching and research," organized by Arvey and cosponsored by the Organization of Biological Field Stations.[76] The field station directors were not immediately successful. Not until FY 1977 did NSF's support for stations jump tenfold from a meager $80,000 to $805,000.[77]

BMS began to fund systematic collections on a larger scale in 1972, providing $950,000 to eight institutions. In 1973, the Division formally established the Biological Research Resources (BRR) Program with Sievers as program director.[78] BRR deliberately directed awards not to institutions as a whole but to particular collections of national importance, generally to enhance existing strength. For each area—botany, insects, birds, vertebrate paleontology—Sievers solicited a report of an ad hoc committee to aid him in making funding choices. In this way, the program was able to provide some support to a larger number of institutions than those represented by the Steere report. Awards initially ranged from $50,000 to $150,000 for five years. By FY 1975, BRR had a budget of $4.0 million, $2.8 million of which went to systematic collections. The program grew slowly but steadily through the remainder of the decade.[79]

The Plant Sciences

Following two decades of relative neglect of the plant sciences, BMS staff came to realize that NSF had a special role to play in plant research. After 1970, when NIH and AEC drastically cut their commitments to basic plant research, the plant sciences found themselves in serious financial difficulty. NIH had dropped many of its grantees in photosynthesis, plant physiology, and pathology, while AEC had shifted its emphasis to more practical research related to the environment. Plant physiologists tried without success to convince the USDA to set up a competitive grants program or to increase the amount of basic research funded at agricultural experimental stations.[80]

It was the current issue of global food needs that convinced two key staff members at NSF, first Elijah Romanoff and then Mary Clutter, to focus on plants. Romanoff, an animal physiologist, came to the Metabolic Biology Program as a rotator from the Worcester Foundation for Experimental Biology and, finding an opportunity to make a difference through administration, stayed until retirement. He recalled becoming personally concerned with overpopulation (an issue receiving widespread attention at the time) and being convinced that plant physiology would provide the basic knowledge necessary to feed more people. When he investigated the place of plants within BMS and in the federal government generally, he found that most BMS programs were funding little plant research and that other granting agencies' support was also insufficient.[81]

About the same time, Robert Burris, chairman of the National Academy of Sciences' Section of Botany, probably with his friend Romanoff's encouragement, wrote to NSF in late 1971 or early 1972 to protest the agency's comparatively poor support of basic research in the plant sciences and their inadequate representation on panels. A delegation of plant scientists came to NSF to argue that plants were being shortchanged.[82]

With this ammunition, Romanoff launched an informal initiative in plant physiology. He increased the number of plant scientists on his panel and let it be known that his program would support plant research. He encouraged prospective grantees to use plant systems. His goal, he recalled, was to support the best botanical proposals even if they were not competitive with proposals in microbiology or biochemistry. He wanted to make the Metabolic Biology Program the focal point for plant physiology at NSF. In 1972 and 1973, BMS featured research on photosynthesis and nitrogen fixation in the NSF annual reports.[83]

By 1976, NSF presented the plant sciences as a budget initiative for the new Division of Physiology, Cellular and Molecular Biology of the Directorate for Biological, Behavioral and Social Sciences (BBS, as BMS was renamed in 1975). The budget sent to Congress claimed that NSF had "an obligation to increase understanding of biological phenomena." As the mission agencies increasingly focused on their own mandates, "the role of NSF in extending the knowledge base becomes more prominent," especially in "the interface areas between, biology, chemistry, physics and mathematics, and in the general field of plant sciences." The division "is a major supporter of basic plant science research and will continue to encourage fundamental analytic and quantitative endeavors in this area." The division's requested budget increase of $9.6 million was to support "selected research areas," particularly "photosynthesis, nitrogen-metabolism, plant cell culture and plant genetics." In FY 1977, metabolic biology devoted some 57 percent of its $7.45 million budget to plant research, the highest proportion of any NSF program.[84]

A few years later, Mary Clutter launched a larger and more formal initiative in "plant biology." A former plant scientist from Yale, Clutter arrived at NSF in 1976 to take charge of developmental biology. She found few plant proposals submitted, and those not competitive. As a result of attending an NSF-sponsored symposium on world food and nutrition a month after her arrival, at which she met representatives of a number of federal agencies, Clutter decided to organize an Interagency Committee on Plant Sciences. This committee conducted a thorough survey of federal support for research in the plant sciences, by agency unit, the first such data collected. As plant research had the potential for increasing world food production, providing alternate and renewable sources of energy, and preserving the environment—all pressing issues in the 1970s—a strong case was made for expansion of basic research and training in the plant sciences. NSF was the second largest supporter of plant research, after USDA. In FY 1977, the year that the committee surveyed, the NSF Directorate for Biological, Behavioral and Social Sciences was spending $22 million on plants, to NIH's $1 million. Through directorate-wide initiatives—which in the 1980s included a targeted fellowship program—NSF steadily increased its commitment to "plant biology." Plants had at last become a funding category.[85]

While in the 1950s and 1960s, NSF biologists had largely ignored USDA, in the 1970s they worked to strengthen ties. As NSF turned toward basic research in areas of social relevance, the agency was inevitably led to agriculture, the key to feeding a rapidly growing world population. For example, in 1975, an NSF grant procured the aid of a panel of experts to identify research areas "in which increased funding could be expected to increase food production." Their report, which appeared in early 1976, provided backing for Clutter's initiatives to increase BBS and other government funding of plant research.[86] NSF's biology staff provided considerable assistance to USDA in the establishment of its national grants program in 1977–78. NSF has nevertheless remained the largest single federal sponsor of competitive grants for basic plant research in academic institutions.[87]

The International Biological Program

Big biology at NSF reached its zenith in the funding of the International Biological Program, which headed the list of "divisional emphases" for fiscal years 1972–74. Favored by Congress, IBP budgets within NSF rose quickly, reaching a level of $4 million by 1970, the first year IBP had a separate line in the budget, and a high of $9.4 million in 1972 when BMS was funding thirteen IBP "integrated research projects" in environmental studies and four in human adaptability, involving some six hundred scientists (see table 9.1). Three of the biome projects (Eastern Deciduous Forest, Grasslands, and Tundra) received more funds ($22.4 million) between 1968 and 1974 than all of the individual projects supported by the General Ecology Program during this same period.[88]

NSF presented the IBP, especially the biome projects, to the public in 1969–72 as aiming to predict the consequences of natural or man-made perturbations to the ecosystem and improve capability for rational management of resources and control of environmental quality. The "new and essential aspect of the U.S. approach to the IBP," NSF spokespersons claimed as late as 1972, was the attempt to link as many ecological processes as possible in a single computer model so as to simulate the entire ecosystem. As ecosystems were highly complex, large integrated research programs were needed. This overselling of IBP by ecologists and NSF made it an easy target for critics.[89]

In 1969, BMS formally established the Ecosystem Analysis Program to handle the IBP projects and other awards in ecosystem ecology. At the same time, the old Environmental Biology Program was renamed General Ecology since, it was argued, the previously unfamiliar term "ecology" had become "well recognized" by Congress and the public. It was headed by a series of university ecologists with continuity provided by James T. Callahan as associate program director and Josephine Doherty as assistant program director. To coordinate the biome and human adaptability projects, NSF set up management offices at Pennsylvania State University and the University of Texas.[90]

In the United States, IBP had become a program in ecology dominated by five large biome projects. The U.S. program operated essentially independently of other nations' programs, though information was disseminated internationally through IBP-supported workshops, conferences, and publications. By 1970 the U.S. National Committee for the IBP (USNC/IBP) ceased to certify unsolicited projects as "IBP-related" and concentrated its attention on the integrated programs initiated through the committee. But the USNC/IBP soon found that it had lost much of its control of the integrated programs, first to NSF, the chief source of support, and then to the large projects themselves. As many organizations dependent on outside funding have discovered, the parts of the program that got carried out were those that meshed with the funding agencies' priorities. The USNC/IBP had hoped to attract a variety of federal funders, but in the end most of the money came from NSF. Projects within NSF's scope such as the biome analyses were well financed, whereas those under the rubric of Human Adaptability received more modest support. A number of projects that the USNC/IBP advocated were never funded.[91]

The human adaptability projects were intended to investigate how people have responded to environmental stress. Because of the complexity of modern industrialized societies, they were deliberately focused on "simpler social systems in which genealogical, cultural, linguistic, and nutritional patterns have been maintained." BMS supported projects on the genetics of the Yanomama Indians of Venezuela under veteran human geneticist James Neel of the University of Michigan; the nutrition of Eskimos headed by H. H. Draper of the University of Illinois; historical migration patterns in the Aleutian Isles under William Laughlin of the University of Connecticut; and the causative factors of migration within the continental United States directed by Everett Lee of the University of Georgia.[92] If not for IBP, BMS would not likely have supported such studies, because they fell into the domains of anthropology, social sciences, or clinical studies (two projects involved clinical teams), rather than basic biology.

Among the ecosystem projects of IBP, the most significant by far were the biome studies, the epitome until the Human Genome Project of big biology. Of the six biomes originally envisioned, all but the Tropical Biome were inaugurated. NSF made large block grants to the project directors who then distributed them among various field sites and subprojects. Of the biomes, the largest were Eastern Deciduous Forest and Grasslands.

The Eastern Deciduous Forest Biome (EDFB), being studied by a team under Stanley Auerbach at Oak Ridge National Laboratory, represented NSF's first major support of scientists at an AEC facility. Auerbach arrived at Oak Ridge in 1954 to begin an ecology program within the Health Physics Division. Starting with a focus on radiation ecology in the 1950s, Oak Ridge became a major center for ecosystem ecology in the 1960s with the Oak Ridge Reservation as field site. Auerbach's team of scientists numbered over fifty by 1967. Working in an AEC laboratory dominated by physical scientists and headed by Alvin Weinberg, promoter of the term "big science," ecologists at Oak Ridge were used to working in a team. The laboratory acquired a reputation for ecological theory and for computer modeling of ecosystems. There, in the 1960s, George Van Dyne, Jerry Olson, and Bernard Patten developed systems ecology, which would later be incorporated into the IBP biome studies. NSF funding enabled this basic research to continue at Oak Ridge at a time when the AEC was turning its resources toward environmental impact statements and other forms of applied environmental science. The EDFB was less centralized than Grasslands; each of the five field sites, corresponding to the five participating institutions (Oak Ridge, Rensselaer Polytechnic Institute, Duke University, and the universities of Georgia and Wisconsin) functioned autonomously. Although it produced a few ecosystem models, the EDFB concentrated on numerous smaller models of ecosystem processes. It received a total of $7.5 million from 1969 to 1974.[93]

The Grasslands study, under the leadership of George Van Dyne, who had gone from Oak Ridge to Colorado State University, was the first, largest, and most hierarchically managed of the biome projects and the only one to retain the goal of mathematically modeling an entire ecosystem. Given the inexperience of some of the cooperating institutions—smaller schools in Colorado and Wyoming—and the newness of ecological research at the sites, it was an exceedingly ambitious undertaking. The project administrators centrally controlled the choice of subprojects; field research was linked to the production of a single ecosystem model. In the period 1968–74, NSF spent $10.4 million on Grasslands, its budget reaching a high of over $2 million a year.[94]

The other three biome projects were the Coniferous Forest Biome under Stanley P. Gessel of the University of Washington ($5 million), the Desert Biome headed by David W. Goodall of Utah State University ($6.6 million), and the Tundra Biome under George C. West of the University of Alaska ($4.5 million). Related to the latter was a cooperative project by investigators of seven institutions to study potential ecological effects of the proposed trans-Alaska oil pipeline and development of Alaska's North Slope.[95] Dieter Mueller-Dombois directed the Hawaii Terrestrial Program to investigate ecology and evolution in the islands, and a sixteen-institution cooperative project headed by Otto Solbrig of Harvard and Harold Mooney of Stanford researched Mediterranean scrub in California and Chile and desert shrub in Arizona and Argentina in an effort to test a fundamental biogeographic question: "Will two very similar physical environments acting on phylogenetically dissimilar organisms in different parts of the world produce ecosystems that are similar in their overall structure and organization?"[96]

These large projects, especially the biomes, remained a source of contention within and without NSF. Those at NSF most closely associated with them—Callahan and Doherty—were convinced that the IBP studies had opened up and consolidated an entire new subfield, ecosystem ecology. W. Frank Blair, in his participant's history of IBP, claimed that "IBP has been responsible for a quantum advance in ecological science as a basic science" and for "significant steps" toward the application of ecology to environmental problems.[97] Many other NSF program staff, especially those in molecular and genetic biology, felt that the IBP projects were too large and overblown. Consolazio, perhaps voicing the opinion of a number of his colleagues at NSF, described the IBP in a memorandum to McElroy in 1972 as "more political than scientific—contains a large element of mediocrity."[98]

Even ecologists were critics. One of the IBP's most outspoken opponents was Nelson G. Hairston, population ecologist and director of the University of Michigan's Museum of Zoology and the last chairman of the BMS Advisory Committee. In 1972, the committee called for a group of scientists—from which NSF staff and anyone involved in IBP funding would be barred—to evaluate the effectiveness of the biomes. The science policy newsletter *Science & Government Report,* identifying Hairston's NSF connection, quoted him as saying that he feared the IBP was producing some "pretty crappy stuff" and that it had become a "boondoggle." Hairston wrote to NSF director Stever that projects such as IBP "give authority to persons outside the Foundation to award what are in essence subgrants. These have sometimes been awarded after a group of the applicant's peers have overwhelmingly recommended rejection of the proposal."[99]

Because of such conflicting assessments, NSF funded two evaluations of the IBP in the mid 1970s, both made available to scientists and the public through the National Technical Information Service. The first was an overview of the entire U.S. IBP effort by an ad hoc committee of the National Academy of Sciences, chaired by long-time BMS advisor Paul J. Kramer of Duke. The committee admitted the biome studies had failed in the goal set before Congress, namely, to produce large ecosystem models that could predict the effects of natural and human changes in the environment. The projects were much more successful in modeling smaller systems such as transpiration, nutrient cycles, and photosynthesis. Despite significant problems of management and coordination and the failure of promised data banks, the Academy committee judged the projects a worthwhile investment. The U.S. IBP "showed field biologists the potential of organized, multidisciplinary, and quantitative study of complex problems and promoted these methods and concepts all over the world."[100]

A group at Battelle Columbus Laboratories, in Columbus, Ohio, conducted the second in-depth review, to compare the Grasslands, Eastern Deciduous Forest, and Tundra biomes with the Hubbard Brook Ecosystem Study—an innovative project funded by NSF since 1963—and with individual research efforts. Recognizing that large integrated projects could produce papers that differed from other types of ecological research in breadth and focus, the Battelle report found that the biome studies—in part because they had "exploded on the research scene powered by massive funding, ambitious goals, and short deadlines"—had not been able to achieve the expected level of integration and efficiency. It suggested that gradually evolving, longer-term, smaller-scale integrated studies such as Hubbard Brook, were less costly and likely to be more productive. When the substance of the Battelle report was published in *Science,* the directors of the biome projects and W. Frank Blair were quick to find fault with the study and come to the defense of big biology.[101]

The Hubbard Brook ecosystem study, seen by some analysts as an alternative model to the biome studies, began as a collaboration between F. Herbert Bormann, a forest ecologist at Dartmouth College and later Yale University, and Gene E. Likens, a Dartmouth limnologist later at Cornell, using the small watersheds of the U.S. Forest Service–managed Hubbard Brook Experimental Forest in the White Mountains of New Hampshire. They were assisted by Robert Pierce, a Forest Service hydrologist who provided access to the forest and later participated in the research. Bormann and Liken's first NSF grant in 1963 was a modest $59,400 for three years. Their ingenious idea was to treat each watershed as a self-contained ecosystem in which inputs and outputs of water and nutrients could be monitored. In 1965, they began a major experiment in which all the trees in one watershed were cut down and left on the ground and an herbicide applied for three years to prevent plants from growing. They and their colleagues then measured the changes in stream-flow, erosion, and chemical composition of the stream, recognizing the process of nitrification as an important factor in the loss of nutrients from the ecosystem. Later experiments investigated the effects of conventional clear-cutting.

Bormann and Likens gradually built up a group of researchers, giving them latitude to pursue their own projects. By the mid 1970s, BMS was funding them at an annual rate of about $200,000. Based in part on field data and a computer simulation model of tree growth, the Hubbard Brook researchers developed an influential Biomass Accumulation Model, which predicted stages of recovery of a forest after clear-cutting. Another major achievement was their contribution to the understanding of acid precipitation. The Hubbard Brook study has since become a staple of ecology textbooks.[102]

Historians who have studied the IBP in its relationship to ecosystem or systems ecology have recognized some of its achievements but have tended to be critical of IBP's failure to live up to its rhetoric. Among the positive legacies of IBP were that it helped to consolidate ecosystem ecology and trained a generation of practitioners; it increased cooperation among American and foreign ecosystem ecologists; it resulted in a permanent increase in funding for the field; it greatly stimulated the use of computer modeling in ecology; and it successfully produced smaller-scale models of productivity, nutrient cycling, and energy flows.[103]

However, IBP turned out to have had only limited applicability to practical problems of environmental management in the 1970s. These problems tended to focus on the economic or health risks of particular pollutants, not the health of the ecosystem in general. A new field of environmental science developed simpler and more practical techniques with greater predictive power than the models produced by academic ecologists.[104] In addition, IBP failed as an experiment in new forms of managed organization of research, because biologists resisted big science in favor of traditional modes of research.[105] Finally, IBP was too big, too structured, and too expensive for optimum effectiveness. Smaller projects such as Hubbard Brook were said to have been a more effective approach to ecosystem ecology than the big-science biome studies.[106]

The question of whether the IBP money could have been better spent on ecology or on other areas of biology is, however, somewhat beside the point. If it were not for the effective selling of IBP to Congress, the funds would not have been available to ecology or biology at all. Both the biome studies and Hubbard Brook served as models for future NSF support of ecosystem ecology. The U.S. IBP officially ended on 30 June 1974, but BMS staff convinced Congress to transfer the line-item funds to the Ecosystem Analysis Program, renamed Ecosystem Studies. This enabled some of the biome projects, including Grasslands and Eastern Deciduous Forest, to be continued for three more years as operations were gradually wound down and a series of synthesis volumes written.[107] Otherwise, projects formerly part of the biomes had to compete for funds on their own merit. After IBP, NSF shifted to smaller-scale ecosystem projects as the Battelle report had advocated. Funds originally supporting IBP were used in 1978 to begin a new program, Long Term Ecological Research (LTER), to study ecosystem structure and function over time at a limited number of field sites. Hubbard Brook was cited as one of the chief models of a successful long-term project.[108]

Funding Biology by Initiatives

In the 1970s, the focus for acquiring increased funding shifted from individual BMS programs covering all of biology to a series of initiatives designed to gain favor with the NSF hierarchy, OMB, and Congress. Those initiatives succeeded best when they combined program officer commitment, the enthusiasm of the NSF hierarchy, the independent lobbying of an identifiable group of biologists, and a cogent argument for social relevance. New emphases in BMS were keyed to evolving social and economic problems. By 1974, following the Arab oil embargo, energy had supplanted the environment as a major national concern. In response, BMS increased support for biological catalysis (enzyme action), the actions of microorganisms in converting materials, photosynthesis, and nitrogen fixation.[109]

The seeming politicization of NSF patronage of biology in the early 1970s strained and brought to a head friction within BMS and between BMS and the scientific community. Moreover, with increased competition for funding regular programs in all the sciences, NIH became a problem in a way that it had not been before. At the same time, social and political groups calling for more openness and fairness in NSF operations led to administrative overhauls in BMS and in NSF generally. Many of the assumptions that had guided NSF patronage of biology in the 1950s and 1960s were no longer valid by 1975. These changed assumptions and relationships in the period 1970–75 are explored in the conclusion of this book.

CHAPTER 10

End of an Era, 1972–1975

The disbanding of the Division of Biological and Medical Sciences in 1975 and the reconstitution of biology at NSF as a part of the new Directorate of Biological, Behavioral, and Social Sciences can be said to mark the end of an era. Challenges to the idealistic founding vision for biology at NSF had taken place over many years, but it was especially in the period 1968 to 1975 that the assumptions underlying funding of biology at NSF had to be reevaluated. The changes that biology at NSF underwent in this period were not all negative, for what emerged was a fairer and more open agency, one that took a more realistic view of its role as one of several agencies funding biology within a larger federal system.

NSF had begun as an agency solely concerned with undirected basic research, which it claimed to be the foundation upon which applied science and technology rested. This tenet, which had been fundamental to the agency's sense of identity, was severely tested in the period after 1968. After the Daddario-Kennedy revision of the NSF Act in 1968, NSF launched a large-scale program of applied research. The research divisions, while continuing to fund only basic research, were forced to justify new funds by focusing on "emphases" that were based upon specific practical needs. The history of biology at NSF after 1975 would therefore incorporate an entirely new element, namely NSF's relation to the growth of new industrial biotechnology.

The distinction between basic research and applied research, which NSF had long depended upon, had become increasingly murky. Retrospective studies of the genesis of technological breakthroughs, such as the military's Project Hindsight reported in 1966, showed that practical achievements did not always depend on prior basic research but often rested on prior technology.[1] Melvin Kranzberg, a founder of the new discipline of history of tech-

nology in America argued in *American Scientist* in 1968 that science and technology were in a complex dialectical relationship. Technology could redirect
science as well as the other way around. At one end of the spectrum of varied relationships existing between science and technology, invention could
progress quite independently of scientific advance. Edwin T. Layton, Jr., also
a historian of technology, referred to science and technology in 1971 as
"mirror image twins."[2] Even NSF Director William McElroy took notice of
Kranzberg's insight to argue for supplementing support of "the autonomous
enterprise of basic science," which remained one route to technological
progress, by a "selective emphasis in new directions to help our nation solve
complex social and environmental problems," another route to practical benefit.[3] In any case, the ideology of basic research—the stockpiling of basic
science as the sole basis for technological advance—was revealed as much
too simplistic.

The elitist belief that NSF should focus only on funding the best science, which was held by the early program directors, had already been challenged by the debates over geographical distribution and science development in the 1960s. It was subjected to a new form of criticism from NSF's
constituents in the early 1970s. The termination of the BMS Advisory
Committee in late 1972 was one casualty of a multifaceted movement by
segments of the scientific community and others to make NSF operations
more fair and open. For two decades, BMS had functioned as an informal
network in which program officers—almost entirely white men—cultivated
friendly relations with grantees and advisors. With little constraint, they had
selected acquaintances as panel members—also, with few exceptions, white
males from major universities. Applicants whose proposals were rejected
were given no information unless they requested it, and then they received
only a statement written by the program officer. Reviews of proposals remained strictly confidential.

In the early 1970s, for the first time, women, African Americans, and
Hispanics, protested directly about NSF's old-boy-network modus operandi.
Why were there scarcely any females or minorities on the professional staff
or on panels and advisory committees? Was peer review fair to grantees, or
was it marred by cronyism? Was it proper for NSF to identify candidates for
professional positions by word of mouth rather than through openly advertised searches? The NSF hierarchy found itself on the defensive.[4]

The issue of women and minorities came before the BMS Advisory
Committee for the first time in 1971, one of many manifestations at this
time of the growing women's movement in science and engineering. In the
same year, a group of feminist scientists founded Association for Women in
Science (AWIS), an organization dedicated to promoting the interests of

women scientists. When confronted, the BMS Advisory Committee was unwilling to admit a problem. The carefully worded minutes stated: "The general sense of the discussion was that, while maintaining particular sensitivity to the need not to discriminate against minority groups or women, the prime consideration should continue to be competence to discharge the required advisory functions."[5] No woman or African American was ever appointed to this committee in its twenty years of existence.

Especially challenged in the years preceding 1975 were the twin beliefs that NSF could foster the unity of biology and that staff and advisors could cooperate effectively for the advance of biological research. By 1970, the original functional organization of BMS had so evolved in practice that it consisted of a jumble of disciplinary programs (e.g., biochemistry, biophysics) and functional programs (e.g., regulatory, metabolic) with overlapping jurisdictions arranged in sections that roughly corresponded to levels of biological organization. BMS staff were agreed that this subdividing of biology needed a thorough rethinking. The organization of BMS by levels did not reflect contemporary biology, they argued, because "today many biologists are working on problems and using approaches at several levels of biological complexity." Much effort went into trying to design a new system based on "biological problem areas," since problems "are more likely to be approached at several levels." They cited neurobiology, a program that was created, as an exemplar. "Research on evolution," which did not become a program until much later, was offered as another problem area. However, advisory committee members were concerned that any reordering might be detrimental to academic biology. In the end nothing was done.[6]

Much was hoped from a National Academy of Sciences Committee on Science and Public Policy report on the life sciences. A major undertaking, *The Life Sciences* and its companion volume, *Biology and the Future of Man,* edited by Academy President Philip Handler, appeared in 1970. But while useful for enticing students to enter biology, the volumes neither provided a unified view of biology nor set priorities for its needs.[7]

At the same time, relations of staff and advisors continued to deteriorate. The BMS Advisory Committee became increasingly dissatisfied with the rapid changes taking place in NSF and its lack of power to do anything about them. In the 1950s and 1960s the staff had been able to work together in harmony with the BMS Divisional/Advisory Committee. BMS directors John Wilson and Harve Carlson were able to call upon these academic advisors as support for staff initiatives. But by the 1970s, the interests of the staff—itself deeply divided—and the committee had diverged. The semiannual meetings turned into unpleasant confrontations as frustrated committee members objected to each new program and emphasis that the staff placed before it. Far less willing to accommodate to altered political realities than the BMS staff, the committee members could not see the need for emphases and earmarking when the regular divisional programs were poorly funded. While the staff pressed forward with initiatives tailored to fit the new constraints, committee members, with few exceptions, advocated a return to the old post-Sputnik consensus. With the creation of the Program Review Office, BMS staff found internal presentations had became more important than those to academic advisors. By the time the BMS Advisory Committee was disbanded in late 1972, it was nearly at cross-purposes with the staff.[8]

The immediate occasion for the demise of the BMS Advisory Committee was the Nixon administration's proposed "sunshine" legislation that threatened to open agency advisors' meetings to the public. (Given the hostile tenor of the last advisory committee meeting in February 1972, no one on the NSF staff was likely to have welcomed public scrutiny.) In order to protect the sanctity of panels to hold closed meetings, NSF replaced the division-level committees with a new advisory committee structure at the level of the assistant director for research whose meetings could safely be made public. Thus, after two decades, there was no longer an NSF advisory group representing all of biology.[9]

The year 1972 also saw the departure of Harve Carlson. The continuing budget crisis had done little to quell the growing division between BMS's "experimental" and "field" biologists. Herman Lewis had devoted his 1969 annual report to what he called "the biology gap" in BMS: "On one hand biology is evolving into a more unified science, which is desirable; on the other hand, our Division is being fragmented by a widening gap which the writer views as a major problem."[10] Through maneuvering by program directors who were concerned that Carlson showed too much favoritism to ecology and systematic biology, the BMS director was asked to retire from the Foundation earlier than he had planned.[11] With Carlson's departure, followed in 1975 by the retirement of the articulate and well-liked deputy director John Mehl, the division lost a broad-minded leadership that had attempted to plan for biology as a whole.

After a long search in which the position was offered to several outside people, the BMS directorship was given in 1973 to Eloise (Betsy) Clark, who had come to NSF in 1969 from Columbia University to be a rotator in developmental biology.[12] She stayed on as program director in biophysics and as head of the Molecular Biology Section before promotion to division director. Eventually, she became the first presidentially appointed Assistant Director for the Biological, Behavioral, and Social Sciences.

Respected for her intelligence, integrity and conscientiousness, Clark faced many difficulties in managing BMS. At a time when women scientists had only recently become vocal, Clark had become the Foundation's first female division director. Communicating on an equal basis with the men in the upper ranks of the NSF hierarchy would have been difficult for any woman at the time. Moreover, oriented by her own research to biochemistry and biophysics, Clark seemed to those in systematics and ecology to lack a broad appreciation of the spectrum of the biological sciences. Unlike Mehl, a biochemist who was admired by ecologists as well as molecular biologists, Clark was unable to bridge the ongoing "biology gap." Many BMS colleagues felt that she lacked the style of forceful leadership that might have compensated for a period of budget decline. Where boldness of vision and action was arguably called for, Clark was perceived as a micromanager who had difficulty delegating authority or making quick decisions. In this she differed radically from her predecessors, Harve Carlson and John Wilson.[13]

In the period 1968–73, BMS lost many of the structures and practices that had contributed to its former sense of unity and mission. After 1968, the division director no longer submitted synthetic annual reports. Program directors stopped writing annual reports after 1972. That same year, advisory committee meetings ceased; no longer was there a focal point for staff or ad hoc committee reports on funding issues. After the departure of McElroy in 1972, even the internal program reviews of biology and medicine, which also contributed to a sense of cohesion, were soon ended. Thus, the whole system of planning for biology at the program and division levels with the aid of advice from the scientific community had been dismantled, and the teamwork that had characterized BMS since 1952 disappeared. Without these institutional structures for joint planning, the division lost an overall sense of purpose. Each program became a separate entity without integration into a larger whole.

In the 1970s, BMS experienced mounting difficulty clarifying its mission in the face of NIH domination of biology. Divisional spokespersons found themselves increasingly called upon to defend NSF's role in biology both to physical scientists and administrators at NSF and to members of Congress. After a few years of stagnant or declining budgets, NIH appropriations began to rise again.[14] From an outsider's perspective, it appeared that the biological sciences were well supported compared to the physical sciences. John Mehl reported that in 1971 HEW supported "a rather surprising 46 to 47 percent of all research obligations to colleges and universities." The life sciences, including clinical medicine, represented over 50 percent of federal funds to universities. The problem, he pointed out while serving as acting director of BMS in 1973, was that "Congress, the Office of Management and Budget, or scientists other than biologists are likely to contrast fields of science using this total base of the life sciences." Even if consideration were limited to academic research, with clinical medicine excluded, NIH's share in 1971 was 63 percent compared to NSF's 12 percent. NIH's primacy, Mehl claimed, had the unfortunate effect of strongly biasing biology toward medical schools and research on humans and vertebrate models, while large areas, including plants, invertebrates, and ecology, were neglected. The question of whether NSF should concentrate its resources on areas of biology with little biomedical concern or whether it should continue to cofund with NIH major investigators in molecular and cellular biology became increasingly acute as the ratio of proposals funded to proposals submitted continued to drop.[15]

In the 1970s, for the first time, NSF biologists perceived NIH as a serious problem. Especially after Senator William Proxmire became chair of the appropriations subcommittee overseeing NSF in late 1972, NSF biology was placed on the defensive. Proxmire, who won notoriety for his Golden Fleece Awards, subjected NSF programs to severe scrutiny for supposed wasting of public money and for "duplication."[16] Why should NSF fund biology at all if NIH was already providing so much money for the life sciences? Each year, NSF had to defend before Congress the distinctiveness of NSF support of biology in relationship to NIH's funding.[17]

In fact, NIH biology *was* different from NSF biology. Dominance by NIH in the biological sciences, while less problematic than military dominance of the physical sciences in the 1950s and 1960s, has nonetheless had the consequence of skewing biological research toward areas that are health-related. NIH, because of its mission to promote research on health and disease, funded some areas of biology heavily while ignoring others. Those areas supported by NIH were relatively better funded overall than those areas—plants, invertebrates, ecology, and systematics, for example—that relied heavily on NSF. Historically, as this book has shown, the two agencies have had very different organizational structures and funding philosophies, which also contributed to differential support of the biological sciences.

Funding patterns have likely had a part in the alignments and realignments of life science departments on university campuses. Both agencies, by bypassing the boundary between botany and zoology in their grants programs, contributed to the mergers of botany and zoology to form departments of biology in the 1960s. Since the 1960s, there has emerged a new split in academic biology in large research universities: that between molecular and cellular biology on the one hand and systematics, ecology, and evolutionary biology on the other. Molecular and cellular biology have called for large laboratories that receive correspondingly large grants year after year, primarily from NIH. Systematics and ecology have survived with smaller and more sporadic individual grants and rely heavily on NSF and agencies other than NIH. The two areas are unable to coexist peacefully in united biology departments, because molecular and cellular biology would always dominate in terms of resource allocation. Since the late sixties, universities have been creating separate departments for what has been variously called organismic biology, integrative biology, or evolution and ecology. This realignment of boundaries of the life sciences in academia thus corresponds not just to differences in biological philosophy and practice, but also to differences in federal funding of biology.[18]

In 1975, the Foundation underwent a major organizational reformation. In place of basic research divisions under an assistant director for research, NSF was reconfigured into directorates headed by presidentially appointed assistant directors who would have direct and equal access to the director of NSF, then H. Guyford Stever. While the reasons for the change were complex, one major factor was the feeling that the Office of the Assistant Director for Research, which had overseen the research divisions, had acquired too much power. In the turf battles that accompanied the successive years of tight budgets, it was blamed for favoring some sciences (particularly the physical sciences) over others. In the reorganization, biology became part of the new Biological, Behavioral, and Social Sciences Directorate (BBS). "Medical" was absent from the title; many thought the term should have been removed long ago. As the social sciences represented a small and politically vulnerable part of NSF, they were placed with the biological sciences in part to protect them. The history of biology at NSF for the next sixteen years would be linked to that of the social sciences.[19]

To many biologists, the demise of the assistant director for research and the splitting of BMS into three divisions (Cellular and Molecular Biology, Environmental Biology, and Behavioral and Neural Sciences) was a welcome sign of empowerment and expansion.[20] However, the new organization gave little identity to biology at NSF. Biology was now represented by two-and-a-half divisions out of four in a directorate that also included nonbiological sciences (anthropology and social psychology were placed in the Division of Behavioral and Neural Sciences). The biological divisions did not cooperate well with each other let alone with the politically suspect and very heterogeneous Social Sciences Division. Although the social sciences intersected with the biological sciences at some points, especially psychobiology and physical anthropology, for the most part such fields as economics, law and social sciences, linguistics, political sciences, social and developmental psychology, and sociology were unrelated to the biological sciences in terms of both graduate training and research methodology. Joining them to biology was a final blow to the former goal of a unified biology at NSF. From 1976 on, there was no Foundation entity corresponding to biology. With the creation of BBS, biology at NSF entered into a new set of organizational re-lationships and a very different era.

In 1991, after the recommendation of a task force on BBS convened to advise on priorities for the directorate, the social sciences were split off to form a separate directorate.[21] With this transformation, biology at NSF has in a sense come full circle from where it began in 1950 as the Division of Biological Sciences. Now that there is a Directorate for Biological Sciences, shorn of past links to either the medical sciences or the social sciences, a renewed era of planning at the level of biology is possible. With the appointment of Rita Colwell as director of NSF in 1998—the first woman and second biologist to head the agency[22]—biology is once more at the forefront at NSF.

Program Officers, 1951–1975, Division of Biological and Medical Sciences

Assistant Director for Biological and Medical Sciences, 1951–1964
Division Director, Division of Biological and Medical Sciences, 1964–1975

1951–52	John Field II (Biological Sciences and Acting Director, Medical Research)
1952–53	Fernandus Payne
1953–54	H. Burr Steinbach
1954–55	Lawrence R. Blinks
1955–61	John T. Wilson
1961–72	Harve J. Carlson
1972–73	John Mehl (acting)
1973–75	Eloise Clark

Deputy Assistant Director/Deputy Division Director

1956–60	Louis Levin
1960–61	Harve J. Carlson
1961–67	David D. Keck
1967–68	Eugene Hess (acting)
1968	Herman Lewis (acting)
1968–75	John W. Mehl

Special Assistant

1960–62   Dixy Lee Ray (for Oceanography)
1963–64   M. Dale Arvey (for Inland Field Stations)

Planning Officer

1965–75   William J. Riemer

Program Directors, 1952–1964

For much of this period, BMS was organized into eight programs. They are listed below with their predecessors.

MOLECULAR BIOLOGY

Molecular and Genetic Biology, 1952–1954
Molecular Biology, 1954–1963

1952–59   William V. Consolazio (joined BMS in 1951)
1959–60   Francis Reithel (Consolazio on sabbatical)
1960–61   William V. Consolazio
1962–63   John W. Mehl
1963–64   Samuel Aronoff

DEVELOPMENTAL BIOLOGY

Developmental, Environmental, and Systematic Biology, 1952–1954
Genetic and Developmental Biology, 1954–1956
Developmental Biology, 1957–1964

1952–53   Frank H. Johnson
1953–54   Hubert B. Goodrich
1954–56   Margaret C. Green (acting)
1956–57   Vernon Bryson
1957–58   Edgar Zwilling
1958–59   Nelson T. Spratt
1959–60   A. C. Clement
1960–62   Meredith N. Runner
1962–64   Philip Grant

GENETIC BIOLOGY

Molecular and Genetic Biology, 1952–1954 (see above)
Genetic and Developmental Biology, 1954–1956 (see above)
Genetic Biology, 1957–1964

1957–59	George Lefevre Jr.
1959–60	Dwight Miller
1960–62	Dean R. Parker
1962–64	Herman Lewis

REGULATORY BIOLOGY

Regulatory Biology and Microbiology, 1952
Regulatory Biology, 1953–1964

1952–57	Louis Levin
1958–59	Arthur W. Martin Jr.
1959–60	Roy P. Forster
1960–61	Paul J. Kramer
1961–64	David B. Tyler

ENVIRONMENTAL BIOLOGY

Developmental, Environmental, and Systematic Biology, 1952–1954
(see above)
Environmental Biology, 1954–1964

1954–56	George Sprugel Jr.
1957–58	Edgar Zwilling
1958–64	George Sprugel Jr.

SYSTEMATIC BIOLOGY

Developmental, Environmental, and Systematic Biology, 1952–1954
(see above)
Systematic Biology, 1954–1963

1954–56	William C. Steere
1956–57	Rogers McVaugh
1957–59	A. C. Smith
1959–62	David D. Keck
1962–64	Walter H. Hodge

METABOLIC BIOLOGY

Metabolic Biology, 1958–1964 (created from parts of Molecular
and Regulatory Biology)

1958–59	Lewis Levin (acting)
1959–60	Samuel J. Ahl
1960–61	Daniel Billen
1961–62	Aubrey W. Naylor
1962–64	Howard J. Teas

PSYCHOBIOLOGY, 1952–1964

1952–59	John T. Wilson
1959–64	Henry S. Odbert

Programs in BMS outside the functional organization of biology:

ANTHROPOLOGY AND RELATED SCIENCES, 1953–1957 (TEMPORARILY IN
BMS BEFORE CREATION OF SOCIAL SCIENCES IN 1957)

1953–57	Harry Alpert

BIOLOGICAL FACILITIES AND SPECIAL PROGRAMS, 1960–1964

1959–60	Harve Carlson
1961–64	Jack T. Spencer

Section and Program Officers, 1964–1975

The reorganization of BMS in 1964 created four sections, each of which contained a number of programs.

SECTION HEAD FOR CELLULAR BIOLOGY

1964	vacant
1965	vacant
1966–75	Herman W. Lewis

Developmental Biology Program

1964–66	Philip Grant
1967–68	Leonard Nelson
1968–69	Ursula K. Abbott

1969–71	Eloise E. Clark
1971–73	Richard W. Siegel
1973–74	E. W. Hanly
1974	William A. Jensen
1975	Antonie W. Blackler
1975	Melvin Spiegel

Genetic Biology Program

1964–73	Herman W. Lewis
1973–74	Margaret Lieb
1974–75	Rose M. Litman
1975	Laurence Berlowitz

Human Cell Biology

1973–75 Herman W. Lewis

SECTION HEAD FOR ENVIRONMENTAL AND SYSTEMATIC BIOLOGY, 1964–1969

Section Head for Ecology and Systematic Biology, 1969–1975
Section Head for Ecology and Population Biology, 1975

1964–65	Walter H. Hodge
1966–67	Robert K. Godfrey
1967–68	Edward S. Deevey Jr.
1968–69	Bostwick H. Ketchum
1969–70	John F. Reed
1971–72	Walter H. Hodge
1973–74	vacant
1975	John L. Brooks

Environmental Biology Program, 1964–1968
General Ecology Program, 1969–1975
Ecology Program, 1975

1964–65	John S. Rankin Jr.
1965–66	John E. Cantlon
1966–67	Robert F. Inger
1967–68	Edward S. Deevey Jr. (acting)

1968–69   Philip L. Johnson
1969   Bostwick H. Ketchum (acting)
1969–75   John L. Brooks

Systematic Biology Program, 1964–1975

1964–65   Walter H. Hodge (acting)
1966–67   Robert K. Godfrey (acting)
1967–68   Kenton L. Chambers
1968–69   Richard F. Johnston
1969–71   John O. Corliss
1971–72   William E. Sievers
1972–74   Donovan S. Correll
1974–75   William E. Sievers
1975   Jack R. Schultz

Biological Oceanography Program, 1968–1970
(continues outside BMS)

1968–69   Edward Chin (acting)
1969–70   Malvern Gilmartin

Ecosystem Analysis Program, 1969–1975
Ecosystem Studies Program, 1975

1969–72   Charles F. Cooper
1972–73   John M. Neuhold
1973–74   William E. Hazen
1974–75   James T. Callahan
1975   vacant

Biological Research Resources Program, 1974

1975   William E. Sievers

SECTION HEAD FOR MOLECULAR BIOLOGY, 1964–1974
BIOCHEMISTRY AND PHYSIOLOGY SECTION, 1975

1964–65   vacant
1966–69   Eugene L. Hess
1970–71   Abraham Eisenstark (acting)

1971–72   Sigmund R. Suskind
1972–74   Eloise E. Clark
1975   vacant

Biochemistry Program

1964–65   Walter L. Koltun
1965–66   R. Bruce Martin (acting)
1966–67   David W. Krogmann
1967–68   Albert Siegel
1968–69   Richard Y. Morita
1969–71   Abraham Eisenstark
1971–72   Sigmund R. Suskind (acting)
1972–73   Stuart W. Tanenbaum
1973–74   Roy L. Kisliuk
1974–75   Roy Repaske
1975   Walter D. Bonner

Biophysics Program

1964–65   Walter L. Koltum, acting
1965–66   R. Bruce Martin
1966–69   Eugene L. Hess (acting)
1970   vacant (asst dir Brenda C. Flam)
1971   Eloise E. Clark
1972–73   Eloise E. Clark (acting)
1974–75   Martin P. Schweizer

SECTION HEAD FOR PHYSIOLOGICAL PROCESSES, 1964–1974
BIOCHEMISTRY AND PHYSIOLOGY SECTION, 1975 (SEE ABOVE)

1964–72   David B. Tyler
1973–74   vacant

Regulatory Biology Program

1964–72   David B. Tyler (acting)
1973–74   Frank P. Conte
1974–75   James W. Campbell
1975   Jack W. Hudson Jr.

Metabolic Biology Program

1964–65	John M. Ward
1965–66	Eugene L. Hess
1966–67	John E. Nellor
1967–68	Seymour Katsh
1968–69	Sidney Solomon
1969–75	Elijah B. Romanoff

OTHER BMS PROGRAMS NOT INCLUDED IN SECTIONS

Psychobiology Program

1964–66	Henry S. Odbert
1966–67	John F. Hall
1967–74	Henry S. Odbert
1974–75	Jacob Beck
1975	David Birch

Neurobiology Program

1971–75	James H. Brown

Facilities and Special Programs

1964–68	Jack T. Spencer

Members of Divisional and Advisory Committees, Biological and Medical Sciences, 1952–1972

Divisional Committee for Biological Sciences, 1952–1953

Marston Bates
George Beadle
Donald P. Costello
Wallace Fenn, Chairman 1953
Jackson W. Foster
Frank A. Geldard, added in 1953
Theodor K. Just
John S. Nicholas
Hubert B. Vickery
Douglas N. Whitaker, Chairman 1952

Divisional Committee for Medical Research, 1953

Frank Brink
Bernard D. Davis
Edward W. Dempsey
Ernest W. Goodpasture
Severo Ochoa
Dickinson W. Richards
George Wald
Arnold D. Welch

Divisional Committee for Biological and Medical Sciences,
1954–1965 (in order of date joined)

Edgar Anderson, 1954–58
Marston Bates, 1954–58; Chairman, 1956–58
Frank Brink Jr., 1954–60
Bernard D. Davis, 1954–58
Jackson Foster, 1954–55
Frank A. Geldard, 1954–60
Ernest Goodpasture, 1954–55
George Wald, 1954–58; Chairman, 1954–55
Arnold D. Welch, 1954–55
H. Albert Barker, 1955
William F. Hamilton, 1956–58
C.N.H. Long, 1956–62; Chairman, 1960–61
C. Phillip Miller, 1956–60
Esmond E. Snell, 1956–58
Lincoln Constance, 1958–63
William D. McElroy, 1958–60
C. P. Oliver, 1958–62
Frank W. Putnam, 1958–63
Kenneth V. Thimann, 1958–63; Chairman, 1961–63
H. Burr Steinbach, 1958–62; Chairman, 1958–59
Horace Davenport, 1960–62
Philip Handler, 1960–62
Lyle H. Lanier, 1960–62
Lawrence R. Blinks, 1962–66
James D. Ebert, 1962–66
William D. Lotspeich, 1962–66
Conrad G. Mueller Jr., 1962–66
Marcus Rhoads, 1962–66
René Dubos, 1962
E. A. Evans Jr., 1962–66
Theodore H. Bullock, 1963–65
Harry Eagle, 1963–67
Ernst Mayr, 1963–66; Acting Chairman, 1965; Chairman, 1966
Paul J. Kramer, 1963–66; Chairman, 1964–65, 1966

Advisory Committee for Biological and Medical Sciences,
1965–1972

Paul J. Kramer, 1965–66; Chairman, 1965

Lawrence R. Blinks, 1965–66

Theodore H. Bullock, 1965

Harry Eagle, 1965–67; Chairman, 1967

James D. Ebert, 1965–66

E. A. Evans Jr., 1965–66

William R. Lotspeich, 1965–66

Ernst Mayr, 1965–66; Chairman, 1966

Conrad G. Mueller Jr., 1965–66

Charles Olmsted, 1965–67

Marcus Rhoads, 1965–66

Chauncey G. Goodchild, 1966–67

William D. Neff, 1966–68

Arthur Geoffrey Norman, 1966–67

Ray D. Owen, 1966–68; Chairman, 1968

Edgar Zwilling, 1966–68

W. Frank Blair, 1967–69

Edward W. Fager, 1967–69

Carl Gottschalk, 1967–69; Chairman, 1969

David M. Gates, 1967–70

J. Woodland Hastings, 1968–70; Chairman, 1970

Charles G. Sibley, 1968–70

Anton Lang, 1968–70

John W. Saunders, 1969–71; Chairman, 1971

Richard L. Solomon, 1969–71

Norton D. Zinder, 1969–71

Nelson G. Hairston, 1970–72; Chairman, 1972

Lawrence L. Pomeroy, 1970–72

George Sayers, 1970–72

L. Sterling Wortman, 1970–72

Robert S. Bandurski, 1971–72

David W. Krogman, 1971–72

Vernon B. Mountcastle, 1971–72

Donald D. Brown, 1972

Michael R. D'Amato, 1972

Charles Yanofsky, 1972

Notes

Unless otherwise stated, all references to manuscript holdings in the National Archives and Records Administration (NARA) are to Record Group 307, The National Science Foundation. Accession numbers and box numbers of frequently cited records, such as program annual reports, are in the Note on NSF Primary Sources. These records, formerly in the Washington National Records Center, were recently moved to NARA. The accession and box numbers are those in effect at the time I consulted this material.

Abbreviations

Ad hoc Comm. Repts.	Notebook, "Ad hoc Committee Reports, 1954-61," Box 3, 76-172, NARA.
Adv. Comm.	Notebook, "Minutes, BMS Advisory Committee Meetings, 1965-2/1972," Box 1, 76-172, NARA.
Adv. Comm. Ann. Repts.	Notebook, "Advisory Committee Annual Reports to NSB, 1965-71," Box 1, 76-172, NARA.
Adv. Panel, 1952	"Advisory Panel Meetings, January & April 1952," Box 20, 70-2191, NARA.
APS Archives	American Physiological Society Archives, Bethesda, Md.
ATW Diary Notes	Alan T. Waterman Diary Notes [filed chronologically], NSF Historian's Files, NARA.
ATW Notes	Alan T. Waterman Notes [filed chronologically], NSF Historian's Files, NARA.
BMS, DSF 1964	"Biological and Medical Sciences," [Director's Subject Files, 1964], Box 3, 70A-3621, NARA.
BMS, DSF 1965	"Biological and Medical Sciences—1965," [Di-

	rector's Subject Files, 1965], Box 14, 70A-3621, NARA.
BMS, DSF 1966	"Biological & Medical Sciences (BMS) 1966," [Director's Subject Files, 1966], Box 22, 70A-3621, NARA.
BMS, DSF 1968	"Biological & Medical Sciences (BMS) 1968," [Director's Subject Files, 1968], Box 1, 74-038, NARA.
BMS, DSF 1969	"BMS," [Director's Subject Files, 1969], Box 1, 75-052, NARA.
BMS, DSF 1970	"BMS," [Director's Subject Files, 1970], Box 1, 75-053, NARA.
BMS, DSF 1971	"BMS," [Director's Subject Files, 1970] Box 1, 77-056, NARA.
BMS, DSF 1972	"BMS–Biological & Medical Sciences," [Director's Subject Files, 1972], Box 1, 77-080, NARA.
BMS, DSF 1974	"BMS," [Director's Subject Files, 1974], Box 1, 79-010, NARA.
BMS Faculty Salaries	"BMS Faculty Salaries," Box 15, 72A-1808, NARA.
Bush Papers	Vannevar Bush Papers, Library of Congress.
Div. Comm., 1952-53	"Biological Sciences Divisional Committee 1st–7th Meetings (1952–53)," Box 20, 70A-2191, NARA.
Div. Staff Papers I & II	Two notebooks, "Division Staff Papers," 1953–70, Box 3, 76-172, NARA.
HF	NSF Historian's Files, NARA.
Levin Diary Notes	Louis Levin Diary Notes, 1952–59, Box 6, 70A-3195, NARA.
NAHC Minutes	Minutes of Meetings of the National Advisory Health Council, 1945–60.
NAS Archives	National Academy of Sciences Archives, Washington, D.C.
NLM	Manuscripts, National Library of Medicine, Bethesda, Md.
NRL HF	Naval Research Laboratory Historian's Office Files.
NSB Minutes	National Science Board Minutes, copies in the National Science Board Office, NSF, and in the NSF Library.

NSF Act of 1950	National Science Foundation Act of 1950, Public Law 507, 81st Cong., 2d sess.
NSF Annual Report	These reports were published for each fiscal year by the GPO, Washington, D.C.
ODSF	Office of the Director Subject Files [Alan T. Waterman], NARA.
OTS Papers	Organization for Tropical Studies Papers, Duke University Archives.
Prog. Ann. Repts., 1952–57	Notebook "Division and Program Annual Reports (FY 52–FY 57)," Box 1, 76-172, NARA.
Prog. Ann. Repts., 1958–60	Notebook "Division and Program Annual Reports (FY 58–FY 60)," Box 1, 76-172, NARA.
Prog. Ann. Repts, 1961–68	Notebooks "Division and Program Annual Reports," Box 2, 76-172, NARA.
RAC	Rockefeller Archive Center, North Tarrytown, N.Y.
RG	Record Group.
Rodman Files	Files lent to author by James Rodman, NSF Program Director.
Steinbach Diary Notes	H. Burr Steinbach Diary Notes, 1953–54, Box 5, 72A-1808, NARA.

Introduction   Envisioning a Federal Patron for Biology

1. Alan T. Waterman, "Federal Support of Fundamental Research in the Biological Sciences," *AIBS Bulletin* 1(5) (October 1951): 11–17, esp. 11, 14. Waterman's purpose was also to garner support for securing from Congress a budget that would get the grants program off the ground.

2. See chap. 5, n. 4.

3. See notes to chaps. 5, 7, 8, 9.

4. See chap. 1, nn. 7 and 10, and also Lily E. Kay, *The Molecular Vision of Life: Caltech, the Rockefeller Foundation, and the Rise of the New Biology* (New York: Oxford University Press, 1993).

5. Nathan Reingold, "Science and Government in the United States Since 1945," *History of Science* 32 (1994): 361–86, esp. 367–69; Roger L. Geiger, *Research and Relevant Knowledge: American Research Universities Since World War II* (New York: Oxford University Press, 1993), 16–17, 168–73.

6. Waterman, "Federal Support of Fundamental Research," 13.

7. Vannevar Bush, *Science—The Endless Frontier* [1945] (Washington, D.C.: National Science Foundation, 1990).

8. Waterman, "Federal Support of Fundamental Research," 14.

9. Daniel J. Kevles, *The Physicists: The History of a Scientific Community in Modern America* (New York: Knopf, 1977).

Chapter 1   *Making a Place for Biology at the "Endless Frontier," 1945–1950*

1. On the 1930s debate over federal patronage of science, see Roger L. Geiger, *To Advance Knowledge: The Growth of American Research Universities, 1900–1940* (New York: Oxford, 1986), 253–64; Carroll W. Pursell Jr., "The Anatomy of a Failure: The Science Advisory Board, 1933–1935," *Proceedings of the American Philosophical Society* 109 (1965): 342–51; Lewis E. Auerbach, "Scientists in the New Deal: A Pre-War Episode in the Relations Between Science and Government in the United States," *Minerva* 4 (1965): 457–62; and A. Hunter Dupree, *Science in the Federal Government: A History of Policies and Activities to 1940* (New York: Harper & Row, 1964 [c. 1957]), 350–61.

2. Daniel J. Kevles, *The Physicists: The History of a Scientific Community in Modern America* (New York: Knopf, 1977).

3. Charles E. Rosenberg, "Rationalization and Reality in Shaping American Agricultural Research, 1875–1914," in Nathan Reingold, ed., *The Sciences in the American Context: New Perspectives* (Washington: Smithsonian Institution Press, 1979), 143–63; Charles E. Rosenberg, *No Other Gods: On Science and American Social Thought* (Baltimore: Johns Hopkins University Press, 1976), 135–95; Margaret Rossiter, "The Organization of the Agricultural Sciences," in Alexandra Oleson and John Voss, eds., *The Organization of Knowledge in Modern America, 1880–1920* (Baltimore: Johns Hopkins University Press, 1979), 211–48; and Diane B. Paul and Barbara A. Kimmelman, "Mendel in America: Theory and Practice, 1900–1919," in *The American Development of Biology*, 281–310.

4. Pursell, "Anatomy of a Failure," 347–48.

5. John P. Swann, *Academic Scientists and the Pharmaceutical Industry: Cooperative Research in Twentieth-Century America* (Baltimore: Johns Hopkins University Press, 1988).

6. George W. Gray, "The Rockefeller Foundation and the Biological Sciences," *AIBS Bulletin* 4(1) (January 1954): 13–15.

7. Robert E. Kohler, *Partners in Science: Foundations and Natural Scientists, 1900–1945* (Chicago: University of Chicago Press, 1991); Kohler, "Science, Foundations, and American Universities in the 1920s," *Osiris*, 2nd series, 3 (1987): 135–64; Kohler, "A Policy for the Advancement of Science: The Rockefeller Foundation, 1924–29," *Minerva* 16 (1978): 480–515; Kohler, "Warren Weaver and the Rockefeller Foundation Program in Molecular Biology: A Case Study in the Management of Science," in Reingold, ed., *The Sciences in the American Context,* 249–93; Gray, "Rockefeller Foundation," 13–15; Geiger, *To Advance Knowledge,* 160–67.

8. Warren Weaver to Chester I. Barnard, "NS Program," 1 January 1952, Folder 14, Box 2, Series 915, RG 3, RAC.

9. Gray, "Rockefeller Foundation," 13–15.

10. Kohler, "Warren Weaver," 250; Warren Weaver, "Molecular Biology: Origin of the Term," *Science* 170 (6 November 1970): 581–82. For a contesting view on the Rockefeller Foundation and the development of "molecular biology," see Pnina Abir-Am, "The Discourse of Physical Power and Biological Knowledge in the 1930s: A Reappraisal of the Rockefeller Foundation's 'Policy' in Molecular Biology," *Social Studies of Science* 12 (1982): 341–82, and responses in vol. 14 (1984): 225–63.

11. Kohler, "Warren Weaver." Weaver's diary notes are in bound volumes at RAC.

12. Bush first headed the National Defense Research Committee in 1940; it became the more inclusive OSRD in 1941. On the OSRD, see Kevles, *The Physicists,* 293–301; Irvin Stewart, *Organizing Scientific Research for War: The Administrative History of the Office of Scientific Research and Development* (Boston: Little, Brown, 1948); and Daniel Lee Kleinman, *Politics on the Endless Frontier: Postwar Research Policy in the United States* (Durham, N.C.: Duke University Press, 1995), 52–73.

13. Stewart, *Organizing Scientific Research for War,* 98–119; and Senate Committee on Military Affairs, Subcommittee, *Hearings on Science Legislation (S. 1297 and Related Bills),* 79th Cong., 1st sess., 22 October 1945, 455–67.

14. See Toby A. Appel, "Organizing Biology: The American Society of Naturalists and its 'Affiliated Societies,' 1883–1923," in Ronald Rainger, Keith Benson, and Jane Maienschein, eds., *The American Development of Biology* (Philadelphia: University of Pennsylvania Press, 1988), 87–120.

15. Gordon Alexander, "The College Curriculum in Wartime and Introductory Courses in Biology," *Science* 99 (28 January 1944): 78–80; C. A. Shull, "General Biology," *Science* 99 (10 March 1944): 199; Leland H. Taylor, "Is Biology a Science?" *Science* 99 (5 May 1944): 364–65; and Maurice B. Visscher, "Basic Biology and General Education," *Science* 99 (12 May 1944): 383–84.

16. On biochemistry, see Robert E. Kohler, *From Medical Chemistry to Biochemistry: The Making of a Biomedical Discipline* (Cambridge: Cambridge University Press, 1982), 158–92. On systematics, see David L. Hull, *Science as a Process: An Evolutionary Account of the Social and Conceptual Development of Science* (Chicago: University of Chicago Press, 1988); and V. B. Smocovitis, "Unifying Biology: The Evolutionary Synthesis and Evolutionary Biology," *Journal of the History of Biology* 25 (1992): 1–65; Joseph Allen Cain, "Common Problems and Cooperative Solutions: Organizational Activity in Evolutionary Studies, 1936–1947," *Isis* 84 (1993): 1–25; and V. B. Smocovitis, "Organizing Evolution: Founding the Society for the Study of Evolution (1939–1950)," *Journal of the History of Biology* 27 (1994): 241–309.

17. Robert F. Griggs, "The Organization of Biology and Agriculture," *Science* 96 (18 December 1942): 545–51, esp. 545; Douglas M. Whitaker to Frank B. Jewett, "Memorandum on increasing the effectiveness of the Biological Sciences in the war," 13 May 1943, Jewett to Vannevar Bush, 24 May 1943, Bush to Jewett, 1 June 1943, 50.713 N.R.C., Divisions of NRC, NAS Archives.

18. Senate Committee, *Hearings on Science Legislation,* 24 October 1945, 602.

19. Robert F. Griggs, "Shall Biologists Set Up a National Institute?" *Science* 105 (30 May 1947): 559–65, esp. 560, 564–65.

20. Griggs to Members of the Division of Biology and Agriculture, "B&A/Future Activities & Needs: Discussion," 19 October 1945, NAS Archives.

21. "Editorial," *American Naturalist* 80 (1 January 1946): 5–16; Jane M. Oppenheimer, "John Spangler Nicholas, March 10, 1895–September 11, 1963," in National Academy of Sciences, *Biographical Memoirs* 40 (1969): 239–89, esp. 273; "Constitution of the American Biological Society," 23 December 1940 draft, and A. P. Hitchens to Nicholas, 8 January 1941, Folder 44, Box 6, Series II, MS 929, John S. Nicholas Papers, Manuscripts and Archives, Yale University; D. H. Wenrich to Griggs, 23 August 1945, "B&A/Inst./Union of American Biological Societies: General, 1940–1946," NAS Archives; "A Society for All Biologists," *Science* 104 (4 October 1946): 325–26; and "American Society of Professional Biologists," *Science* 105 (11 April 1947): 384.

22. Griggs, "Organization of Biology and Agriculture," 547; Robert Chambers and J. S. Nicholas, "Proposal for an American Institute of Biology," *Science* 103 (7 June 1946): 692; Griggs, "Shall Biologists Set Up National Institute?"; Oppenheimer, "Nicholas," 273–74; "NRC News," *Science* 107 (12 March 1948): 267; Clarence J. Hylander, "The American Institute of Biological Sciences: A Historical Resume," *AIBS Bulletin* 1(1) (January 1951): 6–7, and 1(2) (April 1951): 13–15; and Griggs, "Report to the Executive Board of the Activities of the Division of Biology and Agriculture, November 18 to December 20, 1946," B&A/General/1946, NAS Archives.

23. Vannevar Bush, *Science—The Endless Frontier* [1945] (Washington, D.C.: NSF, 1990). On the background and interpretation of the Bush report, see J. Merton England, "Dr. Bush Writes a Report: 'Science—The Endless Frontier,'" *Science* 191 (9 January 1976): 41–47; England, *A Patron for Pure Science: The National Science Foundation's Formative Years, 1954–57* (Washington, D.C.: NSF, 1982), 3–23; Daniel J. Kevles, "The National Science Foundation and the Debate over Postwar Research Policy, 1942–1945: A Political Interpretation of *Science—The Endless Frontier,*" *Isis* 68 (1977): 5–26; Kevles, *The Physicists,* 343–48; Kevles, "Principles and Politics in Federal R&D Policy, 1945–1990: An Appreciation of the Bush Report," in Bush, *Science—The Endless Frontier,* ix–xxv; and Nathan Reingold, "Vannevar Bush's New Deal for Research: or The Triumph of the Old Order," *Historical Studies in the Physical Sciences* 17 (1987): 299–344; Kleinman, *Politics on the Endless Frontier,* 74–99. Roosevelt's letter is reprinted in Bush, *Science—The Endless Frontier,* 3–4.

24. Bush, *Science—The Endless Frontier,* 6, 18–19. The Bowman Committee preferred the term "pure research" to Bush's "basic research" (81–82).

25. Bush, *Science—The Endless Frontier,* 13–40, proposed organization chart, 36. The statement often attributed to Bush, that applied research drives out basic, was

made by Isaiah Bowman's Committee on Science and the Public Welfare: "Under the pressure for immediate results, and unless deliberate policies are set up to guard against this, *applied research invariably drives out pure*" (83).

26. Ibid., 34–40.

27. Ibid., 2, 46–69, esp. 51, 61.

28. Ibid., 1, 37, 24, 43–45, 56, 64n; F. R. Moulton, "Discussion: The Bush Report and Senate Bills," *Science* 102 (12 October 1945): 382–83.

29. On the legislative debate over NSF, see England, *A Patron for Pure Science,* 2–110; Kevles, *The Physicists,* 343–48, 356–58; Kevles, "The National Science Foundation and the Debate over Postwar Research Policy"; and Kleinman, *Politics of the Endless Frontier,* 100–144. Kleinman, in discussing the delay in creating NSF, emphasizes the "permeability" of the state, division between legislative and executive power, and the undisciplined nature of political parties, which allowed individuals to have a greater impact on policy. See also Lyman Chalkley, "Prologue to the National Science Foundation" [1951], HF, especially for chronology. On the social sciences, see Otto N. Larsen, *Milestones and Millstones: Social Science at the National Science Foundation, 1945–1991* (New Brunswick, N.J.: Transaction Publishers, 1992), esp. 1–18; and Mark Solovey, "The Politics of Intellectual Identity and American Social Science, 1945–1970," (Ph.D. diss., University of Wisconsin–Madison, 1996), esp. 50–104.

30. For the Magnuson bill, see S. 1285, 79th Cong., 1st sess., 19 July 1945; for the Kilgore bill, see S. 1297, 79th Cong., 1st sess., 23 July 1945, in Notebook, "Legislative History N.S.F., 78th, 79th, 80th Congress, 1943–1947 (Committee prints and bills)," HF.

31. Griggs, "Shall Biologists Set Up a National Institute?" 561.

32. Bowman's Committee Supporting the Bush Report included such notable biologists as George Beadle, Detlev Bronk, Edward A. Doisy, Linus Pauling, A. N. Richards, Homer W. Smith, Lewis H. Weed, and Warren Weaver. Among the biologists who supported the Committee for a National Science Foundation were Robert Chambers, E. G. Conklin, Max Demerec, Theodosius Dobzhansky, L. C. Dunn, Ralph W. Gerard, Harry Grundfest, Michael Heidelberger, Barbara McClintock, H. Burr Steinbach, Maurice B. Visscher, and Selman Waksman. England, *Patron for Pure Science,* 36–38; Howard A. Meyerhoff, "Science Legislation and the Holiday Recess," *Science* 103 (9 January 1946): 10–11; "Committee for a National Science Foundation," *Science* 103 (11 January 1946): 45, 62–63.

33. Senate Committee, *Hearings on Science Legislation,* 23 Oct. 1945, 472–73, 477, 503–4, 506–10, 1036, esp. 510. For testimony of medical scientists, see 455–537, 1032–38. (Weed's testimony was not given until 2 November.)

34. Ibid., 543. On Dunn, see Bentley Glass, *A Guide to the Genetics Collections of the American Philosophical Society* (Philadelphia: American Philosophical Society, 1988), 38–40.

35. John S. Nicholas and Robert Chambers, "Biological Cooperation," *American Naturalist* 80 (1 February 1946): 113–15; Robert Chambers, "Further Notes on Science Legislation," *Science* 103 (8 February 1946): 160–61; Senate Committee, *Hearings on Science Legislation,* 24 October 1945, 539–609. Among the biologists testifying on 24 October were Dunn, Bronk, Edmund W. Sinnott, Wendell Stanley, and H. Burr Steinbach.

36. Senate Committee, *Hearings on Science Legislation,* 24 October 1945, 540, 541.

37. Ibid., 563–70, esp. 567, 568.

38. Paul and Kimmelman, "Mendel in America," 296–302. Recent historians have pointed out that hybrid corn was a mixed blessing. Forcing farmers to purchase new seed each year, it chiefly benefited the large seed companies. See Deborah Fitzgerald, *The Business of Breeding: Hybrid Corn in Illinois, 1890–1940* (Ithaca: Cornell University Press, 1990), 223.

39. Senate Committee, *Hearings on Science Legislation,* 24 Oct. 1945, 549–50, 559, 561, 569–70, 573, esp. 549, 559, 569–70; Chambers, "Further Notes on Science Legislation," 160–61. See also Chambers, "Recent Developments in Science Legislation," *American Naturalist* 80 (1946): 17–18.

40. See Kilgore's revision, S. 1720, 79th Cong., 1st sess., 21 December 1945, and joint Kilgore-Magnuson bill, S. 1850, 79th Cong., 2nd sess., 21 February 1946.

41. Chambers, "Further Notes on Science Legislation"; Chambers and Nicholas, "Pending Legislation for Federal Aid to Science," *Science* 102 (30 November 1945): 545–48; Chambers, "A United Front for S. 1850," *Science* 103 (21 June 1946): 726.

42. Committee Supporting the Bush Report, "Statement Concerning S. 1850," *Science* 103 (3 May 1946): 558; Howard A. Meyerhoff, "Obituary: National Science Foundation, 1946," *Science* 104 (2 August 1946): 97–98.

43. Public Law 588, 79th Cong., 2nd sess., approved August 1, 1946. In June 1946, Weaver noted that H. M. MacNeille, head of ORI's London office, asked for advice on the premise that "O.R.I. is, in effect, acting as a National Research Foundation." See Warren Weaver, Diaries 1941–46, 200, RAC. The connection was made in print by the end of 1946. See Philip N. Powers (Scientific Personnel Branch, ONR), "A National Science Foundation?" *Science* 104 (27 December 1946): 614–19.

44. Harvey M. Sapolsky, *Science and the Navy: The History of the Office of Naval Research* (Princeton, N.J.: Princeton University Press, 1990), esp. 37–56.

45. Sapolsky, *Science and the Navy,* 9–36, 129, 132 (Table A-1); Harvey M. Sapolsky, "Academic Science and the Military: The Years Since the Second World War," in Reingold, ed., *Sciences in the American Context,* 379–99; "Presentation of ONR Program to the Bureau of Ships," 2 June 1947, NRL HF.

46. Warren Weaver, Diaries 1941–46, 201, RAC.

47. Sapolsky, *Science and the Navy,* 37–56; John E. Pfeiffer, "Office of Naval Research," *Scientific American* 180(2) (February 1949): 11–15.

48. William V. Consolazio, "On Supporting Fundamental Research," December 1949, reprint from *Research Report of the Office of Naval Research,* 1 February 1950, 4, HF.

49. Minutes, Naval Research Advisory Committee, 1st meeting, 14 October 1946, 61, NRL HF.

50. Ibid., 50.

51. Pfeiffer, "Office of Naval Research," 14; Sapolsky, *Science and the Navy,* 53.

52. Minutes, Naval Research Advisory Committee, 1st meeting, 14 October 1946, 48–52.

53. Orr E. Reynolds, "Support of Biological Sciences by the Office of Naval Research," *AIBS Bulletin* 2(2) (April 1952): 18–20, esp. 20.

54. Minutes, Naval Research Advisory Committee, 1st meeting, 14 October 1946, 51; Pfeiffer, "Office of Naval Research," 14; Reynolds, "Support of Biological Sciences," 20.

55. Orr E. Reynolds, interview by author, 12 July 1988. Consolazio explained the need for panels: "We admit that the fast moving frontiers of science require the best scientific minds to keep abreast of developments within their own fields. By selecting an advisory panel of well qualified and objective specialists, this inadequacy can be partially corrected." Consolazio, "On Supporting Fundamental Research," 4.

56. Orr E. Reynolds to John Olive, 17 December 1971, Reynolds membership file, APS Archives, Bethesda, Md.

57. Sapolsky, *Science and the Navy,* 43; Wayne Gruner (former ONR physical sciences program officer who transferred to NSF), interview by author, 2 and 9 April 1991; Consolazio, "On Supporting Fundamental Research," 4.

58. Reynolds interview; England, *Patron for Pure Science,* 262.

59. Sapolsky, *Science and the Navy,* 52–55; Warren Weaver, Diaries 1950, 235, RAC; Pfeiffer, "Office of Naval Research," 15.

60. On the AEC program in biology and medicine, see Richard G. Hewlett and Francis Duncan, *A History of the United States Atomic Energy Commission,* vol. 2, *Atomic Shield, 1947–1952* (University Park, Penn.: Pennsylvania State University Press, 1969), 112–15, 251–55; Paul B. Pearson, "The Biological Program of the Atomic Energy Commission," *AIBS Bulletin* 3(3) (July 1953): 17–19; Kevles, *The Physicists,* 349–52.

61. Hewlett and Duncan, *Atomic Shield,* Appendix 7.

62. Pearson, "Biological Program of AEC," 17.

63. *Science* 111 (7 April 1950): 372. I am indebted to John Beatty for information on the functioning of AEC peer review.

64. Public Law 507, 81st Cong., 2d sess., Sec. 15(a), 8; Annual Report, Genetic Biology, FY 1957, Prog. Ann. Repts., 1952–57.

65. "Quotations: Research," *Science* 102 (14 September 1945): 282–83; Stephen P. Strickland, *Politics, Science, and Dread Disease: A Short History of United States Medical Research Policy* (Cambridge, Mass.: Harvard University Press, 1972), 21–22, 26, 40–41, 48–49. On the founding of the NIH grants program, see also Victoria A. Harden, *Inventing the NIH: Federal Biomedical Research Policy, 1887–1937* (Baltimore: Johns Hopkins University Press, 1986); Donald C. Swain, "The Rise of a Research Empire: NIH, 1930–1950," *Science* 138 (14 December 1962): 1233–37; and Daniel M. Fox, "The Politics of the NIH Extramural Program, 1937–1950," *Journal of the History of Medicine* 42 (1987): 447–66. Valuable oral histories of grants program pioneers by Harlan Phillips in 1963 for historian of medicine George Rosen include Ernest M. Allen, Rolla E. Dyer, David E. Price, Leonard Scheele, and Cassius J. Van Slyke, NLM. I am indebted also to the unpublished work of Harry Marks, Institute of the History of Medicine, Johns Hopkins University.

66. Nathan Reingold, "Choosing the Future: The U.S. Research Community: 1944–1946," *Historical Studies in the Physical and Biological Sciences* 25 (1995): 301–28; Strickland, *Politics, Science, and Dread Disease,* 18–19.

67. Minutes of Meetings of the National Advisory Health Council, 1945–60 (hereafter NAHC Minutes), 19–20 June 1945, 11, 12, 14, in Box 2, RG 443, NIH Office of the Director, NARA; Ernest M. Allen, interview by Harlan Phillips, 3 April 1963, 13–14, NLM. An application from the New York State Health Department was not approved.

68. NAHC Minutes, 28 September 1945, 1–2.

69. "National Advisory Health Council and National Advisory Cancer Council of the Public Health Service, Joint Report on Proposals for a National Science Foundation," 28 September 1945, included in NAHC Minutes, 28 September 1945; S. 1850, 79th Cong., 2d Sess., 21 February 1946.

70. Senate Committee, *Hearings on Science Legislation,* 23 October 1945, 513–20, esp. 513, 514, 516.

71. Strickland, *Politics, Science, and Dread Disease,* 23; NAHC Minutes, 28 September 1945, 2. Dyer noted his acquisition of CMR contracts in his Senate testimony. Senate Committee, *Hearings on Science Legislation,* 519.

72. Strickland, *Politics, Science, and Dread Disease,* 29–30.

73. C. J. Van Slyke, "New Horizons in Medical Research," *Science* 104 (13 December 1946): 559–67, esp. 559; Charles V. Kidd, "The Federal Government and the Shortage of Scientific Personnel," *Science* 105 (24 January 1947): 84–88, esp. 86–87.

74. Strickland, *Politics, Science, and Dread Disease,* 53, 86.

75. Weaver to Barnard, 11 January 1952, RAC.

76. NAHC Minutes, 28 September 1945, and 8–9 March 1946; Van Slyke, "New Horizons," 561. Some study sections were oriented to specific diseases or groups of diseases, surgery, disease organisms, antibiotics, or medical specialties.

Others concerned the preclinical sciences found in medical schools such as physiology, biochemistry and nutrition, and metabolism and endocrinology.

77. Fox, "The Politics of the NIH Extramural Program."

78. Van Slyke, "New Horizons," 561.

79. Ernest M. Allen, interview, 19, NLM.

80. Consolazio, "On Supporting Fundamental Research," 3.

81. Martin Frank, "The History of the Physiology Study Section, 1946–1984," n.d., APS Archives; Richard Mandel, *Division of Research Grants, National Institutes of Health: A Half Century of Peer Review, 1946–1996* (Bethesda, Md.: NIH, 1996).

82. NAHC Minutes, 8–9 March 1946, and 10–11 May 1946.

83. Douglas M. Whitaker to Division of Biological Sciences, 12 August 1951, "Funds Needed by the NSF for Support of the Biological Sciences in the U.S.," Folder: "Division of Biological and Medical Sciences," Box 2, ODSF; Warren Weaver to Chester I. Barnard, 27 September 1951, "NS Program," Folder 14, Box 2, Series 915, RG 3, RAC. NSF estimated that in FY 1952, over three times as much was spent on intramural and extramural "basic research" in the "physical sciences" ($90.5 million) as in the "life sciences" ($27.0 million). See *Federal Funds for Science, II* (1953), 30–33.

84. "Proposed National Science Foundation Act, 1947: S. 526," *Science* 105 (14 February 1947): 191–94; Howard A. Meyerhoff, "The Truman Veto," and "News and Notes" *Science* 106 (12 September 1947): 236–39.

85. "S. 526, National Science Foundation Act of 1947," Legislative History N.S.F., 1943–47, HF; England, *Patron for Pure Science,* 76, 79, 80; George E. Wakerlin, "Comments by Readers," *Science* 105 (14 February 1947): 175. Reingold suggests that the commissions were intended to operate through the NRC and existing private foundations. See Reingold, "Vannevar Bush's New Deal for Research," 337–38.

86. United States. President's Scientific Research Board, John R. Steelman, Chairman. *Science and Public Policy. A Report to the President,* 5 vols. Washington: GPO, 1947, esp. I: 3–9, 29–36.

87. S. 2385, 80th Cong., 2d sess., 25 March 1948 [same as H.R. 6007], "Legislative History, N.S.F., 80th, 81st Congress, 1948–1950 (Committee prints and bills)," HF; Dael Wolfle, "Inter-society Committee for a National Science Foundation. Report of the Meeting of December 28, 1947," *Science* 107 (5 March 1948): 235–36; "News and Notes," *Science* 107 (2 April 1948): 339; England, *Patron for Pure Science,* 91. The Senate eventually dropped the disease commissions but they remained in the bill reported out of the House committee. See S. 2385 [Report No. 1151], 80th Cong., 2d sess., 20 April 1948, HF; H.R. 6007 [Report No. 2223], 80th Cong., 2d sess., 4 June 1948, HF. On the legislative debate from 1948 to 1950, see England, *Patron for Pure Science,* 83–106.

88. *Congressional Record,* House, 81st Cong., 2d sess., 27 February 1950: 2414,

2429–30. A bound volume, *Congressional Record on the National Science Foundation, 1946–1950,* is in the NSF Library.

89. William D. Carey to Elmer Staats, "Status of NSF Legislation," 2 March 1950, "Legislative History Files, BOB—Series 47.1b—NSF Act of 1949," RG 51, NARA; *Congressional Record,* House, 81st Cong., 2d sess., 27 February 1950: 2416–17, 2427, 2436–39; Strickland, *Politics, Science, and Dread Disease,* 77–83.

90. *The Hoover Commission Report on Organization of the Executive Branch of Government* (New York: McGraw-Hill, 1949), 501–3. The creation of an NSF was the report's final recommendation. Its grant awarding function, limited to basic research in "fields not adequately covered" by other agencies, was clearly subordinated to its coordinating and evaluating functions. John T. Wilson credited the Hoover Commission report with establishing NSF. "Notes of Med. Div. Comm.," 23 May 1953, Div. Comm., 1952–53.

91. *Congressional Record,* House, 27 February 1950: 2437, 2439; House, *Conference Report on National Science Foundation Act of 1950,* 81st Cong., 2d sess., 26 April 1950, House Report 1958, 20–21.

92. Frederick C. Schuldt and James L. Grahl to William D. Carey, "Conference Report of the National Science Foundation," 27 April 1950, RG 51, Bureau of the Budget, Series 47.1b—NSF Act of 1949, NARA.

93. James B. Conant, "An Old Man Looks Back. Science and the Federal Government: 1945–1950," *Bulletin of the New York Academy of Medicine* 47 (1971): 1248–51; Reingold, "Vannevar Bush's New Deal for Research."

Chapter 2   *Fashioning a New Federal Patron for Biology, 1950–1952*

1. Milton O. Lee to "Dear Sir" [survey form], 6 October 1949, Lee to Ralph E. Cleland, 20 February 1950, Divisions of the NRC, AIBS, NAS Archives.

2. See Robert E. Kohler Jr., "Warren Weaver and the Rockefeller Foundation Program in Molecular Biology: A Case Study in the Management of Science," in Nathan Reingold, ed., *The Sciences in the American Context: New Perspectives* (Washington, D.C.: Smithsonian Institution Press, 1979), 249–93.

3. NSF Act of 1950, sec. 4.

4. "National Academy of Sciences, Council of the Academy," 18 November 1949, Folder 1912, Box 87; Vannevar Bush to H. Alexander Smith, 3 May 1950, Folder 2462, Box 104; Bush to Bethuel M. Webster, 15 May 1950, Folder 2806, Box 117; and Bush to Guy Martin, 2 October 1950, Folder 1705, Box 69, all in Bush Papers.

5. Lee A. DuBridge and William H. Fowler, interview by Vernice Anderson and Judith Goodstein, 20 October 1981, HF, 21.

6. On the selection and makeup of the board, see J. Merton England, *A Patron for Pure Science: The National Science Foundation's Formative Years, 1945–1957* (Washington, D.C.: NSF, 1982), 113–21.

7. Sophie Bledsoe Aberle, interview by Vernice Anderson, 13 January 1982, HF, 1; DuBridge, interview, 21.

8. James Edward Hammer, *The University of Tennessee, Memphis 75th Anniversary—Medical Accomplishments* (Memphis: University of Tennessee, 1986), 40.

9. Patrick Henry Yancey, S.J., *To God Through Science: Reminiscences of Patrick Henry Yancey, S.J.* (Mobile, Ala.: Spring Hill College Press, 1968), 198–99.

10. NSB Minutes, 2nd Meeting, 3 January 1951, 3–4. On the selection of the Director, see England, *Patron for Pure Science,* 121–27.

11. England, *Patron for Pure Science,* 124–26; NSB Minutes, 3rd Meeting, 13–14 February 1951, 1–3.

12. Bush to Charles S. Garland, 1 April 1948, Folder 1382, Box 57, Bush Papers.

13. England, *Patron for Pure Science,* 127; NSB Minutes, 4th Meeting, 8–9 March 1951, 5–6.

14. On the selection of staff from ONR, see England, *Patron for Pure Science,* 127–40. Orr Reynolds was among those who stayed at ONR. He declined a feeler by Field because, as a strong believer in pluralism, he wanted to assure that ONR would continue to fund biology. Orr E. Reynolds, interview by author, 12 July 1988. See also Reynolds, "Support of the Biological Sciences by the Office of Naval Research," *AIBS Bulletin* 2 (April 1952): 18–20.

15. 12 April 1951, and "Telephone call to Dr. Edwin B. Fred," 16 April 1951, ATW Diary Notes, HF; England, *Patron for Pure Science,* 133.

16. Waterman, "Organization of the National Science Foundation for the Biological and Medical Sciences," 4 August 1952, ATW Notes, HF; NSB Minutes, 15th Meeting, 7–8 August 1952; England, *Patron for Pure Science,* 134. ATW Notes are filed in HF as a separate series from ATW Diary Notes. It should be noted that the assistant directors ("ADs") in areas of science in the 1950s, downgraded to the level of division directors in the 1960s, were different from the presidentially-appointed associate directors ("ADs") for areas of science beginning in 1975.

17. "Interview with John Field . . . ," 16 April 1951, ATW Diary Notes, HF; Ralph R. Sonnenschein, "John Field (1902–1983)," *The Physiologist* 26 (1983): 268; Staff Meeting Notes, 22, 24, 29, 31 May 1951, HF; Waterman, "Federal Support of Fundamental Research in the Biological Sciences," *A.I.B.S. Bulletin* 1(5) (October 1951): 11–17 (Waterman appeared on the cover of the issue).

18. "Staffing of Division of Biological Sciences," 3 July 1951, ATW Diary Notes, HF; Reynolds, interview; Bertha (Bel) Rubinstein, interview by author, 29 September 1988, and Estelle (Kepie) Engel, interviews by author, 29 June, 13 July, and 14 November 1988; Steven M. Horvath and Elizabeth C. Horvath, *The Harvard Fatigue Laboratory: Its History and Contributions* (Englewood Cliffs, N.J.: Prentice-Hall, 1973), esp. 36. Two of Consolazio's grantees persuaded Tufts University to award him an honorary degree in 1966.

19. Waterman, "Federal Support of Fundamental Research in the Biological Sciences," 11–12.

20. Field and Consolazio suggested Lawrence Blinks and Daniel Merriman, who declined the offers, along with Wilson. Staff Meeting Notes, 3 July 1951, HF.

21. John T. Wilson, interview by Frank Edmonson, 12 November 1980, HF, 1–2. Rubinstein, interview.

22. Engel, Reynolds, Rubinstein, interviews; interviews of former staff by author Eugene Hess, 5 June 1989, Joyce Hamaty, 15 November 1988, Lois Hamaty, 15 November 1988 and 13 March 1989, Mildred C. Allen, 5 October 1989, Josephine K. Doherty, 10 April 1990, and Mary Parramore, 17 January 1990.

23. "Organization of NSF for the Biological and Medical Sciences," 4 August 1952, ATW Notes, HF; John Field to Waterman, "Organization of the Division of Biological Sciences and the Division of Medical Research," Prog. Ann. Repts., 1952–57; Waterman to Edmund W. Sinnott, 11 June 1952, ATW Notes, HF.

24. Rubinstein, Engel, interviews; J. Merton England, interview by author, 8 March 1990. Levin ended his career as an administrator at Texas Technological University.

25. NSF Act of 1950, Sec. 8(d).

26. NSB Minutes, 2nd Meeting, 3 January 1951, 3; NSB Minutes, 3rd Meeting, 13–14 February 1951, 3–5. Yancey and Reyniers tried to prevent usurpation of the power of the full board by either the executive committee or the staff. See James A. Reyniers to Detlev W. Bronk, 6 February 1952, Folder 7, Reyniers to Bronk, 20 October 1952 and enclosure, "Preliminary Draft of Memo to the Director, NSF, Relative to Powers and Influence of the Board," Folder 8, and "Comments on the Functions of the Executive Committee submitted by P. H. Yancey, S.J." n.d., Folder 10, Detlev W. Bronk Papers, RG 303-U, RAC.

27. Staff Meeting Notes, 10 May 1951, HF.

28. NSB Minutes, 8th Meeting, 7 September 1951, 7–8; England, *Patron for Pure Science,* 171–73.

29. Yancey, *To God Through Science,* 206–7; NSB Minutes, 9th Meeting, 13 October 1951, 4, 8.

30. NSF Act of 1950, Sec. 1, Sec. 3(a)(2).

31. [Report] "An Interim Divisional Committee on Medical Research," n.d., Div. Comm, 1952–53.

32. "Telephone call from Mr. Frederick C. Schuldt, Bureau of the Budget," 10 May 1951, ATW Diary Notes, HF; England, *Patron for Pure Science,* 147–48.

33. NSB Minutes, 6th Meeting, 11 May 1951, 5–6; NSB Minutes, 7th Meeting, 28 July 1951, 1; "Telephone call from Mr. Schuldt, Bureau of the Budget," 18 May 1951, ATW Diary Notes, HF.

34. Staff Meeting Notes, 23 July 1951, HF; NSB Minutes, 7th Meeting, 28 July

1951, 15–17. A copy of H.R. 3371, 82nd Cong., 1st sess., 20 March 1951, is in the Detlev W. Bronk Papers, Folder 8, RG 303-U, RAC.

35. "Telephone call to Dean Wilbert Davisson, School of Medicine, Duke University, Durham, North Carolina, to discuss H.R. 3371," 25 July 1951, ATW Diary Notes, HF; John Field to Waterman, 24 July 1951, Folder "Legislation Pertaining to Medical Research," Box 12, ODSF.

36. NSB Minutes, 7th Meeting, 28 July 1951, 16–17; Stephen P. Strickland, *Politics, Science, and Dread Disease: A Short History of United States Medical Research Policy* (Cambridge, Mass.: Harvard University Press, 1972), 55–74.

37. "Comments on Federal Support in Life Sciences," 28 January 1953, Louis Levin Memos, 1954–58, HF; NSB Minutes, 5th Meeting, 6 April 1951, 9; Robert F. Loeb to James B. Conant, 18 April 1951, Folder 8, Detlev W. Bronk Papers, RG 303-U, RAC; House Committee on Appropriations, *Supplemental Appropriation Bill for 1952, Hearings before the House Subcommittees of the Committee on Appropriations,* 82nd Cong., 1st sess., Part 2, 1951, 556.

38. House Committee, *Supplemental Appropriation Bill for 1952,* 538–72; Senate Committee on Appropriations, *Supplemental Appropriations for 1952: Hearings before the Senate Committee on Appropriations,* 82nd Cong., 1st sess., 1106, 1109–10. See also England, *Patron for Pure Science,* 141–60.

39. Waterman to NSB, 27 August 1951, ATW Notes, HF; Senate Committee, *Supplemental Appropriations for 1952,* 1125.

40. Field's draft of a biology program, heavily weighted toward biochemistry, had covered even less of biology and led Reyniers to send a letter of complaint to Waterman. Field also proposed more explicit military applications than appeared in the Senate justification, including biological and chemical warfare. "Biological Science Programs," 31 August 1951, and Douglas M. Whitaker, "Funds Needed by NSF for the Support of the Biological Sciences in the U.S.," 12 August 1951, in folder "Division of Biological and Medical Sciences," Box 2, ODSF; Waterman to NSB, 27 August 1951, and "Phone Conversation with Reyniers, Colby College," 27 August 1951, ATW Notes, HF.

41. Senate Committee, *Supplemental Appropriations for 1952,* 1126–29. See analogous listing for physical sciences which included biophysics, 1129–37.

42. "Telephone call from Dr. James B. Conant," 23 October 1951, ATW Diary Notes, HF.

43. "Comments on candidates for Assistant Director for Medical Research," 17 July 1951, and "Telephone call from Dr. D. W. Bronk," 18 July 1951, ATW Diary Notes, HF.

44. "Organization of the National Science Foundation for the Biological and Medical Sciences," 4 August 1952, ATW Notes, HF; *NSF Annual Report 1950–51,* 25.

45. "Telephone call from Dr. David E. Price, National Institutes of Health," 15

January 1952, and "Telephone call from Dr. Leonard Scheele," 16 January 1952, Diary Notes, Box 4, ODSF; Scheele to Waterman, "Prospective Personnel, 1951–1956," 29 January 1952, Box 14, ODSF.

46. 10 March 1952, and "Telephone call from Mr. George Merck," 9 April 1952, Diary Notes, Box 4, ODSF.

47. Waterman to A. Baird Hastings, 12 May 1952, Hastings to Waterman, 19 May 1952, and "Prospective Personnel," Box 14, ODSF; "Organization of NSF for the Biological and Medical Sciences," 4 August 1952, ATW Notes, HF.

48. Waterman, "Federal Support of Fundamental Research in the Biological Sciences," 15. See also John T. Wilson to H. Burr Steinbach, "National Science Foundation Policy with Reference to the Medical Sciences," 23 September 1953, in Mildred C. Allen, Policy Book III, HF.

49. Waterman, "A Message from the National Science Foundation," Dedication of the Cancer and Pathology Research Laboratories, University of Tennessee College of Medicine, Memphis, 4 October 1951, Waterman Papers, Library of Congress.

50. NSB Minutes, 8th Meeting, 7 September 1951, 5; House Committee on Appropriations, *Independent Offices Appropriations for 1953, Hearings before the House Subcommittee of the Committee on Appropriations,* 82nd Cong., 2d sess., Part 1, 190.

51. Staff Meeting Notes, 31 July 1952, HF; "Telephone call from Dr. D.W. Bronk," 4 August 1952, Diary Notes, Box 4, ODSF.

52. NSB Minutes, 15th Meeting, 7–8 August 1952, 2, 11–12; "Conference with Dr. Fernandus Payne, Indiana University," 16 July 1952, and "Telephone call from Dr. Fernandus Payne," 2 August 1952, Diary Notes, Box 4, ODSF.

53. Waterman, "Organization of NSF for the Biological and Medical Sciences," 4 August 1952, ATW Notes, HF; NSB Minutes, 15th Meeting, 7–8 August 1952, 11; "Telephone call to Mr. Chester I. Barnard," 24 June 1952, Diary Notes, Box 4, ODSF; Waterman to Sunderlin, Klopsteg, Kelly, Wilson, and Harwood, 16 July 1952, ATW Notes, HF; Waterman to E. W. Goodpasture, 30 June 1952, Waterman to Distribution List, 22 July 1952, "Visit of Dean E. W. Goodpasture," 22 July 1952, and "Prospective Personnel, 1951–1956," Box 14, ODSF. See also Staff Meeting Notes, 31 July 1952, HF.

54. "Minutes of the First Meeting of the Divisional Committee for Medical Research," 23 May 1953, HF; "Notes of Med. Div. Comm.," 23 May 1953, Div. Comm., 1952–53; "Minutes of the Meetings of the Divisional Committees for Biological Sciences and Medical Research," 19–20 November 1953, HF. The Medical Research Committee included Dempsey, Goodpasture, Frank Brink Jr., Bernard D. Davis, Severo Ochoa, Dickinson W. Richards, George Wald, and Arnold D. Welch.

55. James B. Conant to Wallace O. Fenn, 17 October 1951, Waterman to Fenn, 30 November 1951, and "NSF 1951–53," Folder 5, Box 30, Wallace O. Fenn Papers, Edward G. Miner Library, University of Rochester. Appointment letters signed by

NSB chairman Conant maintained the fiction that the board named the committees and that the committees would primarily advise the board and director. NSB Minutes, 10th Meeting, 3 December 1951, 7.

Other members included entomologist Marston Bates, University of Michigan; future Nobel prize plant geneticist George Beadle, Caltech; cytologist Donald P. Costello, University of North Carolina; bacteriologist Jackson W. Foster, University of Texas; biochemist Hubert B. Vickery, Connecticut Agricultural Experiment Station; botanist Theodor K. Just, Chicago Natural History Museum; and embryologist John S. Nicholas, Yale.

56. 9 March 1951, ATW Diary Notes, HF; NSB Minutes, 3rd Meeting, 13–14 February 1951, 3; England, *Patron for Pure Science,* 161–164.

57. "Guide for Submission of Research Proposals" is reprinted in *NSF Annual Report FY 1952,* 50–52. See copy of the original mimeographed version, ATW Notes, 1951, HF. Information on amounts requested is found in Adv. Panel, 1952.

58. John Field to __, [27 December 1951], "National Science Foundation, Consultants Present, January 18–19, 1952," Adv. Panel, 1952. The members of the first panel were Lawrence Blinks (Stanford), John M. Buchanan (Penn.), Elmer G. Butler (Princeton), Britton Chance (Penn.), Ralph E. Cleland (Indiana), David R. Goddard (Penn.), Irwin C. Gunsalus (Illinois), Sterling B. Hendricks (USDA), Fritz A. Lipmann (Harvard), Tracy Sonneborn (Indiana), and H. Burr Steinbach (Minnesota).

59. "'Ground Rules' Suggested for Evaluation of Proposals for the National Science Foundation," n.d., Folder 5, Box 30, Fenn Papers. Information on the proposals submitted, the range of ratings given each, and the reasons for acceptance or decline is found in Adv. Panel, 1952.

60. Lee Anna Embrey to John Field, "Impressions of the Meeting of the Panel of Consultants to the Division of Biological Sciences, January 18, 1952," 21 January 1952, Adv. Panel, 1952.

61. Statements on ratings are based on perusal of grants folders for FY 1952 in Boxes 1 and 2, 60A-262, NARA; Embrey to Field, "Impressions," 21 January 1952.

62. Ratings of panel members are found in "Total Scores of Advisory Committee on Proposals," Adv. Panel, 1952. Conflict of interest rules were introduced at the NSB meeting. Board members left the room during discussion of and did not vote on proposals from their own institutions. See England, *Patron for Pure Science,* 167.

63. See individual grants folders and untitled, undated documents "Attached herewith are rating summaries of proposals considered at the Advisory Committee Meeting held on 18–19 January 1952," and "The following proposals have been evaluated and rated by the Advisory Panel for Biological Sciences at a meeting held on 18–19 January 1952," Adv. Panel, 1952. Those who received grants are listed in *NSF Annual Report FY 1952.*

64. "Tentative agenda," n.d., Folder 5, Box 30, Fenn Papers; "Biological Sciences Divisional Committee Meeting," 25–26 January 1952, Div. Comm, 1952–53, NARA. Examples of summary sheets can be found in Box 20, 70A-2191, NARA. Comparison to the Rockefeller Foundation is based on study of individual grants files of the Division of Natural Sciences in RAC.

65. James A. Reyniers to Bronk, 6 February 1952, Folder 7, Detlev W. Bronk Papers, RG 303-U, RAC; NSB Minutes, 11th Meeting, 1 February 1952, 4–9. The most controversial were two institutional awards, one to aid *Biological Abstracts,* and the other to provide operating funds for the Pacific Science Board of the National Academy of Sciences.

66. NSB Minutes, 12th Meeting, 29 February 1952, 2.

67. Ibid., 6–7; NSB Minutes, 14th meeting, 13 June 1952, 8, 10. See Adv. Panel, 1952.

68. See BMS grants listed in *NSF Annual Report FY 1952.* Other well-known (or soon to be well-known) grantees in FY 1952 included Max Delbruck, Frits Went, James Bonner, and Arthur Galston (all of Caltech), James D. Ebert (Indiana), Michael Pelczar Jr. (Maryland), Theodore H. Bullock (UCLA), Albert Blakeslee (guest professor at Smith College), and G. Evelyn Hutchinson (Yale). The data is from Fernandus Payne to Director, "Interim Report," 27 January 1953, filed with BMS annual reports, HF.

69. "Summary of Proposals Declined Support/Divisions of Biological Science and Medical Research," 8 May 1952, Div. Comm., 1952–53; Embrey to Field, "Impressions," 21 January 1952. In June, Reyniers' committee specifically discussed the reasons given for declining proposals in the biological sciences. See NSB Minutes, 12th Meeting, 29 February 1952, 7; NSB Minutes, 14th Meeting, 13 June 1952, 3–4.

70. Embrey to Field, "Impressions," 21 January 1952.

71. Robert H. Knapp and Hubert B. Goodrich, "The Origins of American Scientists," *Science* 113 (11 May 1951): 543–45; Knapp and Goodrich, *Origins of American Scientists: A Study Made Under the Direction of a Committee of the Faculty of Wesleyan University* (Chicago: University of Chicago Press, 1952); Staff Meeting Notes, 3 July 1951, HF. Despite their general title, the authors omitted women from the study, thus revealing the prevailing lack of concern for women scientists and the role of women's colleges in producing them. See Margaret W. Rossiter, *Women Scientists in America: Before Affirmative Action, 1940–1972* (Baltimore: Johns Hopkins University Press, 1995), 213–14.

72. See note 63.

73. Staff Meeting Notes, 23 February 1952, HF.

74. "Biological Sciences Division Consultants Advisory Panel, February 18, 1952, Merit versus Locality," Div. Comm., 1952–53.

75. "Quarterly Summary of Proposals Received in Biological and Medical Sci-

ences Division through 30 June 1952 (Geographic Distribution)," Div. Comm., 1952–53.

76. "G35 Bryn Mawr College B50," Box 7, 59A-2463, NARA; "G941 Yale University B780," Box 2, 60A-262, NARA. The first panel rejected Herman Branson, Howard University biophysicist, and women from Ursuline College and Washburn Municipal University, schools with little reputation for research.

77. Minutes, BMS Divisional Committee, 16 May 1952, Box 2, ODSF. Minutes of the BMS Divisional Committee, taken by the staff and distributed to committee members, are variously called "Staff Notes," "Notes," or "Minutes." They have been normalized to "Minutes."

78. See note 63.

79. *NSF Annual Report FY 1953,* 78. ONR funded Michener and Sokal's project on resistance of insects to DDT. See *Federal Grants and Contracts for Unclassified Research in the Life Sciences, Fiscal Year 1954* (Washington, D.C.: NSF, 1955), 59.

80. Minutes, BMS Divisional Committee, 16 May 1952, Box 2, ODSF; Waterman, memorandum for files, "Distribution of Research Funds between MPE and B&M Sciences," 23 December 1952, Wilson F. Harwood to Waterman, "Distribution of Research Funds . . . ," 22 December 1952, and Paul Klopsteg to Waterman, 15 December 1952, ATW Notes, HF; *NSF Annual Report FY 1952,* 44. Klopsteg called for a sixty-forty distribution in favor of the physical sciences.

81. Field to Waterman, "Organization of the Division of Biological Sciences and the Division of Medical Research," May 1952, Prog. Ann. Repts., 1952–57.

82. Paul Weiss, "Medicine and Society: The Biological Foundations" [1953], reprinted in Weiss, ed., *Within the Gates of Science and Beyond: Science and Its Cultural Commitments* (New York: Hafner, 1971), 79–95; Paul Weiss, "Memorandum on the Proposed Establishment of a Committee Council in the Division of Biology and Agriculture of the National Research Council of the National Academy of Sciences," 24 January 1952, and "NRC Governing Board," 3 February 1952, "B&A: Biology Council, Beginning of Program, 1952–54, NAS Archives.

83. Wilson interview, 10–11.

84. Minutes, 4th meeting, BMS Divisional Committee, 1 October 1952, HF.

85. William V. Consolazio to Fernandus Payne, "The State of the Biological Sciences Division" [January 1953], BMS annual reports, 1. See also Frank H. Johnson to Payne, "Meaning and Significance of Developmental, Environmental and Systematic Biology" [January 1953], ibid.

86. Wilson, interview, 11; Alan T. Waterman, "The National Science Foundation and the Life Sciences," *Public Health Reports* 69 (1954): 378–84.

87. Pnina Abir-Am, "The Discourse of Physical Power and Biological Knowledge in the 1930s: A Reappraisal of the Rockefeller Foundation's 'Policy' in Molecular Biology," *Social Studies of Science* 12 (1982); Robert C. Olby, "The Molecular Revolution in Biology," in Olby et al, eds., *Companion to the History of Modern*

Science (London: Routledge, 1990), 503–20, esp. 503–5; Doris T. Zallen, "Withdrawing the Boundaries of Molecular Biology: The Case of Photosynthesis," *Journal of the History of Biology* 26 (1993): 65–87; Consolazio to Payne, State of the Biological Sciences Division, 1, 3. In 1952, Consolazio emphasized protein synthesis, a long-standing biochemical problem area, rather than the study of nucleic acids. An example of the breadth of the Molecular Biology Program is its funding in 1952 of ecologist G. Evelyn Hutchinson's research on amino acid analyses of lake sediments.

88. Louis Levin to Fernandus Payne, "Comments on Federal Support of the Life Sciences" [January 1953], filed with BMS annual reports.

89. Consolazio to Payne, "The State of the Biological Sciences Division," 1.

90. Johnson, "Meaning and Significance of Developmental, Environmental and Systematic Biology". I am grateful to Scott F. Gilbert for discussion of developmental biology.

91. Johnson, "Meaning and Significance of Developmental, Environmental and Systematic Biology."

92. *NSF Annual Report FY 1952*, 18–19. See David Hull, *Science as a Process: An Evolutionary Account of the Social and Conceptual Development of Science* (Chicago: University of Chicago Press, 1988); Joseph Allen Cain, "Common Problems and Cooperative Solutions: Organizational Activity in Evolutionary Studies, 1936–1947," *Isis* 84 (1993): 1–25; and V. B. Smocovitis, "Organizing Evolution: Founding the Society for the Study of Evolution (1939–1950)," *Journal of the History of Biology* 27 (1994): 241–309.

93. Wilson to Payne, "Information concerning the Psychobiology Program for inclusion within material supporting the 1954 budget," 19 January 1953, BMS annual reports. On the term "psychobiology," see Donald Dewsbury, "Psychobiology," *American Psychologist* 46 (1991): 198–205. Dewsbury argues that the term was most often used "to combat excessive reductionism in areas overlapping psychology and biology as traditionally defined" (p. 203).

94. Frank H. Johnson, interview by author, 6 June 1989.

95. H. B. Steinbach to Waterman, "Tentative Staff Paper on Divisional Committees for Biological Sciences and Medical Research," 23 November 1953, Div. Comm., 1952–53.

96. Wilson, interview, 11.

Chapter 3   Expanding and Experimenting in the 1950s

1. J. Merton England, *Patron for Pure Science: The National Science Foundation's Formative Years, 1945–57* (Washington, D.C.: NSF, 1982), pp. 224, 274–75. John T. Wilson, interview by J. Merton England and Milton Lomask, 21 May 1974, HF, 2, 24. England has noted that NSF's biologists were considerably more experimental in their grant giving than the physical scientists.

2. NSF 88-3, *Report on Funding Trends and Balance of Activities: National Science Foundation, 1951–1988,* 1988, 18.

3. Wilson's ultimate goal was for NSF to fund all aspects of research and higher education. Wilson interview by England and Lomask, and John T. Wilson, interview by Frank Edmondson, 12 November 1980, HF. See also John T. Wilson, *Academic Science, Higher Education, and the Federal Government, 1950–1983* (Chicago: University of Chicago Press, 1983).

4. BMS Annual Report FY 1957, 21; BMS Annual Report FY 1960, 4. Copies of all BMS annual reports are located in HF. From 1953 on the division also housed a program in anthropology and related sciences under Harry Alpert; in 1958 this program was moved to the newly founded Office of the Social Sciences.

5. Wallace O. Fenn to Fernandus Payne, 2 October 1952, and Payne to Fenn, 8 October 1952, Folder 5, Box 30, Wallace Osgood Fenn Papers, Edward G. Miner Library, University of Rochester.

6. Wilson interview by Edmondson, 11.

7. BMS Annual Report FY 1955; BMS Annual Report FY 1957, 21; BMS Annual Report FY 1958, 6; BMS Annual Report FY 1959, 6–7.

8. BMS Annual Report FY 1959, 10.

9. Material on Waterman's searches for candidates can be found in annual folders in ATW Notes and ATW Diary Notes in HF and in ODSF, especially "Diary Notes," Box 4, and "Prospective Personnel," Box 14.

10. "Conversation with Dr. D. W. Bronk with reference to staffing of the Division of Biological Sciences," 26 June 1952, and "Conference with Dr. Fernandus Payne, Indiana University," 16 July 1952, ATW Diary Notes, HF. On Payne's career, see William R. Breneman, Tracy M. Sonneborn, and Ruth V. Dippell, "Memorial Resolution, Fernandus Payne (February 13, 1881–October 13, 1977)," Indiana University Archives. Waterman had first approached William J. Robbins, Director of the New York Botanical Garden, and Edmund Sinnott, Dean of the Sheffield Scientific School, Yale University.

11. H. Burr Steinbach to Director, Annual Report of the Assistant Director for Biological and Medical Sciences, Limited Distribution," 31 August 1954, HF (this is a longer and restricted version of BMS Annual Report, FY 1954); "H. B. Steinbach," *Biological Bulletin* 163(1) (1982): 47–48.

12. Winslow Briggs, Arthur Giese, and David Epel, "Lawrence R. Blinks, 1900–1989," Stanford University *Campus Report,* 28 February 1990, 14–15; Lawrence Blinks, "The Hopkins Marine Station, Stanford University and its Connections with L. R. Blinks (or vice-versa): A Golden Anniversary," n.d. [ca. 1983], manuscript autobiography, Hopkins Marine Station, Pacific Grove, Calif.

13. Wilson interview by Edmondson, 12–14; Wilson interview by England and Lomask, 48–49.

14. Field to Waterman, "Organization of the Division of Biological Sciences

and the Division of Medical Research" [May 1952]," filed with Prog. Ann. Repts., 1952–57.

15. Levin's discussion of such issues as loss of balance between teaching and research, "empire building," and the tendency to create a second class group of investigators wholly dependent on federal funds, can be found in the folder: "Louis Levin—Memos: 1954–58," HF. The work of program directors is documented in Prog. Ann. Repts, 1952–57, and summarized in BMS annual reports.

16. BMS Annual Report FY 1960, 26; "Guide for Submission of Research Proposals," December 1951, reprinted in *NSF Annual Report FY 1952*, 50–52; Estelle (Kepie) Engel, interviews by author, 29 June, 13 July, and 14 November 1988.

17. See Waterman's Diary Notes, 1951–63, in ODSF. Copies of many of these notes are in ATW Diary Notes, HF. Louis Levin and H. Burr Steinbach also left "diary notes." See Abbreviations, above, and Note on NSF Primary Sources. By the late 1950s a program director might visit annually some twenty institutions in all parts of the country. See BMS Annual Report FY 1959, 25. Site visits were listed in program annual reports.

18. Consolazio to Waterman, 6 May 1963, in William V. Consolazio Memos, 1963–64, HF. Consolazio acquired a reputation for always being able to work out a grant if it supported good science. For example, a 1961 award to future Nobelist Rita Levi-Montalcini, an émigré Italian scientist at Washington University, St. Louis, enabled her to spend part of the year in Italy, where she established a collaborative research unit with WU. See Rita Levi-Montalcini, *In Praise of Imperfection: My Life and Work* (New York: Basic Books, 1988), 188, 193.

19. See, for example, Wilson interview by Edmondson, 15.

20. Consolazio to Steinbach, "Rough Minutes on Transactions of Divisional Committee for Medical Research—Meeting of November 20," 23 November 1953, and "Notes of Med. Div. Comm.," 23 May 1953, Div. Comm., 1952–53. On the flexibility of the role of the NSF program director in biology, see the following interviews by Norman Kaplan: Herman Lewis, ca. 1967, Jack T. Spencer, August 17, 1967, and Eugene Hess, May 9, 1967, Norman Kaplan Papers, American Philosophical Society, Philadelphia. I am grateful to Margaret Rossiter for calling these interviews to my attention.

21. See BMS staff lists in NSF and BMS annual reports. On the importance of rotators, see Waterman Memorandum, 7 December 1954, ATW Notes.

22. BMS Annual Report FY 1954. Her husband, Earl L. Green, was as this time program manager for genetics at AEC. The Greens left Washington in 1956 for the Jackson Memorial Laboratory in Bar Harbor, Maine, where she was a senior scientist and longtime NSF grantee and he served as director.

23. Hubert B. Goodrich, "Draft Memorandum to the Division of Biology and Medicine of the National Science Foundation on Undergraduate Research," n.d., and Goodrich, "A Preliminary Report for DES for Estimate of Budget for FY 56,"

8 April 1954, in BMS: "Program - Div., En., & Sys., 1952–1954," HF; Frank H. Johnson, interview by author, 6 June 1989.

24. Interviews by author: Engel; Bertha (Bel) Rubinstein, 29 September 1988 and 9 February 1989; Brenda Flam, 9 August 1989; Mary Parramore, 17 January 1990; Josephine Doherty 10 April 1990 and 28 August 1990; Holly Schauer, 5 July 1990; Cecilia Spearing, 13 September 1990 and October 1990. A significant conduit from NIH was the laboratory of a female physiologist, Willie White Smith.

25. Wilson, interview by England and Lomask, 49.

26. Interviews by author: Mildred C. Allen, 5 October 1989; Lois Hamaty, 13 March 1989; Joyce Hamaty, 15 November 1988; also Engel, Rubinstein, Doherty, Flam, Spearing, Parramore, Schauer. Wilson was said to have treated his female employees exceptionally well.

27. Wilson to Payne, "Notes on Day-to-Day Activities within the Division of Biological and Medical Sciences," 12 September 1952, filed with Prog. Ann. Repts., 1952–57. Names of panel members were listed in NSF annual reports. After the late 1950s, the physical sciences' panels acted as general advisory committees at the discipline level and did not evaluate projects. Wayne Gruner, interview by author, 2 and 9 April 1991.

28. See Levin, "Talk to NSF on Regulatory Biology Program," 21 October 1954, Louis Levin Memos, 1952–58, HF.

29. On the minutes, taken by staff, of divisional committee meetings, 1952–65 and staff and ad hoc committee reports, see Abbreviations, above, and Note on NSF Primary Sources. On divisional committee support of Wilson, see minutes for 21 January 1958 and 12–13 March 1958. For the elimination of the divisional committee's role in approving grants, see "Conference with Dr. Sunderlin, Mr. Harwood and Mr. Callender to discuss the Grants Procedure," ATW Diary Notes, 2 November 1953, HF.

30. In FY 1970, Ursula K. Abbott and Rita Colwell became members of the developmental biology and oceanography panels, respectively. See NSF annual reports for divisional committee members. When the divisional (later advisory) committee was disbanded in 1972, it had never had a female or a black member.

31. BMS Annual Report FY 1960, 5. See NSF annual reports for lists of awards. See BMS annual reports for statistics on what proportion of sums requested were awarded.

32. An exception was Sprugel in environmental biology.

33. See, for example, George Sprugel Jr., Annual Report, Environmental Biology Program FY 1955, 2, and Annual Report, Environmental Biology Program FY 1956, 1, 3, Prog. Ann. Repts., 1952–57.

34. BMS Annual Report FY 1959, 1; John T. Wilson, "Support of Research by the NSF," *AIBS Bulletin* 11 (December 1961): 21–24, 31. Success rates are now tabulated somewhat differently as the ratio of awards to all actions taken.

35. BMS Annual Report FY 1959, 6; BMS Annual Report FY 1956, 12; BMS Annual Report FY 1963, 6a.

36. BMS Annual Report FY 1956, 6.

37. Ibid. For an example of a scientist objecting to summer salaries, see Fenn to Payne, 11 August 1953, Folder 5, Box 30, Fenn Papers.

38. See Louis Levin, "The Role of the National Science Foundation in Biological Science," *AIBS Bulletin* 4 (October 1954): 419–21.

39. Minutes, BMS Divisional Committee, 10th Meeting, 21 January 1958, 2, and 12th Meeting, 20–21 October 1958, 3, HF.

40. BMS Annual Report FY 1960, 1–2; NSF-60-111, press release 21 February 1960; William V. Consolazio, Annual Report, Molecular Biology Program FY 1961, 2, in Prog. Ann. Repts., 1958–60.

41. See NSF annual reports. Recipients of awards over $200,000 in the period 1960–63 included Shinya Inuoue (Dartmouth), Robert B. Corey and Linus Pauling (Caltech), Paul Doty (Harvard), Boris Ephrussi (Western Reserve), Seymour Benzer (Purdue), and Robert M. Bock and Harlyn O. Halvorson (Wisconsin).

42. Wilson, interview by England and Lomask, 14–15. On the Johnson Foundation, see George W. Corner, *Two Centuries of Medicine: A History of the School of Medicine, University of Pennsylvania* (Philadelphia: J. B. Lippincott, 1965), 264, 268, 308.

43. Waterman to Senior Staff, "NSF Policy with Respect to the Support of Foreign Basic Research," 19 November 1959, ATW Notes, HF. See NSF annual reports for names of foreign grantees. Consolazio, who took a sabbatical to travel to European laboratories in 1960, discussed his views of funding European biologists in "A Summary of Some Observations on European Biology," June 23, 1960, Norman Kaplan Papers, American Philosophical Society, Philadelphia.

44. "Biological Sciences Divisional Committee Meeting," 25–26 January 1952, Div. Comm., 1952–53; BMS Annual Report FY 1954; *NSF Annual Report FY 1954*, 24–26; NSF-90, press release 27 March 1954.

45. Rubinstein, interview; Levin, "Remarks to NSB Committee on Biological and Medical Sciences," 14 October 1954, Louis Levin Memos, HF; "Biological Sciences Division, Consultants Advisory Panel, February 18, 1952," Div. Comm., 1952–53.

46. "Topics for Discussion at Divisional Meetings with Regard to National Science Policy," n.d. [ca. 1953], Louis Levin Memos, 1952–58, HF.

47. Fernandus Payne, "Interim Report of the Division of Biology & Medicine to the Director," 27 January 1953, Prog. Ann. Repts., 1952–57. The House subcommittee overseeing the NSF budget, however, typically requested a list of awards by state. Figures for distributions in 1952–54 and 1957 are based on tabulation of grants in NSF annual reports.

48. BMS Annual Report FY 1960, 6.

49. Margaret W. Rossiter, *Women Scientists in America: Before Affirmative Action, 1940–1972* (Baltimore: Johns Hopkins University Press, 1994). Among women who were regularly funded in the 1950s were Grace E. Pickford (Yale), Mildred Cohn (Washington University and University of Pennsylvania), Ruth Patrick (Academy of Natural Sciences, Philadelphia), Barbara Low (Harvard and Columbia) and Dorothy Wrinch (Smith).

50. James H. M. Henderson (Tuskegee), Edward G. High (Meharry), Herman Branson and Lawrence M. Marshall (Howard), Mary Reddick (Atlanta University), and Robert James Terry (Texas Southern) were among black biologists who received grants. Identification of black biologists is based on Vivian Ovelton Sammons, *Blacks in Science and Medicine* (New York: Hemisphere, 1990).

51. "G-1692, Sarah Lawrence College, Collins and Wylie," Box 1, 60A-262, NARA.

52. Wilson, interview by England and Lomask, 2.

53. "Biological Sciences Divisional Committee Meeting, 25 and 26 January, 1952," Div. Comm., 1952–53.

54. Minutes, Biological Sciences Divisional Committee, 24 November 1952, and 23 March 1953, HF; Payne to Staff, "Recommendations of Divisional Committee for Biological Sciences," 6 January 1953, and Waterman to Board, "Recommendations of the Divisional Committee for Biological Sciences," 21 January 1953, Div. Comm, 1952–53.

55. BMS Annual Report FY 1961, 10–11. On the broad scope of the early BMS grants, see also Levin, "Role of the National Science Foundation in Biological Science."

56. See BMS annual reports and NSF annual reports for lists of conferences funded.

57. NSB Minutes, 7th Meeting, 28 July 1951, 9; Payne to Waterman, "Material and Statements Concerning the Work of the Divisions of Biology and Medicine; forwarding of," 17 October 1952, Prog. Ann. Repts., 1952–57; "C6 National Academy of Sciences," Box 6, 59A-2463, NARA.

58. See Levin Diary Notes, 8 September 1952, 2, 23, and 27 October 1952, 17 November 1952, 1 and 26 January 1953. The conference is described briefly in *NSF Annual Report FY 1953*, 20, 23–24. See also NSF-54, press release 28 May 1953.

59. See A. C. Smith, "Annual Report, Systematic Biology Program FY 1958," Prog. Ann. Repts., 1958–60.

60. NSF-25, press release 6 June 1952; NSF-51, press release 16 April 1953; Levin Diary Notes, 17 November 1952.

61. Senate Committee on Military Affairs, Subcommittee, *Hearings on Science Legislation (S. 1297 and Related Bills),* 79th Cong., 1st sess., 1945, 598; Staff Meeting Notes, 1 August 1951, 22 December 1952, HF; NSB Minutes, 11th Meeting, 1 February 1952, 8. On *Biological Abstracts,* see William Campbell Steere, *Biological Ab-*

stracts / BIOSIS, The First Fifty Years: The Evolution of a Major Science Information Service (New York: Plenum Press, 1976).

62. BMS Annual Report FY 1955, 5; BMS Annual Report FY 1956, 9.

63. BMS Annual Report FY 1958, 20–21; BMS Annual Report FY 1959, 27.

64. *NSF Annual Report* FY 1955, 124; BMS Annual Report FY 1955, 4.

65. BMS Annual Report FY 1955, 2; *The Emory Alumnus* 40(1) (February 1964): 24; "Yerkes Marks Its 60th Year," *Inside Yerkes* (Spring 1990): 5. I am grateful to the Emory University Archives and the Yerkes Regional Primate Center for these latter two sources.

66. *Twenty-Five Years of High-Altitude Research: White Mountain Research Station* (Berkeley: University of California Press, 1973), esp. Nello Pace, "A History," 9–17; Warren Weaver Diaries, 26 May 1952, 144–45, RAC; Minutes, Divisional Committee for Biological Sciences, 16 May 1952, HF; Levin Diary Notes, 13 July 1954; NSF-29, press release 30 July 1952; Ralph H. Kellogg, interview by author, 30 April 1990. See also NSF annual reports for 1953, 1955, 1957, and 1959. The White Mountain Research Station has sponsored research in a wide variety of fields including physiology, neurobiology, endocrinology, metabolic studies, ecology, physics, geology, meteorology, and astronomy.

67. Joel B. Hagen, "Problems in the Institutionalization of Tropical Biology: The Case of the Barro Colorado Island Biological Laboratory," *History and Philosophy of the Life Sciences* 12 (1990): 225–47, esp. 240–43. On later NSF support of the Smithsonian facility, see Chapter 7. I am grateful to Joel Hagen for additional information on the significance of NSF's role in funding the laboratory.

68. On LOBUND, see "Division of Biological and Medical Sciences. LOBUND and Related Matters," Box 2, ODSF, especially Wilson F. Harwood to Waterman, 26 February 1953, C. E. Sunderlin, Memorandum for the file, "Germfree Animal Production," 2 April 1954, Blinks to Waterman, "Further CONFIDENTIAL Remarks and Opinions concerning LOBUND," 9 June 1955, Levin, Memorandum for the file, 2 September 1955, and Theodore Hesburgh, "LOBUND Memorandum," 19 June 1956. See also "Telephone call to Dr. James A. Reyniers," 26 January 1953, ATW Diary Notes, HF; Levin Diary Notes, 6 July 1955, and 19 October 1955; Levin to Conrad A. Elvehjem, et al., "Comments on Report of Ad Hoc Committee on Microbiological Facilities," 23 July 1956, ATW Notes, HF; and Minutes, BMS Divisional Committee, 4th Meeting, 25–26 November 1955, 12–13 October 1956, 5, 5–6 April 1957, 5–6, HF

69. A copy of the report has not been located.

70. Edmund P. Joyce to Waterman, 14 January 1958, in "Division of Biological and Medical Sciences," Box 23, ODSF; Theodore M. Hesburgh with Jerry Reedy, *God, Country, Notre Dame* (New York: Doubleday, 1990), 66–67. Reyniers became president and director of the Germfree Life Research Center in Tampa, Florida. Not until FY 1964, after a new LOBUND director had been chosen and a research pro-

gram was underway, did BMS finally provide a facilities grant of $616,000 for a new laboratory building. Harve J. Carlson to Waterman, "Resubmission of a proposal from Lobund Institute, University of Notre Dame," 7 March 1963, and Frederick D. Rossini to Carlson, 6 May 1963, in "Div. of Biological & Medical Sci. 1963," Box 70, ODSF; *NSF Annual Report FY 1964,* 111–12.

71. BMS Annual Report FY 1956, 7.

72. BMS Annual Report FY 1958, 9–10; Marine Biological Laboratory, Annual Report, 1954–60, *Biological Bulletin* 111 (1955): 1–44, esp. 6; 113 (1957): 1–43, esp. 6, 8; 115 (1958): 1–52, esp. 6–7; 117 (1959): 1–53, esp. 6–7; 119 (1960): 1–56, esp. 6–8; 121 (1961): 1–56, esp. 6–8.

73. See BMS annual reports, especially 1957–61.

74. *Federal Grants and Contracts for Unclassified Research in the Life Sciences, Fiscal Year 1952* (Washington, D.C.: NSF, 1954). The last year of this series, a set of which is in the NSF Library, was FY 1958. See also William V. Consolazio and Margaret C. Green, "Federal Support of Research in the Life Sciences," *Science* 124 (21 September 1956): 522–26, and William V. Consolazio and Helen Jeffrey, "Federal Support of Research in the Life Sciences," *Science* 126 (26 July 1957): 154–55. How these reports were used in practice is not known.

75. Wilson to Steinbach, "National Science Foundation Policy with Reference to the Medical Sciences," 23 September 1953, Mildred C. Allen Policy Book III, HF.

76. *NSF Annual Report FY 1959,* 226. Life science fields included agriculture, anthropology, biochemistry, biophysics, botany, general biology, genetics, medical sciences, microbiology, psychology, and zoology.

77. BMS Annual Report FY 1961, 7; BMS Annual Report FY 1956, 6; Consolazio, Annual Report, Molecular Biology Program FY 1954, 10, Prog. Ann. Repts., 1952–57.

78. BMS Annual Report FY 1954, 11. This discussion is found only in the "Limited distribution" version. See note 11.

79. BMS Annual Report FY 1956, 7; Minutes, BMS Divisional Committee, 8th Meeting, 5–6 April 1957, HF; BMS Annual Report FY 1957, 18; Consolazio, Annual Report, Molecular Biology Program FY 1957, 6, Prog. Ann. Repts., 1952–57; Waterman to Wilson, 15 January 1957, in "Division of Biological and Medical Sciences," Box 23, ODSF, copy in ATW Notes, HF. According to a recent biography, the award fell through "because Szilard—in a typical fit of 'independence'—refused to tell a National Science Foundation grants officer exactly how much time he might spend at each institution." William Lanouette with Bela Szilard, *Genius in the Shadows: A Biography of Leo Szilard, the Man Behind the Bomb* (New York: Charles Scribner's Sons, 1992), 397.

80. BMS Annual Report FY 1955, 3.

81. BMS Annual Report FY 1959, 15.

82. BMS Annual Report FY 1955, 5; BMS Annual Report FY 1957, 10, 12.

The last project was supplemented by funds from NIH and AEC. See John E. Brobeck, Orr E. Reynolds, and Toby A. Appel, eds., *History of the American Physiological Society: The First Century, 1887–1987* (Bethesda, Md.: American Physiological Society, 1987), 118–19.

83. BMS Annual Report FY 1956, 10; BMS Annual Report FY 1957, 12.

84. George Wald to Waterman, 1 April 1955, in "Divisional Committee for Biological and Medical Sciences," Box 2, ODSF. See also *NSF Annual Report FY 1955*, 62–63. Levin's staff report has not been located. One later BMS project related to medical schools was the sponsoring of a conference on two-year medical schools in 1962. See Minutes, BMS Divisional Committee, 22nd Meeting, 25–26 May 1962, 4, and 23rd Meeting, 12–13 October 1962, 5, HF.

85. BMS Annual Report FY 1955, 6–7; BMS Annual Report FY 1956, 6; BMS Annual Report FY 1957, 9–10.

86. Waterman, Memorandum, 19 May 1955, ATW Notes, HF; "Conference with National Institute of Health Administrators," 16 November 1955, Levin Diary Notes. In 1961, when NIH was able to provide eight student research stipends to any medical school that applied, BMS discontinued its medical student program despite the strong recommendation of the advisory panel that it be continued "on the competitive merit system, which is the distinguishing feature of NSF's approach as compared to that of NIH." Because several members of the divisional committee wanted the program to continue, Harve Carlson requested the education division to consider taking the program over, but nothing came of the idea. BMS Annual Report FY 1961, 10; BMS Annual Report FY 1962, 11.

87. BMS Annual Report FY 1954, 9; Wilson, "Memorandum for the File: Grants for Equipment and Laboratory Assistance to Departments of Psychology at under-supported Institutions, including Small Colleges," 1 April 1955, and Wilson to Waterman, "Proposals B-1333 (Univ. of Wichita); B-1443 (Sarah Lawrence); B-1611 (Howard Univ.) . . . ," 18 May 1955, in "G-1692, Sarah Lawrence College, B-1443, Collins and Wylie," Box 1, 60A-262, NARA.

88. Wilson, interview by England and Lomask, 13–14.

89. BMS Annual Report FY 1956, 6; Minutes, BMS Divisional Committee, 4th Meeting, 25–26 November 1955, HF; *NSF Annual Report FY 1956*, 144–45.

90. BMS Annual Report FY 1957, 11; Waterman to Wilson, 13 November 1957, ATW Notes, HF.

91. The Caltech grant was for $20,700 for three years. *NSF Annual Report FY 1957*, 158; BMS Annual Report FY 1957, 10–11. On Beadle's views concerning dissertations, see "Biological Sciences Division Consultants Advisory Panel, February 12, 1952," Division of Biological and Medical Sciences, Box 2, ODSF.

92. Harry C. Kelly to Waterman, "Background for Consideration of Locally-Administered Fellowship Programs," 23 January 1957, and attachment, Bowen C.

Dees to Kelly, "Request for Reconsideration of Two Proposals," 22 January 1957, ATW Notes, HF.

93. BMS Annual Report FY 1958, 1, 12; Minutes, BMS Divisional Committee, 11th Meeting, 12–13 March 1958, 3; Waterman to AD's for BMS, MPES, and SPE, 5 March 1957, ATW Notes, HF; and Waterman to AD's for BMS, MPES, and SPE, "Responsibility for Programs Involving Education in the Sciences," 26 April 1957, in folder "Office of the Director Letters (1954–63) for Historical Purposes," HF. I was unable to find Waterman's 8 September memorandum mentioned by Wilson.

94. Minutes, BMS Divisional Committee, 9th Meeting, 11–12 October 1957, HF.

95. Minutes, BMS Divisional Committee, 11th Meeting, 12–13 March 1958, 6.

96. BMS Annual Report FY 1958, 12.

97. Minutes, BMS Divisional Committee, 10th Meeting, 21 January 1958, 2; Minutes, BMS Divisional Committee, 11th Meeting, 12–13 March 1958, 5; BMS Annual Report FY 1958, 23.

98. Minutes, BMS Divisional Committee, 11th Meeting, 12–13 March 1958, 2; NSB Minutes, 54th meeting, 28–30 June 1958, 13–14, 22–23.

99. *NSF Annual Report FY 1958,* 53; *NSF Annual Report FY 1959,* 72, 226.

100. Waterman to Senior Staff, "NSF Policies and Programs for Support of Graduate Training Through Fellowships and Research Grants," 20 February 1959, ATW Notes, HF.

101. Minutes, BMS Divisional Committee, 12th Meeting, 20–21 October 1958, 5–6.

102. Dees to Kelly, "Thesis Research," 18 May 1959, Kelly to Waterman, "Thesis Research," 21 May 1959, William J. Hoff and Robert B. Brode to Waterman, "Grant to University of Chicago (Park) B&MS #44," 22 May 1959, and Waterman to Wilson, "Proposed Grant to University of Chicago for Dr. Thomas Park," 31 March 1959, in Division of Biological and Medical Sciences, Box 23, ODSF. The award was for $32,000 for three years. Waterman noted that only a few laboratory directors objected to the common practice of graduate students completing their dissertations while employed as research assistants on a grant. See Waterman to Senior Staff, 20 February 1959.

103. See BMS annual reports and divisional committee minutes, e.g., BMS Annual Report FY 1958, 1, 12, 17–19, 23.

104. NSF-59-122, press release 29 March 1959; NSF-59-138, press release 22 June 1959.

105. BMS Annual Report FY 1959, 28.

106. Consolazio, Annual Report, Molecular Biology Program FY 1961, 11 July 1961, 19.

107. BMS Annual Report FY 1960, 8, 28; BMS Annual Report FY 1961, 2, 20. BMS "logistic" awards were made to the Caltech "phytotron" and the Highlands Biological Station.

Chapter 4   *Government Relations and Policy-making in the Cold War Era*

1. On pluralism, see Jeffrey K. Stine, A *History of Science Policy in the United States, 1940–1985: Report Prepared for the Task Force on Science Policy, Committee on Science and Technology, House of Representatives, Ninety-Ninth Congress, Second Session* (Washington, D.C.: GPO, 1986), 34–35, 37, 40. See also A. Hunter Dupree, *Science in the Federal Government: A History of Policies and Procedures,* 2nd ed. (Baltimore: Johns Hopkins University Press, 1986), vii–xviii.

2. "Comments on Federal Research Support of Life Sciences," 28 January 1953, Louis Levin Memos, 1954–58, HF.

3. See "A Rare Glimpse Inside the Budget Bureau," *Scientific Research* 2 (1967): 29–31.

4. John Wilson, interview by J. Merton England and Milton Lomask, 21 May 1974, 16, HF.

5. See J. Merton England, *Patron for Pure Science* (Washington, D.C.: GPO, 1982), 216–21, esp. table showing appropriations process for 1951–58, 217. House Committee on Appropriations, Subcommittee, *Hearings on Independent Offices Appropriations for 1956,* 84th Cong., lst sess., 1956, 227–343; Senate Committee on Appropriations, Subcommittee, *Hearings on H.R. 5240, Independent Offices Appropriations for 1956,* 84th Cong., 1st sess., 1956, 411–52, esp. 437.

6. "Conversation with Dr. James B. Conant and Dr. D. W. Bronk," 9 March 1951, ATW Diary Notes.

7. See, for example, Senate Committee, *Independent Offices Appropriations for 1956,* 421–22. Agriculture was least frequently noted. On the "ideology of basic research," see above, Introduction, n. 5.

8. Senate Committee, *Independent Offices Appropriations for 1956,* 419, 437; House Committee on Appropriations, Subcommittee, *Hearings on Independent Offices Appropriations for 1957,* 84th Cong., 2nd sess., 1957, 509–621, esp. 529–30; House Committee, *Hearings on Independent Offices Appropriations for 1956,* 242–43. When pressed, Waterman would say that 75 percent of proposals received were meritorious, i.e., deserved funding.

9. House Committee, *Hearings on Independent Appropriations for 1956,* 229.

10. Nicholas DeWitt, *Soviet Professional Manpower: Its Education, Training and Supply* (Washington, D.C.: NSF, 1955); House Committee, *Hearings on Independent Offices Appropriations for 1956,* 522. On Thomas's skepticism, see ibid., 229, 234.

11. Senate Committee on Appropriations, Subcommittee, *Hearings on H.R. 9739, Independent Offices Appropriations for 1957,* 84th Cong., 2nd sess., 1957, 375–89. NSF received $40 million of a requested $41.3 million. Both Waterman and

Bronk were ambivalent about NSF budgets depending on the Cold War mentality.

12. Senate Committee, *Hearings on Independent Offices Appropriations for 1956,* 422.

13. See House Committee, *Hearings on Independent Offices Appropriations for 1956,* 283–89.

14. House Committee, *Hearings on Independent Offices Appropriations for 1956,* 286–87; Waterman to AD's and PD's in B&M and MPE Divs., 6 February 1957, ATW Notes. For an example of research highlights, see NSF *Annual Report FY 1958,* 27–35. Headings, which emphasized the fundamental nature of the research and its possible practical results, included: "Wax-Eating Birds Provide Clue for Control of Tuberculosis," "Nonnerve Tissue Tumor Agents Specifically Induce Nerve Growth," "Protein-like Materials Synthesized Directly from Amino Acids," and "The Alteration of Proteins at Will."

15. House Committee on Appropriations, Subcommittee, *Hearings on Independent Offices Appropriations for 1958,* 85th Cong., lst sess., 1956, 1268–1409, esp. 1286.

16. Senate Committee on Appropriations, Subcommittee, *Hearings on H.R. 4663, Independent Offices Appropriations for 1954,* 83rd Cong., lst sess., 1954, 315–36, esp. 330; Senate Committee, *Hearings on Independent Offices Appropriations for 1957,* 384.

17. On AEC and the issue of duplication versus coordination, see, for example, House Committee, *Hearings on Independent Offices Appropriations for 1958,* 1351–52, 1402–3; Senate Committee on Appropriations, Subcommittee, *Hearings on H.R. 6070, Independent Offices Appropriations for 1958,* 85th Cong., lst sess., 1958, 247–300, esp. 273–74; and House Committee on Appropriations, Subcommittee, *Hearings on Independent Offices Appropriations for 1954,* 83rd Cong., lst sess., 1954, 45–99, esp. 83–91.

18. John T. Wilson to H. Burr Steinbach, "National Science Foundation Policy with Reference to the Medical Sciences," 23 September 1953, Mildred C. Allen Policy Book III, HF.

19. Waterman to Roger W. Jones, 24 October 1955, ATW Notes.

20. See, for example, "Hearings of the Wolverton Committee on Status and Future Prospects of Research Related to Health," 9 October 1953, Levin Diary Notes; Wilson to Waterman, "Comments on Report of the Committee of Consultants on Medical Research to the Subcommittee on Appropriations for Departments of Labor, Health, Education and Welfare (Boisfeuillet Jones Committee)," 8 June 1960, Division of Biological and Medical Sciences 1960, Box 42, ODSF. Hubert Humphrey chaired a subcommittee that prepared a report that made a case for an NSF role in international health research. See Senate Committee on Government Operations, The *National Science Foundation and the Life Sciences,* Committee Print (Washington, D.C.: GPO, 1959).

21. "Telephone call from Representative Albert Thomas (Democrat, Texas)" 16 March 1958, ATW Diary Notes; 2 and 9 April 1958; "Telephone conversations with Dr. Klopsteg re: Proposal of Dr. Flocks, Iowa," 2 April 1958, and "Telephone conversation with Dr. R. H. Flocks (State U. of Iowa), 9 April 1958, Levin Diary Notes; England, *Patron for Pure Science*, 399–400 n. 27; *NSF Annual Report FY 1959*, 173. For another example, see "Telephone call from Mr. John R. Gomien, Administrative Assistant to Senator Dirksen," 1 March 1955, ATW Diary Notes.

22. Marie S. Rogers to Senator Hugh Scott, July 1960, and Waterman to Senator Joseph S. Clark, 9 February 1960, ATW Notes; *NSF Annual Report FY 1960*, 203.

23. For the text of executive order, see *NSF Annual Report FY 1954*, 118–19. Reprinted in England, *Patron for Pure Science*, 353–55; see also ibid., 196–202, 311–20.

24. William V. Consolazio to Fernandus Payne, "The State of the Biological Sciences Division," 15 January 1953, 10, BMS annual reports, HF.

25. Senate Committee, *Hearings on Independent Offices Appropriations for 1954*, 320. See also House Committee, *Hearings on Independent Offices Appropriations for 1954*, 58–60.

26. DuBridge, for one, vigorously opposed the executive order because any curtailment of DOD support (which included payment of full costs of research) would be disastrous for Caltech. See Lee A. DuBridge to Arthur S. Fleming, Office of Defense Mobilization, 12 August 1953, folder 10, Detlev W. Bronk Papers, RAC.

27. England, *Patron for Pure Science*, 311–20.

28. Senate Committee, *Hearings on Independent Offices Appropriations for 1954*, 316, 319; House appropriations hearings, 1954, 92; House Committee on Appropriations, Subcommittee, *Hearings on Independent Offices Appropriations for 1955*, 83rd Cong., 2nd sess., 1955, 750–803, esp. 751; England, *Patron for Pure Science*, 233–35.

29. NSF *Annual Report FY 1953*, 86; *NSF Annual Report FY 1954*, 97; *NSF Annual Report FY 1955*, 134.

30. *Hearings on Independent Offices Appropriations for 1955*, 766, 770; Harry C. Kelly to Waterman, 19 February 1953, ATW Notes.

31. F. C. Sheppard to Acting Assistant Director for Administration, "Proposed transfer of National Institutes of Health general purpose fellowship program to the National Science Foundation," 2 September 1953, Bowen C. Dees to File, "Transfer of the National Institutes of Health Postdoctoral Fellowship Program to the National Science Foundation," 2 October 1953, Dees to File, "Conference with Representatives of the National Institutes of Health Concerning Postdoctoral Fellowships," 12 October 1953, Dees to File, "Second Meeting with Representatives Concerning the NIH Postdoctoral Fellowship Program," 15 October 1953, ATW Notes; "Telephone call to Dr. Robert F. Loeb," 22 October 1953, ATW Diary Notes; Waterman to John T. Edsall, 11 December 1953, ATW Notes. In one in-

stance Waterman expressed willingness to earmark funds, when he proposed to the board that NIH postdoctoral fellowships be transferred to NSF. See ATW Diary Note [no title], 23 October 1953, ATW Notes.

32. Harry C. Kelly, "National Institutes of Health Predoctoral Program (Telephone Conversation with Mr. Charles Kidd)," 5 May 1954, and attached F. C. Sheppard to Waterman, ATW Notes.

33. "Senior Staff Meeting, Monday, June 7, 1954," Steinbach Diary Notes.

34. "Telephone call from Dr. William H. Sebrell," 12 October 1954, ATW Diary Notes; "NIH Fellowship Program," 3 August 1954, Steinbach Diary Notes. Some of the money was also to be used to initiate summer fellowships for medical students in competition with the proposed NSF program.

35. National Science Foundation Act of 1950, 1. On the evaluation function of NSF, see England, *Patron for Pure Science,* 181–96.

36. "Conversation with Dr. James B. Conant and Dr. D. W. Bronk," 9 March 1951, ATW Diary Notes.

37. Wilson, interview by England and Lomask, 51. For an example of interpreting policy making as "information gathering," see Louis Levin, "The Role of the National Science Foundation in Biological Science," *AIBS Bulletin* 4 (October 1954), 1.

38. Payne to Waterman, "Material and Statements Concerning the Work of the Divisions of Biology and Medicine" 17 October 1952, filed with Prog. Ann. Repts., 1952–57.

39. House appropriations hearings, 1954, 55–56; NSF press releases 23, 21 May 1952, 33, 2 October 1952, and 35, 6 October 1952. Fernandus Payne, skeptical of the value of these studies in providing useful information to the Foundation, thought them to be of "direct value to the research workers and to the development of their respective fields of science, which, after all, is one of the principal functions of the Foundation." Payne to Waterman, 17 October 1952.

40. On the APS perspective, see John R. Brobeck, Orr E. Reynolds, and Toby A. Appel, eds., *History of the American Physiological Society: The First Century, 1887–1987* (Washington, D.C.: American Physiological Society, 1987), 97–100, 105.

41. Ralph W. Gerard, *Mirror to Physiology: A Self-Survey of Physiological Science* (Washington, D.C., 1958), esp. 1, 131. The proposal to NSF and cover letter noting the informal discussions with NSF staff are included as Appendix A, 261–65.

42. Orr E. Reynolds, interview by author, 12 July 1988; Brobeck, et al., *History of the American Physiological Society,* 118–19.

43. NSF *Annual Report FY 1953,* 17, 81; NSF Staff Meeting Notes, 5 August 1953, HF; NSF press release 35, 6 October 1952.

44. I am grateful to historians of psychology James Capshew and Donald Dewsbury for help in interpreting the significance of the psychology survey.

45. Sigmund Koch, ed., *Psychology: A Study of a Science,* 6 vols. (New York: Mc-

Graw-Hill, 1959–63). A seventh volume, Sigmund Koch, "Psychology and the Human Agent," was never published.

46. Kenneth E. Clark, *America's Psychologists: A Survey of a Growing Profession* (Washington, D.C.: American Psychological Association, 1957).

47. Levin and Consolazio to Waterman and Eugene Sunderlin, 16 April 1953, ATW Notes; "Visit with Dr. Gerty Cori," 12 January 1953, Levin Diary Notes. See also "Biochemistry Survey (meeting of advisory group, February 6, 1953)," filed under "Biological and Med. Sciences (BMS)," HF.

48. On the Long Report episode, see England, *Patron for Pure Science*, 325–29, and Nathan Reingold, "Choosing the Future: The U.S. Research Community: 1944–1946," *Historical Studies in the Physical and Biological Sciences*, 25 (1995): 301–28.

49. *The Advancement of Medical Research and Education through the Department of Health, Education, and Welfare: Report of the Secretary's Consultants on Medical Research and Education* (Washington, D.C.: HEW, 1958), 60; Stephen P. Strickland, *Politics, Science and Dread Disease: A Short History of United States Medical Research Policy* (Cambridge, Mass.: Harvard University Press, 1972), 91, 97.

50. NSB Minutes (Executive Session), 30th Meeting, 5 November 1954; Strickland, *Politics, Science and Dread Disease,* 97–98.

51. NSB Minutes (Executive Session), 31st Meeting, 6 December 1954; Waterman to Detlev Bronk, Theodore Hesburgh, O. W. Hyman, George Merck, and Robert F. Loeb, "Memo (Private and Confidential)," 23 November 1954, and diary note, 15 December 19S4, ATW Notes.

52. 15 December 1954, and "Memo (Private and Confidential)," 23 December 1954, ATW Notes. BOB's William Carey asked if the request should come from the president but NSF representatives preferred Mrs. Hobby. A copy of the letter can be found in the Long Report (see n. 56 below), 66–68.

53. NSB Minutes (Executive Session), 32nd Meeting, 21 January 1955.

54. A. Baird Hastings and E. W. Goodpasture both declined to chair the committee but agreed to serve on it. The other members were Edward A. Doisy, Charles B. Huggins, Colin M. MacLeod, C. Phillip Miller, and Wendell M. Stanley. All but Doisy and Stanley held M.D. degrees. The Executive Secretary, selected by Waterman, was Joseph Pisani, M.D., assistant dean of the College of Medicine, State University of New York.

55. "Notes for 1st Meeting of the Special Committee on Medical Research to be held at the National Science Foundation on July 22, 1955," 21 July 1955; Waterman to Members of Special Committee on Medical Research and Dr. Pisani, 24 February 1956; and Waterman to National Science Board, 29 February 1956: all in ATW Notes; NSB Minutes (Executive Session), 37th Meeting, 5 December 1955.

56. *Medical Research Activities of the Department of Health, Education and Welfare, Report of a Special Committee on Medical Research Appointed by the National Science*

Foundation at the Request of the Secretary of Health, Education and Welfare, December 1955, [Long Report], mimeograph, 70 p., copy in HF, 20.

57. Ibid., 31.

58. Ibid., 26, 40–41, 45–46, 51.

59. Ibid., 53; "Conversation with Dr. L. T. Cogeshall, special assistant to the secretary of Health, Education and Welfare," 5 January 1956, ATW Diary Notes; Waterman to C. N. H. Long, 6 February 1956, and Waterman to Virgil Hancher, 18 June 1956, ATW Notes. Two decades later, Wilson recalled that Long, "a very able man, who called a spade a spade—wrote a very good report, and it was buried and nobody ever saw it in the light of day." Wilson, interview, 21 May 1974, 22. The BMS staff was not allowed to take part in the Long Committee's deliberations, but the divisional committee discussed the final report at its meeting in April 1956. Long became a member of the divisional committee in 1956. See "Notes for lst Meeting of the Special Committee on Medical Research," 21 July 1955, ATW Notes; Minutes, BMS Divisional Committee, 5th Meeting, 21 April 1956.

60. Philip M. Smith, *The National Science Board and the Formulation of National Science Policy, A Report to Lewis M. Branscomb, Chairman, National Science Board* (Washington, D.C.: NSF, 1981), 13, 16, 23.

61. On NSF's handling of the loyalty issue, see England, *Patron for Pure Science,* 329–37; NSF *Annual Report FY 1956,* 9–11; and Smith, *The National Science Board and the Formulation of National Science Policy.* On McCarthyism as it affected universities, see Ellen Schrecker, *No Ivory Tower: McCarthyism and the Universities* (New York: Oxford University Press, 1986).

62. Consolazio to Waterman, "The Security Problem in the Federal Extramural Basic Research Program," 8 October 1953, ATW Notes; "Review of the Regulatory Biology Program during Fiscal Year 1954,," 6 April 1954, Louis Levin Memos. 1954–58, HF.

63. Consolazio to Waterman, "Security Problem," 8 October 1953.

64. "Minutes of the Meetings of the Divisional Committees for Biological Sciences and Medical Research," 19–20 November 1953, HF.

65. Leonard Scheele to Waterman, 27 April 1954, and attachments, ATW Notes.

66. Russell R. Larmon, acting secretary of HEW, reported to Sherman Adams that there were thirty-nine cases of denial of support up to August 1954 and suggested the matter be placed on the Sub-Cabinet agenda for a government-wide solution. Larmon to Adams, 30 August 1954, file E4-1 (1952–56), series 52.1, RG 51, NARA, copy in HF.

67. Martin D. Kamen, *Radiant Science, Dark Politics: A Memoir of the Nuclear Age* (Berkeley: University of California Press, 1985), 275. Kamen added, "The backing of the National Science Foundation was to continue for the rest of my research career, providing absolutely crucial support in all those years and a reliable backup

source when the National Institutes of Health came out from under the security cloud and were able to take over a major share of the funding I needed." See also *NSF Annual Report FY 1956*, 132.

68. NSF *Annual Report FY 1955*, 113; Elvin A. Kabat, "Getting Started 50 Years Ago—Experiences, Perspectives, and Problems of the First 21 Years," *Annual Review of Immunology* 1 (1983): 1–32, esp. 29, 31–32. Kabat's dating of relations with Levin and Consolazio is off.

69. William V. Consolazio, Annual Report, Molecular Biology Program FY 1954, 2, Prog. Ann. Repts., 1952–57.

70. John T. Edsall, "Government and the Freedom of Science," *Science* 121 (29 April 1955): 615–19, esp. 618.

71. Howard K. Schachman, interview by author, 25 July 1990.

72. Divisional Committee for Biological and Medical Sciences to Director, "Security," 3 May 1954, attached to Minutes, BMS Divisional Committee, 1st Meeting, 3 May 1954.

73. William J. Hoff to Waterman, "To What Extent Should Considerations of Loyalty and Security Apply in Making Grants for Nonclassified Research," 18 May 1954, ATW Notes.

74. BMS made an award of $30,000 for three years for the project, "Configurations of Polypeptide Chains in Proteins," NSF *Annual Report FY 1955*, 112; England, *Patron for Pure Science*, 330, 417 n. 70. See also Thomas Hager, *Force of Nature: The Life of Linus Pauling* (New York: Simon & Schuster, 1995), 439–40. By FY 1958, Pauling had two grants from NIH as well as ONR and NSF funding. See *Federal Grants and Contracts for Unclassified Research in the Life Sciences, Fiscal Year 1958* (Washington, D.C.: NSF, 1961), 4.

75. England, *Patron for Pure Science*, 322, 330; "Special NSF Staff Meeting Notes," 29 April 1954, Senior Staff Meeting Notes—1954, HF.

76. Consolazio, Annual Report, Molecular Biology Program FY 1954, 11.

77. *NSF Annual Report FY 1955*, 5, 18–20; Leonard Scheele to Waterman, 21 March 1956, and attachments, ATW Notes.

78. *NSF Annual Report FY 1956*, 10–11. For the Academy report, see "Loyalty and Research: Report of the Committee on Loyalty in Relation to Government Support of Unclassified Research," Science (20 April 1956): 660–62.

79. Consolazio, Annual Report, Molecular Biology Program FY 1954, 5, 9.

80. Robert F. Griggs to Ross G. Harrison, Chairman, National Research Council, "The Prospects and Program of the Division of Biology and Agriculture," 12 October 1945, esp. 18–21, "Biology & Agriculture / Future Activities & Needs: Discussion," NAS Archives.

81. Wilson, interview, by England and Lomask, 24; Wilson, interview by Frank Edmonson, 12 November 1980, 21, HF.

82. Consolazio to J. Merton England, 13 February 1983, in folder "J. Merton England," HF.

83. See Rexmond C. Cochrane, *The National Academy of Sciences: The First Hundred Years, 1863–1963* (Washington, D.C.: National Academy of Sciences, 1978), 483–90; Harold J. Coolidge, "Biological Research in the Pacific Area," AIBS *Bulletin 4* (1954): 19–20; Annual Report of the Division of Biology and Agriculture, 1951–52, NAS Archives.

84. B&A Annual Report, 1951–52; William V. Consolazio Diary Note, "Meeting at the National Research Council on the Desirability of Convening a Conference on Photosynthesis," ca. 9 January 1952, and John Field to Waterman, "Summary Sheet on Proposal Contract for Activation of the Committee on Photobiology of the National Research Council," 29 April 1952, in "C6 National Academy of Sciences B58," Box 6, 59A-2463, NARA.

85. B&A Annual Report, 1951–52.

86. Field and Consolazio, "Conversation with Doctor Paul Weiss, Chairman of the Division of Biology and Agriculture, NRC," 17 December 1951, in "C6 National Academy of Sciences B58," Box 6, 59A-2463, NARA.

87. Paul Weiss, "Memorandum on the Proposed Establishment of a Committee Council in the Division of Biology and Agriculture of the National Research Council of the National Academy of Sciences," 24 January 1952, and "NRC Governing Board," 3 February 1952, in "B&A: Biology Council, Beginning of Program, 1952–1954," NAS Archives; B&A annual reports, 1951–52, 1952–53, NAS Archives.

88. "NRC Governing Board," 3 February 1952.

89. "Divisional Committee Meeting, May 16, 1952," Box 2, ODSF, copy in HF.

90. Frank H. Johnson to Weiss, 2 December 1952, "B&A: Biology Council, Beginning of Program, 1952–1954," NAS Archives.

91. Warren Weaver to Weiss, 9 June 1952, "B&A: Biology Council, Beginning of Program, 1952–1954," NAS Archives.

92. B&A Annual Report, 1952–53; Weiss to Detlev Bronk, 11 December 1952, "B&A: Biology Council, Beginning of Program, 1952–1954," NAS Archives.

93. Weiss to Bronk, 11 December 1952; Russell B. Stevens to Weiss, 21 December 1956, "B&A: Biology Council, Funding: Dept of Health Educ & Welfare," NAS Archives; Orr Reynolds, interview by author, 12 July 1988.

94. Russell B. Stevens to John T. Wilson, February 1956, "B&A: Biology Council, 1955–1956, Funding: Natl Sc Foundation Support Proposed," NAS Archives. Steinbach approved the NSF award only when committee chairman Dean Phillips of Emory University, not Weiss, was listed as principal investigator. 9 March 1954, 25 March 1954, 26 March 1954, Steinbach Diary Notes; NSF *Annual Report FY 1954*, 91.

95. "Biology Council," NAS-NRC *News Report* 4 (1954): 25–26; Russell B. Stevens, "The Biology Council," NAS-NRC *News Report* 5 (1955): 57–59. The original chairmen of the functional committees were Weiss (developmental), Stanley A. Cain (environmental and group), Ralph W. Gerard (regulatory), Berwind P. Kaufman (genetic and systematic), and F. O. Schmitt (molecular and cellular).

96. B&A Annual Report, 1953–54.

97. 25 January 1954, 16 February 1954, 20 May 1954, 24 May 1954, and 4 August 1954, Steinbach Diary Notes.

98. John S. Nicholas to Weiss, 2 November 1954, "B&A: American Inst of Biological Scs 1954/Separation from NRC," NAS Archives.

99. "Proposed Contract with AIBS for Assistance in Setting Up Committees and for Other Functions" 4 August 1954, Steinbach Diary Notes; *NSF Annual Report FY 1955,* 124.

100. "Excerpt from letter, dated July 19, 1954, from Dr. Paul Weiss to Dr. F. L. Campbell," attached to Campbell to L. A. Maynard, 5 August 1954, "B&A: Amer. Inst of Biological Scs/1954/Separation from NRC," NAS Archives.

101. H. Bentley Glass, "A Letter to All Biologists," *AIBS Bulletin* 4 (October 1954): 5. AIBS was restructured to allow for individual memberships. Detlev Bronk long held a grudge against Glass for taking AIBS out of the Academy. See Bronk to Glass, 27 December 1957, and enclosure, Bronk to Bostwick H. Ketchum, 27 December 1957, "Earth Sciences/Committee on Oceanography: Appointments, 1957," NAS Archives.

102. Apparently B&A requested $70,000 for three years. Weiss to L. A. Maynard, 23 November 1955, and G. D. Meid and Maynard to Waterman, 19 January 1956, "B&A: Biology Council, 1955–1956, Funding: Natl Sc Foundation Support Proposed," NAS Archives.

103. Minutes, BMS Divisional Committee, 5th Meeting, 21 April 1956.

104. Ernest M. Allen to L. A. Maynard, 18 March 1957, "B&A: Biology Council, Funding: Dept of Health Educ & Welfare Support, Long-term Support Proposed, 1956–1957," NAS Archives.

105. Russell B. Stevens to Drs. Maynard, Weiss and Phillips, 10 April 1957, and Weiss to Members of the Biology Council, 28 June 1957, "B&A: Biology Council, End of Program, 1957," NAS Archives.

106. B&A continued to seek support for restoration of the Biology Council into the 1960s. As chairman of B&A in 1959, Steinbach, too, supported the need for a Biology Council to give B&A a more well-rounded program. Frank L. Campbell to Members of Division of Biology and Agriculture, Letter 9, December 1959, Letter 35, 22 January 1963, "News and Views from the Division of Biology and Agriculture to its Members," NAS Archives.

107. In 1961, social scientist C. B. Baldwin, Jr. compared the strategies of NIH and NSF in seeking funds and concluded that "the fundamental cause of NIH's suc-

cess is the practical, humanitarian, and emotional appeal of its programs." NSF goals were, in comparison, more general, more difficult to understand, "and cannot be translated into personal terms as easily." C. B. Baldwin Jr., "Federal Support of Research by the National Institutes of Health and National Science Foundation: A Comparison and Analysis of the Factors Affecting their Appropriations," unpublished manuscript, 1961, p. 9, NIH Historian's Office Files, Manuscripts, National Library of Medicine. I thank Peter Hirtle for bringing this document to my attention.

108. See David H. DeVorkin, "Who Speaks for Astronomy? How Astronomers Responded to Government Funding after Word War II," to be published in *Historical Studies in the Physical and Biological Sciences;* Frank Edmondson, "AURA and KNPO: The Evolution of an Idea, 1952–58," *Journal for the History of Astronomy* 22 (1991): 68–86; and Allan A. Needell, "Lloyd Berkner, Merle Tuve, and the Federal Role in Radio Astronomy," Osiris, 2nd ser. 3 (1987): 261–88.

Chapter 5   Competing within a Pluralist Federal Funding System, 1952–1963

1. Alan T. Waterman, "Federal Support of Fundamental Research in the Biological Sciences," AIBS *Bulletin* 1(5) (October 1951): 14.

2. Louis Levin to Fernandus Payne, "Support of 'Nonprogrammatic' Research by Federal Agencies," 14 October 1952, in folder "Division of Biological and Medical Sciences (Misc.), 1951–1955," HF; Levin, "Comments on Federal Research Support of Life Sciences," 28 January 1953, Louis Levin Memos, 1954–58, HF.

3. Payne to Director, "Interim Report," 27 January 1953, filed with BMS annual reports, HF.

4. On military funding of the physical sciences see, for example, Paul Forman, "Behind Quantum Electronics: National Security as a Basis for Physical Research in the United States, 1940–1960," *Historical Studies in the Physical and Biological Sciences* 18 (1987): 149–229; A. Hunter Dupree, "National Security and the Post-war Science Establishment in the United States," Nature 323 (18 September 1986): 213–16; Daniel J. Kevles, *The Physicists* (New York: Knopf, 1977); Harvey M. Sapolsky, *Science and the Navy: A History of the Office of Naval Research* (Princeton: Princeton University Press, 1990); and Roger L. Geiger, "Science, Universities, and National Defense, 1945–1970," in *Science After '40,* edited by Arnold Thackray, Osiris 7 (1992): 26–48. Forman argues that physics "underwent a qualitative change in its purposes and character" due to military funding (p. 150).

5. Orr E. Reynolds, "Support of the Biological Sciences by the Office of Naval Research," *AIBS Bulletin* 2 (April 1952): 18–20; Orr E. Reynolds, interview by author, 12 July 1988.

6. See *Federal Grants and Contracts for Unclassified Research in the Life Sciences, Fiscal Year 1958* (Washington, D.C.: NSF, 1961), 20, 63. Wilson estimated annual army

expenditures for extramural research in biology and medicine to be about $10 million for FY 1955–58. See John T. Wilson to Deputy Director, "Information on science programs of other federal agencies," 6 February 1957, Div. Staff Papers I.

7. Nick A. Komons, *Science and the Air Force: A History of the Air Force Office of Scientific Research* (Arlington, Va.: Office of Aerospace Research, 1966), 30, 41, 65–66, 144–45. Unfortunately this source provides no information on investigators or projects in the life sciences. Wilson estimated that the air force spent about $5 million annually on extramural grants, FY 1955–58. Air force grants appear not to have been listed in *Federal Grants and Contracts in the Life Sciences, FY 1958*.

8. BMS Annual Report FY 1958, 23–24.

9. BMS Annual Report FY 1959, 26; BMS Annual Report FY 1960, 27.

10. See John A. Pitts, *The Human Factor: Biomedicine in the Manned Space Program to 1980,* NASA History Series (Washington, D.C.: NASA, 1985), esp. 80–82.

11. BMS Annual Report FY 1954, "Limited Distribution" version, 14–15, HF. Steinbach's comments were omitted from the version labeled "General Distribution."

12. Vernon Bryson and Rogers McVaugh, "The Status of Basic Research in Agriculture," April 1956, 1, 39, 44, in Div. Staff Papers I.

13. Minutes, BMS Divisional Committee, 4th Meeting, 25–26 November 1955, 5th Meeting, 21 April 1956, and 6th Meeting, 12–13 October 1956, HF.

14. Minutes, BMS Divisional Committee, 25th Meeting, 24–25 May 1963, 7.

15. Frank H. Kaufert and William H. Cummings, *Forestry and Related Research in North America* (Washington, D.C.: Society of American Foresters, 1955), 19–21.

16. *Forest Science* was launched in early 1955, before NSF funding was received. See BMS Annual Report FY 1955, 5, and E. L. Demmon, "Forest Science: A Quarterly Journal of Research and Technical Progress," *Forest Science* 1 (1955): 3–5.

17. BMS Annual Report FY 1961, 1, 13–14; Paul J. Kramer, "A Survey of Basic Research in Forestry," September 1961, 9, Div. Staff Papers I.

18. Kramer, "Survey of Basic Research in Forestry," 1.

19. John Beatty and Jack M. Holl, personal communications, ca. 1990.

20. Richard G. Hewlett and Jack M. Holl, *Atoms for Peace and War, 1953–1961: Eisenhower and the Atomic Energy Commission* (Berkeley: University of California Press, 1989), 576–77.

21. *Federal Funds for Science, IX,* 1959, 66–67.

22. John Beatty, "Genetics in the Atomic Age: The Atomic Bomb Casualty Commission, 1947–1956," in Keith R. Benson, Jane Maienschein, and Ronald Rainger, eds., *The Expansion of American Biology* (New Brunswick, N.J.: Rutgers University Press, 1991), 284–324.

23. See John Beatty, "Weighing the Risks: Stalemate in the Classical/Balance Controversy," *Journal of the History of Biology* 20 (1987): 289–319, esp. 301–11.

24. William V. Consolazio and Margaret C. Green, "Federal Support of Research in the Life Sciences," *Science* 124 (21 September 1956): 522–26.

25. *Federal Grants and Contracts for Unclassified Research in the Life Sciences, FY 1958,* 87–96. According to this data, AEC gave fewer than forty contracts in "genetic biology" in FY 1958, but since NSF was using its own definitions, molecular genetics would have been counted as "molecular biology."

26. George Lefevre Jr., Annual Report for Genetic Biology FY 1957, Prog. Ann. Repts., 1952–57.

27. On Calvin's laboratory, for which BMS helped to build a new building in the early 1960s, see Chapter 8.

28. Joel B. Hagen, *An Entangled Bank: The Origins of Ecosyste Ecology* (New Brunswick, N.J.: Rutgers University Press, 1992), 100–111, esp. 101. Odum obtained training in radiation ecology at the Nevada Proving Grounds and Hanford through a 1957 NSF senior postdoctoral fellowship (115–16).

29. On AEC support of ecology, see Chunglin Kwa, "Mimicking Nature: The Development of Systems Ecology in the United States, 1950–1975" (Ph.D. diss., University of Amsterdam, 1989); Hagen, *An Entangled Bank,* 100–121; Stephen Bocking, *Ecologists and Environmental Politics: A History of Contemporary Ecology* (New Haven: Yale University Press, 1997), esp. 63–88; and John Beatty, "Ecology and Evolutionary Biology in the War and Postwar Years: Questions and Comments," *Journal of the History of Biology* 21 (1988): 245–63, esp. 259–62.

30. J. Merton England, *A Patron for Pure Science: The National Science Foundation's Formative Years, 1945–57* (Washington, D.C.: NSF, 1982), 292–97. NSF supported design studies of the Midwest Universities Research Association (MURA) and tried to secure a portion of the funding of the Stanford Linear Accelerator (SLAC).

31. Stephen P. Strickland, *Politics, Science, and Dread Disease: A Short History of United States Medical Research Policy* (Cambridge, Mass.: Harvard University Press, 1972), chaps. 6 and 7. After the 1956 departure of Surgeon General Leonard Scheele, Waterman became increasingly involved in nonbiological issues and had little direct contact with NIH. See also James A. Shannon, "The Advancement of Medical Research: A Twenty-Year View of the Role of the National Institutes of Health," *Journal of Medical Education* 42 (1967): 97–118.

32. *The Advancement of Medical Research and Education through the Department of Health, Education, and Welfare: Report of the Secretary's Consultants on Medical Research and Education* (Washington, D.C.: HEW, 1958), 58.

33. BMS Annual Report FY 1956, 4.

34. NSB Minutes (Executive Session), 49th Meeting, 14 October 1957.

35. *Advancement of Medical Research and Education,* 5, 7, 29.

36. Ibid., 7.

37. Bowen C. Dees to Harry C. Kelly, "Apparent Trends in Federal Fellowship Support," 16 May 1956, and Waterman, "Historical Role of the Federal Government in Graduate Research and Education in the Sciences," Table 5, 13 January 1960, ATW Notes.

38. See *Federal Grants and Contracts for Unclassified Research* in *the Life Sciences, FY 1958,* 287–302.

39. Louis Levin to File, "Marquette University Proposal," 18 January 1957, ATW Notes.

40. H. Burr Steinbach, Diary Note, "Meeting of the Chairmen of the Divisional Committees of the National Science Foundation with the Director and certain other officers of the Foundation, November 25, 1958," included with Minutes, 12th Meeting, BMS Divisional Committee, 20–21 October 1958, HF. See also William G. Colson to Waterman, "Meeting of Chairmen of NSF Divisional Committees, November 25, 1958," ATW Notes.

41. Steinbach Diary Note, 25 November 1958; Minutes, 16th Meeting, BMS Divisional Committee, 26 May 1960, 4, HF.

42. Wilson noted these changes in his annual report under the heading "Indirect Policy Influences," as evidence of the beneficial effect of more liberal NSF policies on other agencies. BMS Annual Report FY 1956, 16–17.

43. *1989 NIH Almanac* (Bethesda, Md.: NIH, 1989): 124–25, 175.

44. Estelle ("Kepie") Engel, interview with author, 29 June, 14 July, and 14 November 1988.

45. U.S. NIH Study Committee, *Biomedical Science and Its Administration: A Study of the National Institutes of Health* (Washington, D.C.: The White House, 1965), 2–3.

46. Joseph D. Cooper, "Onward the Management of Science: The Wooldridge Report," Science 148 (11 June 1965): 1433–37.

47. National Science Foundation, *Federal Funds for Scientific Research and Development at Nonprofit Institutions, 1950–1951 and 1951–1952,* 12.

48. *Federal Funds for Science, I,* 14–15.

49. For FY 1953, NSF listed its life-science contribution of $1,159,000 entirely under the biological sciences. In FY 1959, NSF distributed $19,179,000 into $17,844,000 for biological and $1,329,000 for medical sciences. In FY 1963, NSF listed $40,000 under agricultural sciences. How such figures were arrived at is not known.

50. *Federal Funds for Science, IX,* 74. Of a total of $39.4 million for biological sciences, the NSF share was $17.8 million and the NIH share $4.0 million ($4.3 million for all of HEW).

51. NSF's share of "basic biological sciences" decreased from 45.2 percent to 28.9 percent between 1959 and 1963, largely because the AEC and NASA listed large portions of their research under this category. By 1963, the data show NSF becoming the largest single supporter of basic biological sciences, followed closely by the AEC.

52. The first volume appeared in 1954, the last in 1961. Copies of this relatively rare publication are in the NSF library.

53. *Federal Grants and Contracts for Unclassified Research in the Life Sciences, Fiscal Year 1958* (Washington, D.C.: NSF, 1961).

54. Wilson reported for several years NSF's share of federal extramural basic biological *and* medical research. This figure reached a high of 25 percent in FY 1959 (which Wilson declared "a relatively small segment"), declining to 20 percent the following year. How these figures were derived is not clear. See BMS Annual Report FY 1959, 6, and BMS Annual Report FY 1960, 3.

55. John T. Wilson to Deputy Director, "Information on science programs of other federal agencies," 6 February 1957, Div. Staff Papers I.

56. Minutes, BMS Divisional Committee, 12th Meeting, 21–22 October 1958, 2–3, HF.

57. BMS Annual Report FY 1958, 5.

58. Diary Note, "Telephone call from Dr. Graham DuShane, AAAS," 20 November 1959, in "Division of Biological and Medical Sciences," Box 23, ODSF.

59. Minutes, BMS Divisional Committee, 14th Meeting, 22–23 October 1959, 4, 6, HF.

60. Alan T. Waterman, "Introduction," in Vannevar Bush, *Science—The Endless Frontier* (Washington, D.C., NSF: 1960), vii–xxvi, esp. xii.

Chapter 6   Funding Individuals and Institutions in the 1960s

1. This committee was renamed the Committee on Science and Technology in 1975. See U.S. House of Representatives, *Toward the Endless Frontier: History of the Committee on Science and Technology, 1959–79* (Washington, D.C.: GPO, 1980.)

2. On science policy after Sputnik, see Jeffrey Stine, *Science Policy Study Background Report No. 1: A History of Science Polic in the United States, 1940–1985* (U.S. Congress, House, 99th Cong., 2nd sess., Serial R, September 1986), 41–55; Daniel J. Kevles, "Principles and Politics in Federal R&D Policy, 1945–1990: An Appreciation of the Bush Report," preface to Vannevar Bush, *Science—The Endless Frontier* (Washington, D.C.: NSF, 1990), ix–xxxiii, esp. xviii–xx; Roger L. Geiger, *Research and Relevant Knowledge: American Research Universities Since World War II* (New York: Oxford University Press, 1993), esp. 157–97.

3. See, for example, Leland J. Haworth, "Some Problem Areas in the Relationships between Government and Universities," *BioScience* 15 (1965): 339–45.

4. A. C. Leopold, "The Man in the White Lab Coat," *BioScience* 17 (1967): 233–35.

5. Harve J. Carlson, interview by author, 4 November 1988.

6. Herman Lewis, interview by author, 18 March 1991; Carlson interview.

7. Communication from Arthur Cronquist, New York Botanical Garden, ca. 1990; Lewis interview. See Joel B. Hagen, "Experimentalists and Naturalists in Twentieth-Century Botany: Experimental Taxonomy, 1920–1950," *Journal of the History of Biology* 17 (1984): 249–70. Keck was part of the celebrated team, with Jens

Clausen and William Hiesey, that carried out extensive work in experimental plant taxonomy in the 1920s and 1930s under Carnegie Institute of Washington sponsorship.

8. Lewis interview; Herman Lewis, interview by Charles Weiner, 10 November 1975, Recombinant DNA History Collection, 1966–78, Box 10, MC 100, MIT Archives. For Lewis's views on the role of the NSF program director, see Herman Lewis interview by Norman Kaplan, ca. 1967, Norman Kaplan Papers, American Philosophical Societ, Philadelphia.

9. David B. Tyler to Haworth, "Departmental Grants Program," 3 June 1965, BMS, DSF, 1965, and Tyler to Haworth, Wilson, Robertson, and Carlson, "Institutional-type Research Grants Program," 14 December 1966, BMS, DSF 1966. After leaving NSF in 1972, Tyler had a distinguished career as professor of pharmacology at the College of Medicine, University of South Florida.

10. NSF press release 58-145, 2 September 1958. A complete set of press releases is located in the Office of Legislative and Public Affairs, NSF.

11. A few men with Ph.D.s held these positions beginning with Carl A. Kuether, associate program director for molecular biology, in FY 1962. See BMS Annual Report FY 1962, 23, HF.

12. For example, Lewis recalled that program directors routinely joined Carlson for lunch. Lewis interview by author.

13. See, for example, George T. Mazuzan, *The National Science Foundation: A Brief History* (Washington, D.C.: NSF, 1988), 13; Haworth, "Some Problem Areas," 345; Geiger, *Research and Relevant Knowledge,* 173–229. Nathan Reingold questions the widespread belief in a "golden age" and notes that for many scientists, the post-Sputnik era was an age of anxiety. See Nathan Reingold, "Science and Government in the United States Since 1945," *History of Science* 23 (1994): 361–85.

14. For example, in 1962, when requested by the NSF administration to supply and defend a "blue sky" figure, BMS argued for $115 million for FY 1964, which was nearly three times its 1963 budget. Minutes, BMS Divisional Committee, 22nd Meeting, 25–26 May 1962, 6, HF.

15. Minutes, BMS Divisional Committee, 28th Meeting, 2–3 October 1964, 5, HF. BMS Annual Report FY 1968, 1.

16. Wilson recalled that Consolazio once had "a hell of a row" on the issue of overhead with Lee DuBridge, Chairman of the Committee on Sponsored Research of the American Council on Education, and one of the chief exponents of full costs of research. "DuBridge just was furious." John T. Wilson, interview by J. Merton England and Milton Lomask, 21 May 1974, 48, HF.

17. American Council on Education, "Recommendations on Faculty Salaries Charged to Government Contracts: A Statement of the Committee on Sponsored Research Endorsed by the Commission on Federal Relations" (Washington, D.C.: American Council on Education, 1963), copy in BMS Faculty Salaries; President's

Science Advisory Committee, *Scientific Progress, the Universities, and the Federal Government* (Washington, D.C.: The White House, 1960).

18. In 1963 NIH paid full salaries of temporary faculty in medical schools while the school usually still paid permanent faculty salaries. By the end of the decade, NIH also regularly paid salaries of permanent tenured faculty. Division of Biological and Medical Sciences, "Payment of Faculty Salaries," 14 January 1963, and David D. Keck, "Faculty Salaries," 17 January 1964, attached to David Tyler to Members, BMS Divisional Committee, "Program for Support of Faculty Salaries," 3 March 1964, and Harve J. Carlson to Professional Staff, "Robertson Plan on Faculty Salaries," 7 February 1968, BMS Faculty Salaries.

19. William G. Rothstein, *American Medical Schools and the Practice of Medicine: A History* (New York: Oxford University Press, 1987), 224–26; Rosemary Stevens, *American Medicine and the Public Interest* (New Haven: Yale University Press, 1971), 357–73, esp. 359.

20. Waterman, "For the Personal Attention of Presidents of Educational Institutions Applying for Research Support from the National Science Foundation," 3 June 1960, and attachment, "Amendment to National Science Foundation January 1960 edition on brochure on Grants for Scientific Research," 3 June 1960, BMS Faculty Salaries.

21. Minutes, BMS Divisional Committee, 15th Meeting, 21 January 1960, 3, HF; Keck, "Faculty Salaries."

22. Randal Robertson, "AD(R) Instruction 1963.2. Academic Year Faculty Salary Reimbursement," 1 October 1963, BMS Faculty Salaries.

23. Waterman to NSB, "Faculty Salary Procedures," 14 May 1963, Carlson to Professional Staff, "Robertson Plan on Faculty Salaries," BMS Faculty Salaries. Later versions called for payment of 65 percent of costs of estimated time. Robertson to Haworth, "Faculty Salary Plan," 2 April 1965, BMS Faculty Salaries.

24. Tyler, "Program for Support of Faculty Salaries."

25. Eugene Hess, interview by author, 5 June 1989, Lewis interview by author.

26. Wilson interview, 21 May 1974, 32–33.

27. Stephen P. Strickland, *Politics, Science, and Dread Disease: A Short History of United States Medical Research Policy* (Cambridge, Mass.: Harvard University Press, 1972), 169–78.

28. On the AIBS Affair, see J. Merton England, "The National Science Foundation and Curriculum Reform: A Problem of Stewardship," *The Public Historian* 11 (1989): 23–36; Daniel S. Greenberg, "American Institute of Biological Sciences Accused of Misuse of NSF Grant Funds," Science 139 (25 January 1963): 317 21; James D. Ebert, "'Biology on the Cuff'—Is AIBS Worth Saving? An Open Letter from the Institute's President," *Science* 139 (25 January 1963): 321–22; Greenberg, "AIBS: Emergency Meeting of Board Produces Steps Designed to Promote Financial Solvency and Confidence," *Science* 139 (1 February 1963), 392; Greenberg, "AIBS:

Happy Ending in Prospect, But Case Adds to Congressional Skepticism on Support for Science," *Science* 139 (1 March 1963): 814–15; Minutes, BMS Divisional Committee, 23rd Meeting, 12–13 October 1962, 8–11, HF; NSB Minutes, 81st Meeting, 15–17 November 1962, 18; and James D. Ebert, interview by author, 27 March 1988. See additional manuscript sources in folder "AIBS," Box 69, ODSF, copies in ATW Notes.

29. On Rosenthal, see NSF press release, 61-144, 5 October 1961.

30. Greenberg, "AIBS: Happy Ending in Prospect," 814.

31. Consolazio to Waterman, 6 May 1963, Consolazio Memos, HF. See also Consolazio, "Annual Report of Activities of the Program for Molecular Biology for Fiscal Year 1961," Prog. Ann. Repts., 1961–68, NARA.

32. John Mehl, "Annual Report of Activities of the Program Director for Molecular Biology for Fiscal Year 1963," 10 July 1963, Prog. Ann. Repts., 1961–68. Mehl's reflective critique was copied verbatim (and unattributed) by Carlson into his BMS Annual Report FY 1963, 21–26.

33. Wilson interview, by England and Lomask, 2.

34. Carlson to BMS Divisional Committee, 14 October 1964, and enclosure, Howard J. Teas and David D. Keck, "Biological Sciences Assistance Program (BSAP)," 1 June 1964, in folder "BMS Divisional Committee Meeting, 28th- 10/2–3/64," Box 20, 70A-2191, NARA.

35. Ibid.

36. Lewis interview by author; Estelle ("Kepie") Engel, interview by author, 29 June, 13 July, and 14 November 1988.

37. Minutes, BMS Divisional Committee, 28th Meeting, 2–3 October 1964, 4, HF; Paul J. Kramer, "1965 Annual Report to the National Science Board of the Advisory Committee for Biological and Medical Sciences," 20 January 1966, Adv. Comm. Ann. Repts.

38. Wilson to Carlson, "BMS Annual Report for FY 1964," 24 August 1964, filed with BMS annual reports.

39. On Wilson's aspirations, see John T. Wilson, *Academic Science, Higher Education, and the Federal Government, 1950–1983* (Chicago: University of Chicago Press, 1983). Wilson took a dim view of the U.S. Office of Education. He would have transferred the OE Bureau of Higher Education to NSF and limited the OE to providing block grants to states for secondary education. Wilson interview, by England and Lomask, 4, 6, 8–9.

40. Consolazio, for example, in January 1953, suggested that biology, compared to either the physical or medical sciences, could be likened to the youngest child in a large family receiving vacated buildings as cast-off clothing. "The results are that one finds biology today needing desperately more up-to-date laboratory space and new modern equipment and tools." Consolazio to Fernandus Payne, "The State of the Biological Sciences Division," 15 January 1953, 9–10, filed with BMS annual reports.

41. NSF officially supported the 1956 Act. Waterman to Roger W. Jones, 11 January 1956, 24 July 1956, ATW Notes.

42. Minutes, BMS Divisional Committee, 6th Meeting, 12–13 October 1956, 6–7, HF; Minutes, BMS Divisional Committee, 8th Meeting, 56 April 1957, 3–4, HF.

43. Minutes, BMS Divisional Committee, 10th Meeting, 21 January 1958, 5, HF.

44. "Position Paper on the Support of College and University Science Laboratory Refurbishment," 4 March 1958, and attachment, "Laboratory Facility Grants by NIH Falling in NSF Areas of Interest," in folder "Laboratory Refurbishment," Box 31, ODSF.

45. Ibid, 7; Office of the Director, "Position Paper on Alternate Courses of Action for Consideration by the Board Regarding Grants for the Equipping, Renovation or Construction of College and University Science Laboratories," 4 May 1958, 3, in folder "Laboratory Refurbishment," Box 31, ODSF.

46. Waterman to Senior Staff, 28 May 1959, ATW Notes.

47. Joshua Leise, interview by author, 26 April 1991; Howard J. Teas to C. H. Carter, OIP, "UL 1039," 19 August 1963, and Aubrey W. Naylor to Carter, "UL 651," 27 March 1962, in folder "Memoranda–Sept. '61," Box 7, 70A-3915, NARA.

48. Lists of awards are found in NSF annual reports through 1963 and the annual NSF publication, *Grants and Awards,* after 1963.

49. Leise interview.

50. Leise recalled that funding was based on institutional or departmental merit and not on any geographical criterion. Another program director, J. Merton England, noted that much reference was made to an institution's "potential." Leise interview; J. Merton England, interview by author, 24 April 1991.

51. Waterman to Phillip S. Hughes, 20 July 1959, ATW Notes. See also Waterman to Maurice H. Stans, 18 March 1959, ATW Notes.

52. Wilson interview by England and Lomask, 12.

53. See J. Merton England, "Institutional Grants of the National Science Foundation," *Science* 148 (25 June 1965): 1693–96; England, "Investing in Universities: Genesis of the National Science Foundation's Institutional Programs, 1958–1963," *Journal of Policy History* 2 (1990): 131–56, esp. 139–45; and *The NSF Science Development Programs, Vol. 1: A Documentary Report* (Washington, D.C.: NSF, 1977), 7–8, 18–19.

54. England interview.

55. This was done in order to avoid duplicating the base. England, "Investing in Universities"; England interview.

56. Minutes, BMS Divisional Committee, 20th Meeting, 27–28 October 1961, 2, HF.

57. See Waterman, "Notes for Board Meeting, January 19, 1961," Waterman to

NSB, "National Professorship Program," 2 March 1961, and attachment, "Announcement of a Program of National Science Professorships," and Waterman to NSB, "National Science Professorship Program," 23 March 1961, and attachment, Waterman to David E. Bell, 7 July 1961, ATW Notes; England, "Investing in Universities," 145.

58. See James A. Shannon and Charles V. Kidd, "Federal Support of Research Careers," *Science* 134 (3 November 1961): 1399–1402.

59. PSAC, *Scientific Progress, the Universities, and the Federal Government* (n. 20), 14, 28.

60. See England, "Investing in Universities," 147–51; *The NSF Science Development Programs, Vol. 1: A Documentary Report* and *Vol. 2: Budgets, Statements of Goals, and Grantees' Summaries of Grant Impact* (Washington, D.C.: NSF, 1977); and Howard E. Page, "The Science Development Program," in Harold Orlans, ed., *Science Policy and the University* (Washington, D.C.: Brookings Institution, 1968), 101–19.

61. In the education division, the College Science Improvement Program supplied funds for equipment to undergraduate colleges.

62. See National Science Foundation, *The NSF Science Development Programs* (Washington, D.C.: NSF, 1977). The figure for biology included $36 million in "biological and medical sciences" and $2 million in biochemistry which was classified with chemistry (ibid., vol. 1: 68–69).

63. On Georgia, see *The NSF Science Development Programs*, vol. 2: 14–17. This volume consists of university administrators' evaluations of the effect of the awards on their institutions.

64. On Iowa, see ibid, 19–21.

65. These programs and the difficulties of evaluating the SDP are discussed in The *NSF Science Development Programs*, vol. 1.

66. Tyler proposed five-year awards of up to $100,000 a year to be spent at the discretion of the departmental chairman for salaries of junior staff, summer salaries, technical assistance, research equipment, and travel. Tyler to Haworth, "Departmental Grants Program," 3 June 1965, BMS, DSF 1965.

67. Report, BMS Advisory Committee, 4th Meeting, 11–12 May 1967, 2–4, and Report, BMS Advisory Committee, 6th Meeting, 2–3 May 1968, 1, 12, and Appendix D, 4, in Adv. Comm.

68. Wilson to Carlson, "BMS Annual Report for FY 1964," 24 August 1964, filed with BMS annual reports.

69. Geiger, *Research and Relevant Knowledge*, 174, 195–97.

Chapter 7   Promoting Big Biology

1. Derek J. de Solla Price, *Little Science, Big Science* (New York: Columbia University Press, 1963), esp. 2. Price was one of the founders of scientometrics. On the origin of the concept of "big science" and its variety of meanings, see James Cap-

shew and Karen A. Rader, "Big Science: Price to the Present," in Arnold Thackray, ed., *Science After '40, Osiris* 7(1992): 3–25.

2. Alvin M. Weinberg, "Impact of Large-Scale Science on the United States," *Science* 134 (21 July 1961): 161–64; Weinberg, *Reflections on Big Science* (Cambridge, Mass.: MIT Press, 1967), esp. 39. Weinberg warned scientists and federal administrators of the dangers of "Big Science," which he felt should therefore be confined to national laboratories.

3. Robert W. Smith, *The Space Telescope: A Study of NASA, Science, Technology, and Politics* (Cambridge: Cambridge University Press, 1989), 373–98; Peter Galison, "Introduction: The Many Faces of Big Science," in Peter Galison and Bruce Hevly, eds., *Big Science: The Growth of Large-scale Research* (Stanford: Stanford University Press, 1992), 1–7; and Bruce Hevly, "Afterword: Reflections on Big Science and Big History," in ibid., 355–63.

4. *NSF Annual Report FY* 1963, 61–66.

5. See Daniel S. Greenberg, *The Politics of Pure Science: An Inquiry into the Relationship between Science and Government in the United States* [1967] (New York: Plume Books, 1971), 170–206.

6. On astronomy, see Allan A. Needell, "Lloyd Berkner, Merle Tuve, and the Federal Role in Radio Astronomy," *Osiris,* 2nd series, 3 (1987): 261–88; David H. DeVorkin, "Who Speaks for Astronomy? How Astronomers Responded to Government Funding after Word War II," to be published in *Historical Studies in the Physical and Biological Sciences;* and Frank K. Edmondson, "AURA and KNPO: The Evolution of an Idea, 1952–58," *Journal for the History of Astronomy* 22 (1991): 68–86. On meteorology, see George T. Mazuzan, "Up, Up, and Away: The Reinvigoration of Meteorology in the United States: 1958 to 1962," *Bulletin of the American Meteorological Society,* 69 (1988): 1152–63.

7. Minutes, BMS Divisional Committee, 12th Meeting, 20–21 October 1958, 8, HF.

8. BMS Annual Report FY 1959, 11, HF.

9. For example, NSF treated small prefabricated growth chambers, formerly considered specialized facilities, as research equipment. The graduate laboratory program handled small departmental museums, but Special Facilities funded large national or regional museums. BMS Annual Report FY 1959, 27–28; BMS Annual Report FY 1960, 12–13.

10. On the Rockefeller Foundation's initiatives to sponsor cooperative projects in biology, see, for example, Robert E. Kohler, *Partners in Science: Foundations and Natural Scientists, 1900–1945* (Chicago: University of Chicago Press, 1991), and Lily E. Kay, *The Molecular Vision of Life: Caltech, the Rockefeller Foundation and the Rise of the New Biology* (New York: Oxford University Press, 1993).

11. James F. Bonner, oral history, 13–14 March 1980, Caltech Archives.

12. Frits W. Went, "The Phytotron," *Engineering and Science* 12(9) (June 1949):

3–6 (copy in Caltech Archives). See NSF annual reports and Frits W. Went, *The Experimental Control of Plant Growth, with Special Reference to the Earhart Plant Research Laboratory at the California Institute of Technology* (New York: Ronald Press, 1957).

13. H. Burr Steinbach, "Conversation with Dr. F. C. Steward, Cornell University, concerning funds for a phytotron," 18 August 1954, Steinbach Diary Notes.

14. House Committee on Appropriations, Subcommittee, *Hearings on Independent Offices Appropriations for 1957*, 84th Cong., 2nd sess., 1957, 588; Senate Committee on Appropriations, Subcommittee, *Hearings on H.R. 11574, Independent Offices Appropriations for 1959*, 85th Cong., 2nd sess., 1956, 301.

15. "Report of the Committee on Controlled Environment Laboratories," 13 March 1957, Ad hoc Comm. Repts., NARA. BMS often funded ad hoc committees through contracts to scientific societies as a means of using research funds rather than scarce program management funds. Usually the BMS staff named the committee members.

16. Sterling B. Hendricks and Frits Went, "Controlled-Climate Facilities for Biologists," *Science* 128 (5 September 1958): 510–12; "Controlled-Climate Facilities: A Report of a Committee of the Botanical Society of America," February 1959, copy furnished by the staff of The Biotron, University of Wisconsin. This report has not been located in NSF records.

17. The Biotron staff placed the entire initiative with NSF. See "History of the Wisconsin Biotron," n.d., University of Wisconsin–Madison Archives. Wilson told the National Science Board that the *Science* article had said that NSF would entertain proposals. This was by no means clear from the article. In any case, no formal solicitation was made.

18. Transcript, Biotron Conference, University of Wisconsin, 10–12 December 1959, The Biotron, University of Wisconsin.

19. NSB Minutes, 60th Meeting, 24–25 May 1959, 7–8; Louis Levin, Annual Report of the Facilities Program for 1959, 29 June 1959, HF.

20. Transcript, Biotron Conference, 14–15; BMS Annual Report FY 1959, 13.

21. Leland J. Haworth, Memorandum to Members, National Science Board, 10 January 1966, BMS, DSF 1966. Award amounts are found in NSF annual reports, FY 1963–66 and 1968.

22. "SEPEL. The Southeastern Plant Environment Laboratories of Duke University and North Carolina State University. A History of Their Planning and Financing . . . ," n.d., Phytotron Records, 1955–78, Botany Department Records, Duke University Archives. Project Summaries, FY 1968, are found in "Fact Book, Facilities and Special Programs, Division of Biological and Medical Sciences, National Science Foundation" (hereafter Fact Book), April 1968, items N2 and N3, HF. Kramer chaired the universities' joint Phytotron Board from 1962 on.

23. Summary Sheet, University of Wisconsin, Senn, 16 November 1965, Fact

Book, item N4; "History of the Wisconsin Biotron"; Charles Baum and Theodore Tibbitts, interview by author, 30 October 1991.

24. See Harold A. Senn, "University of Wisconsin Biotron, Operation of the Biotron, National Science Foundation Grant GB-4371, Report for the period September 1, 1965 to August 31, 1966," University of Wisconsin–Madison Archives. Senn, a Canadian botanist, directed the Biotron from 1960 to 1977. James Bonner, interview by author, 2 August 1990; John L. Brooks, interview by author, 26 September 1988; Josephine Doherty, interview by author, 10 April and 28 August 1990; Baum and Tibbitts, interview. Now funded by the state of Wisconsin, the Biotron is used for both applied and basic research in areas such as agriculture, veterinary medicine, and pharmacology. Industries also rent space to test products in adverse weather conditions.

25. BMS program director Howard J. Teas sarcastically commented that "The NCS [North Carolina State] writeup sounds as though the applied agricultural people who have been doing greenhouse and field experiments up to now can barely wait to get into a phytotron to solve economic problems of the Reynolds Tobacco Company and other farm and industry groups." Teas to Jack T. Spencer, "Phytotron Proposal from Duke University & North Carolina State—BF-306," 8 January 1964, in "Review of Proposal," Box 7, 70A-3915, NARA.

26. Bonner, interview; *Phytotron Duke University, Durham, North Carolina* (Durham: Duke University [ca. 1983]) [24 page brochure].

27. BMS Annual Report FY 1960, 2.

28. "Research Priorities," 19 October 1959, in folder "Office of the Director Letters (1954–63) for Historical Purposes," HF; BMS Annual Report FY 1960, 2; ibid., FY 1961, 11–14.

29. On computers in the life sciences and forestry, see William D. Neff, "Report on the use of Computers in the Biological Sciences," 20 February 1963, Div. Staff Papers II, and Paul Kramer, "A Survey of Basic Research in Forestry," September 1961, Div. Staff Papers I; BMS Annual Report FY 1960, 27 and FY 1961, 12–14. BMS supported the NAS-NRC Committee on Computers in Life Science Research, organized a conference, and hired Neff as a consultant to survey U.S. computer centers.

30. Minutes, BMS Divisional Committee, 15th Meeting, 21 January 1960, 3, HF.

31. Minutes, BMS Divisional Committee: 12th Meeting, 20–21 October 1958, 2, 8–9, quotations on p. 8, HF; 13th Meeting, 30 April 1959, 1; 14th Meeting, 22–23 October 1959, 5, HF; 20th Meeting, 27–28 October 1961, esp. p. 3, HF.

32. On BMS funding of biological oceanography, see Toby A. Appel "Marine Biology/Biological Oceanography and the Federal Patron: The NSF Initiative in Biological Oceanography in the 1960s," in Keith R. Benson and Philip F. Rehbock, eds., *The Pacific and Beyond: Proceedings of the Fifth International History of Oceanography Meeting* (Seattle: University of Washington Press, in press).

33. Rexmond C. Cochrane, *The National Academy of Sciences: The First Hundred Years, 1863–1963* (Washington, D.C.: NAS, 1978), 497–507; "American Oceanography Survey and Proposals for Ten-Year Program Made by NAS-NRC Committee," *Science* 129 (27 February 1959): 550–51; NAS Committee on Oceanography, *Oceanography 1960 to 1970* (Washington, D.C.: NAS-NRC, 1959–62). Further stimulation for support of oceanography came from the Special Committee on Oceanic Research of the International Council of Scientific Unions, an umbrella group associated with the United Nations.

34. Wilson to Waterman, "Current Controversy Between Physical Oceanographers and Biological Oceanographers," 2 March 1960, in Division of Biological and Medical Sciences, 1960, Box 42, ODSF.

35. George Sprugel to Members, BMS Divisional Committee, "Foundation's Role in Biological Oceanography," 20 May 1960, Div. Staff Papers I; Wilson to Waterman, 2 March 1960; "A Report to the National Science Foundation by the *ad hoc* Committee on Biological Oceanography," August 1961, 2–3, Div. Staff Papers I. I am grateful for discussions with historian of oceanography Eric Mills.

36. Sprugel, "Foundation's Role," 20 May 1960; Bostwick H. Ketchum, "The New Oceanographic Research Vessel," *AIBS Bulletin* 10(2) (April 1960): 19–20; Minutes, BMS Divisional Committee, 16th Meeting, 26 May 1960, 3–4.

37. Sprugel, "Foundation's Role," 20 May 1960, 10–11.

38. Wilson to Dixy Lee Ray, 8 June 1960, enclosed with memorandum, Wilson to Waterman, 10 June 1960, folder "Division of Biological & Medical Sciences," Box 42, ODSF; Minutes, BMS Divisional Committee, 18th Meeting, 3 March 1961, 5–6 and 20th Meeting, 27–28 October 1961, 6, HF. Harve Carlson, who knew Ray through his former position as scientific director of ONR's San Francisco office, claimed he had initiated her appointment. Harve Carlson, interview by author, 4 November 1988.

39. Report, *ad hoc* Committee on Biological Oceanography, 5. See also Harve J. Carlson, "Biological Oceanography," January 1966, 3, and John S. Rankin, "Role of the National Science Foundation in Support of Biological Oceanography FY 1958–1965," January 1966, 11, Div. Staff Papers II.

40. Report, *Ad hoc* Committee on Biological Oceanography, 10. Committee members represented major centers of marine biology and oceanography: Rolf Bolin (Stanford), Ralph Emerson (Berkeley), Erling Ordal (University of Washington), John Ryther (WHOI), H. Burr Steinbach (Chicago and MBL), and Karl Wilbur (Duke). They took a generally elitist view of funding priorities stating that "under no circumstances should the available funds be spread so thin that the best laboratories are inadequately supported" (p. 18).

41. Rankin, "Role of NSF in Support of Biological Oceanography," Appendix.

42. See Rankin, "Role of NSF in Support of Biological Oceanography"; Carl-

son, "Biological Oceanography"; and George Sprugel, Jr., "National Science Foundation," *AIBS Bulletin* 13(5) (October 1963): 33–34 [in an issue on marine biology]. BMS support of oceanography and marine biology represented just one part of the Foundation's commitment. The Division of Mathematical, Physical, and Engineering Sciences, which coordinated the Foundation's participation in IIOE, the Office of Antarctic Programs, which sponsored biological projects on the NSF-owned *U.S.N.S. Eltanin,* and the U.S.-Japan Cooperative Program also funded oceanography projects. Many of BMS's awards were split-funded with other arms of NSF.

43. Jack T. Spencer, Annual Report, Facilities and Special Programs, FY 1968, 7, Prog. Ann. Repts, 1961–68.

44. Carlson, "Biological Oceanography," 11–12.

45. Ibid., 13–14.

46. BMS Annual Report FY 1965, 9. Plans to build the Bodega Laboratory became embroiled in the politics surrounding the ill-fated Bodega Bay Nuclear Plant. See J. Samuel Walker, "Reactor at the Fault: The Bodega Bay Nuclear Plant Controversy, 1958–1964—A Case Study in the Politics of Technology," *Pacific Historical Review* 59(1990): 323–48; and Joel Hedgpeth Papers, Bancroft Library, University of California, Berkeley.

47. *NSF Annual Report FY 1964,* 58; Annual Report, Facilities and Special Programs, 1968, 8–9; *Research Vessel Eastward: Fifteen Years of Service,* compiled by Sharlene and Orrin Pilkey (Durham, N.C.: Duke University, n.d.), copy in Duke University Archives; Hopkins Marine Station of Stanford University, *Bulletin* 1967; "Stanford Oceanographic Expeditions 1969, Biological Oceanography for Graduate Students, Hopkins Marine Station, Stanford University," flyer, copy supplied by Hopkins Marine Station. Information on NSF support of the *Eastward* and *Te Vega* can be found in Box 5, 76-172, NARA.

48. Annual Report, Facilities and Special Programs, 1968, 8.

49. See Elizabeth Noble Shor, *Scripps Institution of Oceanography: Probing the Oceans, 1936 to 1976* (San Diego: Tofua Press, 1978), 167–83; Annual Report, Facilities and Special Programs, 1968, 8–9; *NSF Annual Report FY 1966,* 63–65.

50. Carolyn H. Krooskos, "Guide to the Records of the S.I.O Alpha Helix Program Management Office, 1966–1980, Archival Collection—AC 7," 1985, Scripps Institution of Oceanography Archives; Shor, *Scripps,* 168; Per F. Scholander, "Rhapsody in Science," *Annual Review of Physiology,* 40 (1978): 1–17, esp. 14.

51. *NSF Annual Report FY 1963,* 54.

52. Carlson, "Biological Oceanography," 7. For NSF involvement in IIOE, see *NSF Annual Report FY 1962,* 48–50; *FY 1963,* 54–55; *FY 1964,* 53–55; 1965, 97–98; *FY 1966,* 58–62.

53. David D. Keck, "Report on Asian Trip Made in Order to Gather Information about Marine Biological Research Facilities and the Forthcoming Indian Ocean Expedition, August 18–September 26, 1960," 24 October 1960, Div. Staff Papers I;

BMS Annual Report FY 1961, 5; Minutes, BMS Divisional Committee, 17th Meeting, 21–22 October 1960, 4, HF; *NSF Annual Report FY 1964,* 54.

54. Waterman to Jerome B. Wiesner, 24 April 1961, Waterman to King of Denmark, 5 December 1962, and Waterman to Ragnar Rollefson, 8 December 1962, ATW Notes.

55. Carlson, "Biological Oceanography," 8, 21–24; John H. Ryther, "International Indian Ocean Expedition," *AIBS Bulletin* 13(5) (October 1963): 48–51.

56. Annual Report, Facilities and Special Programs, 1968, 11.

57. M. B. Schaefer to Frederick Seitz, 24 March 1965, and Haworth to Seitz, 8 May 1965, in folder "R/V Anton Bruun 1965," Box 14, 70A-3621, NARA.

58. Annual Report, Facilities and Special Programs, 1968, 12. Figures for program costs were derived from NSF annual reports.

59. M. Dale Arvey and William J. Riemer, "Inland Biological Field Stations of the United States," *BioScience* 16 (1966): 249–54, esp. 251.

60. Ibid.

61. Dale Arvey to John T. Wilson and Randal M. Robertson, "Conference on Inland Biological Field Stations," 26 May 1964, BMS, DSF 1964.

62. *The University of Michigan Biological Station, 1909–1983* (University of Michigan Biological Station, 1983), 21. Data on BMS support of inland field stations is found in Fact Book, items U1–U6.

63. Annual Report, Facilities and Special Programs, 1968, 15–17.

64. David D. Keck, "Tropical Biology," August 1961, 3, Div. Staff Papers I.

65. Ibid., 11.

66. Frank L. Campbell to William J. Robbins, 25 July 25 1959, in folder "Tropical Botany—General," NAS Archives.

67. "Conference on Tropical Botany, Fairchild Tropical Garden, May 5–6–7, 1960," NAS-NRC Publication 822 (Washington, D.C.: NAS-NRC, 1960), pp. iii–iv, 1–16.

68. Keck, "Tropical Biology," 5–7; Donald E. Stone, "The Organization for Tropical Studies (OTS): A Success Story in Graduate Training and Research," in Frank Almeda and Catherine M. Pringle, eds., *Tropical Rainforests: Diversity and Conservation* (San Francisco: California Academy of Sciences and Pacific Division, AAAS, 1988), 143–87, esp. 145–46.

69. Frank L. Campbell to John T. Wilson, 21 December 1960, David D. Keck to William J. Robbins, 19 January 1961, in folder "Tropical Plant Science Committee/Establishment," NAS Archives. Keck suggested that such activities might be fundable under the recently passed American Republics Cooperation Act. He further explained that if funds were appropriated for scientific projects under this act and were administered through NSF, grants would be processed by the Office of Special International Programs, and not by BMS.

70. Minutes, Executive Committee, Division of Biology and Agriculture, March 8, 1961, in folder: "Biology & Agriculture 1961/Executive Com," NAS Archives. By the end of the year the Executive Committee was advocating a Tropical Science Board that "should be concerned with parasitology, climatology, soils, etc. as well as with botany." Minutes, Executive Committee, Division of Biology and Agriculture, December 9, 1961," in ibid.

71. Keck, "Tropical Biology," 11; Walter H. Hodge and David D. Keck, "Biological Research Centers in Tropical America," The Association for Tropical Biology, *Bulletin No. 1* (1962): 107–20, manuscript version in Div. Staff Papers I.

72. William L. Stern, interview by author, 17 September 1992; John W. Purseglove, "Neotropical Botany Conference," *Plant Science Bulletin* 8(3) (September 1962): 4–7; "News and Notes," in ibid., 9. On the Hodge and Keck report, see previous note.

73. University of Michigan, "Preliminary Draft/Second Report of the Committee on a Proposed Center for Tropical Studies," November 10, 1959, and University of Michigan,"Proposal for the Establishment of a Center for Tropical Studies," February 3, 1962, OTS North American Office files, Duke University. I thank Donald Stone for providing access to these records.

74. Stone, "OTS," 145–47; Jay M. Savage, "Final Report on the Conference on Problems in Education and Research in Tropical Biology, Held at the Universidad of Costa Rica, April 23–27, 1962," n.d., folder "OTS-Formative Steps," Box 17, OTS Papers.

75. Norman Hartweg, Chairman, memorandum to Organizing Committee, Organization for Tropical Studies, Inc., "Summary of Coral Gables meeting of institutional representatives, March 8, 1963," and Jay M. Savage, "Organization for Tropical Studies, Minutes of the Board of Directors Meeting, June 25–26, 1963," folder "OTS-Formative Steps," Box 17, OTS Papers.

76. According to Stone, at the time of its formation there was "an explicit understanding or expectation at least that OTS would be designated as a 'national laboratory' comparable to the Kitt Peak National Observatory or the National Center for Atmospheric Research." Stone, "OTS," 182. For a claim of "national laboratory" status, see "The Program of Tropical Studies and Research of the Organization for Tropical Studies, Inc.," 5 August 1966, 9, folder "OTS-Board Minutes/Advisory Council Minutes," Box 17, OTS Papers.

77. Reed C. Rollins to Members of the Board of Directors, OTS, 13 September 1965, folder "OTS—NSF Grant (Original)," Box 17, OTS Papers.

78. Stone, "OTS," 150–54, 184; National Science Foundation, *Grants and Awards for the Fiscal Year Ended June 30, 1968* (Washington, D.C.: GPO, 1969), 220. The NSF education division also supported undergraduate education in the tropics through a grant to Associated Colleges of the Midwest, a consortium of ten colleges,

which, like OTS, operated a program in Costa Rica. See William J. Riemer, "Tropical Science: A Report to the Biological and Medical Sciences Advisory Committee, Orally Presented 29 October 1965," 4, Div. Staff Papers II.

79. Robert F. Inger to James S. Bethel, 11 August 1967, OTS North American Office files, Duke University; Stone, "OTS," 171–72. See also L. D. Gómez and J. M. Savage, "Searchers on that Rich Coast: Costa Rican Field Biology, 1400–1980," in Daniel H. Janzen, ed., *Costa Rican Natural History* (Chicago: University of Chicago Press, 1983), 1–11, esp. 9.

80. *Director's Program Review, Biological Sciences,* 16 January 1973 (Washington, D.C.: NSF, 1973), 24. Copies of these internal printed reports are in HF.

81. *NSF Annual Report FY 1963,* 195; Joel B. Hagen, "Problems in the Institutionalization of Tropical Biology: The Case of the Barro Colorado Island Biological Laboratory," in *History and Philosophy of the Life Sciences* 12 (1990): 225–48, esp. 243.

82. The Associated Colleges of the Midwest and the University College of the West Indies apparently planned field stations and the National Academy of Sciences proposed to establish a Tropical Research Foundation and four research centers. See Stone, "OTS," 144; Keck, "Tropical Biology," 9; Riemer, "Tropical Science," 3, 6.

83. Allan A. Needell, Nuclear Reactors and the Founding of Brookhaven National Laboratory," *Historical Studies in the Physical Sciences* 14 (1983): 93–122; Allan A. Needell, "Lloyd Berkner, Merle Tuve, and the Federal Role in Radio Astronomy," *Osiris,* 2nd series, 3 (1987): 261–88.

84. Gerald Tape, AUI, communication with author, 12 August 1992; "A Conference on Tropical Marine Biology," 24 June 1966, in Fact Book, item I1; Associated Universities, Inc., "Planning Document for Establishing and Operating a Tropical Marine Science Center," final report on work done under NSF Grant 6B-5672, [August] 1967, 1, AUI, Washington, D.C.

85. T. Keith Glennan to Leland J. Haworth, August 31, 1967, Harve J. Carlson to Haworth, "Need to Establish an NSF Position Regarding Establishment of a National Tropical Marine Center," January 2, 1968, Fact Book, item I3.

86. See "Planning Document for Establishing and Operating a Tropical Marine Science Center, Prepared for the National Science Foundation by Associated Universities, Inc., Washington, D.C.," [August] 1967, Associated Universities, Inc., Washington D.C.; The Ralph M. Parsons Company, "Engineering Study, Tropical Marine Science Center, for Associated Universities, Inc., Washington D.C.," 30 June 1967, AUI. I thank Gerald Tape for access to AUI records.

87. Minutes, OTS Executive Committee, 6 December 1966 and 17–18 April 1967, folder "OTS-Board Minutes/Advisory Council Minutes" Box 17, OTS Papers.

88. On Smithsonian aspirations in the tropics, see Helmut K. Bouchner and F. Raymond Fosberg, "A Contribution Toward a World Program in Tropical Biology,"

BioScience 17 (1967): 532–38; and S. Dillon Ripley, "Perspectives in Tropical Biology," *BioScience* 17 (1967): 538–39.

89. Carlson to Haworth, January 2, 1968; Jack T. Spencer to Harve Carlson, "Supplementary Information on a Proposed Tropical Marine Science Center," January 1968, Fact Book, item I4; "An Assessment of University Interest in A Tropical Marine Science Center, prepared for the National Science Foundation by Associated Universities, Inc., Washington, D.C., 1968" [submitted 1 March 1968], AUI. On University of Miami dissent, see also "A Conference on Tropical Marine Biology."

90. T. Keith Glennan, "A Tower of Biology," *BioScience* 17 (1967): 630–31.

91. Annual Report, Facilities and Special Programs, 1968, 14–15; Stone, "OTS," 156–61.

92. AUI, "A Proposal to the National Science Foundation for the Phase One Development of a Tropical Marine Science Center," 15 October 1968, AUI; Franklin Long to Haworth, 17 October 1968, AUI; Harve Carlson to Franklin A. Long, 24 March 2 1969, Carlson to Lloyd Slater, 24 March 1969, BMS, DSF 1969; Gerald Tape, communication with author, ca. 1990.

93. *Director's Program Review,* 3 March 1971, 32; ibid., 11 January 1972, 5; ibid., 16 January 1973, 21–24.

94. BMS/BBS spent $2.9 million on La Selva, FY 1973–FY 1987. See "National Science Foundation Biological Research Resources Program, Categorized Support from FY 72 to FY 87," ca. 1987, copy and information on La Selva kindly provided by James Edwards, NSF.

95. John Kanwisher to Malvern Gilmartin, 14 July 1970, Kanwisher to Theodre H. Bullock, n.d., and Bullock to Kanwisher, 7 August 1970, "Alpha Helix—National Advisory Board 1969–70," Box 28, A. Baird Hastings Papers, National Library of Medicine. The Hastings Papers are a rich source on the controversies surrounding the *Alpha Helix.*

96. NSF press release 59-138, 22 June 1959; Sprugel, "National Science Foundation," 34.

97. Minutes, BMS Divisional Committee, 12th Meeting, 20–21 October 1958, 8–9.

Chapter 8   Allocating Resources to a Divided Science

1. BMS Annual Report FY 1964, 20–23, esp. 21; ibid., FY 1962, 24–26, HF.

2. Herman Lewis, "Perspectives in the Life Sciences," [ca. May 1963], Div. Staff Papers II. Among the more provocative articles in the journal debate were Ernst Mayr, "The New versus the Classical in Science" (Editorial), *Science* 141 (30 August 1963): 765; James Bonner, "The Future Welfare of Botany," *BioScience* 13(1) (February 1963): 20–22; George Gaylord Simpson, "The Crisis in Biology," [1967], reprinted in Simpson, *Biology and Man* (New York: Harcourt Brace Jovanovich,

1969), 3–18; W. Hardy Eshbaugh and Thomas K. Wilson, "Departments of Botany, Passé?" *BioScience* 19 (1969): 1072–74; and Theodosius Dobzhansky, "Biology, Molecular and Organismic," *American Zoologist* 4 (1964): 443–52.

3. BMS Annual Report FY 1962, 24–26, HF.

4. Garland Allen, *Life Science in the Twentieth Century* (Cambridge: Cambridge University Press, 1978), 187.

5. *NSF Annual Report FY 1965,* 78; Herman Lewis, Annual Report for Genetic Biology FY 1966, 1–2, Prog. Ann. Repts, 1961–68. On the history of molecular biology after Watson and Crick, see Horace Freeland Judson, *The Eighth Day of Creation: The Makers of the Revolution in Biology* (New York: Simon and Schuster, 1979), expanded ed. (Plainview, N.Y.: CHSL Press, 1996), and Robert Olby, "The Molecular Revolution in Biology," in Olby et al., eds., *Companion to the History of Modern Science* (London: Routledge, 1990), 493–520.

6. Dobzhansky, "Biology, Molecular and Organismic," 443.

7. NSF press releases: 65-115, "Structure of RNA Determined by Foundation Grantee," 19 March 1965; 65-145, "Foundation Grantee Synthesizes Self-Duplicating RNA," 29 September 1965; 67-121, "Harvard Scientist Demonstrates Control of Genetic Activity at the Molecular Level," 4 June 1967; and 68-162, "Two Nobel Laureates Aided by National Science Foundation Grants," 16 October 1968. Eugene L. Hess to Leland J. Haworth, "Authorization Hearings and Nobel Prizes," 1 November 1968, BMS, DSF 1968. *NSF Annual Report FY 1964,* 28–29; ibid., *FY 1965,* 78–82; ibid., *FY 1969,* 35–36.

8. See Olby, "Molecular Revolution in Biology," 503–4. Eugene L. Hess to Harve J. Carlson, 27 November 1968, copy lent by Brenda Flam, NSF. Budget figures are based on BMS annual reports.

9. Herman Lewis, Annual Report for Genetic Biology FY 1963, 3, Prog. Ann. Repts, 1961–68; Lewis, "Perspectives in the Life Sciences," esp. 3–5. Many biologists shared Lewis's prediction that the next frontiers in biology would be molecular control of development (differentiation) and the molecular basis of thought (neurobiology). Budget figures are based on BMS annual reports.

10. Other leading figures in molecular genetics supported by BMS in the 1960s included Thomas Anderson, Seymour Benzer, Melvin Cohn, Robert Corey, Boris Ephrussi, Walter Gilbert, Heinz Fraenkel-Conrat, Alan Garen, Matthew S. Meselson, Mark Ptashne, Howard K. Schachman, Sol Spiegelman, Charles Yanofsky, and Norton Zinder. See NSF annual reports.

11. *1988 NIH Almanac* (NIH Publication 88-5) (Washington, D.C.: U.S. Department of Health and Human Services, 1988), 153–54. On NIH budget politics, see Stephen P. Strickland, *Politics, Science, and Dread Disease* (Cambridge: Cambridge University Press, 1972).

12. On NIH grants, see NIH, *Research Grants Index* (n. 17 below), published annually in the 1960s. Unfortunately, amounts awarded are not given.

13. Herman Lewis, oral history by Charles Weiner, 10 November 1975, transcript, 19, Recombinant DNA History Collection, Box 10, MC 100, MIT Archives, Cambridge, Mass.

14. See BMS Advisory Committee Book, Tab C, 8th Meeting, 24–25 April 1969, in notebook "Advisory Committee for Biological and Medical Sciences," 9th Meeting, 30–31 October 1969, HF.

15. This and the following paragraphs are based on interviews by the author with former staff members and scientists, especially: Harve J. Carlson, 4 November 1988; Herman Lewis, 18 March 1991: Howard K. Schachman, 25 July 1990; Estelle ("Kepie") Engel, 29 June, 13 July, and 14 November 1988; Eugene Hess, 5 June 1989.

16. BMS Annual Report FY 1967, 1, and Lewis interview. For a good discussion of peer review in molecular biology, see interview of Eugene Hess by Norman Kaplan, 9 May 1967, Norman Kaplan Papers, American Philosophical Society, Philadelphia. On NIH peer review, see Richard Mandel, *Division of Research Grants, National Institutes of Health: A Half Century of Peer Review, 1946–1996* (Bethesda, Md.: NIH, 1996). The staff at NIH eventually came to take a more active role in the peer review process.

17. Of the Nobelists and those listed in note 10 above, scientists funded by NSF but not NIH in 1965 included Benzer, Delbruck, Ephrussi, Fraenkel-Conrat, Gilbert, Jacob, Lipmann, Meselson, Porter, Wald, and Zinder. Neither agency supported Pauling, who had left Caltech, and Ptashne was still a graduate student. See National Institutes of Health, Division of Research Grants, *Research Grants Index, Fiscal Year 1965,* 2 vols. (Bethesda, Md.: Public Health Service, 1965).

18. Lewis interview.

19. John Mehl, Annual Report for Molecular Biology FY 1963, 5–6, Prog. Ann. Repts., 1961–68; Lewis interview.

20. See James H. Cassedy, "Stimulation of Health Research," *Science* 145 (28 August 1964): 897–902.

21. John R. Platt, "The Need for Better Macromolecular Models," *Science* 131 (19 April 1960): 1309–10; Minutes, BMS Divisional Committee, 23rd Meeting, 12–13 October 1963, 3, HF; Walter L. Koltun, "Precision Space-Filling Atomic Molecules: The Corey-Pauling-Koltun Models to Construct Macromolecules Soon Will Be Available for Research and Teaching," July 1965, BMS, DSF 1965; NSB Minutes, 81st Meeting, 15–17 November 1962, 5–6; Engel and Hess interviews; Annual Report for Molecular Biology Section FY 68, Prog. Ann. Repts, 1961–68; Cassedy, "Stimulation of Health Research," 901.

22. "Outdoor Ceremony Marks Opening of Biodynamics Lab," *The Magnet,* 8(4) (April 1964), 1, 10, copy supplied by Lawrence Berkeley Laboratory Archives and Records Office, Berkeley, Cal.; "Melvin Calvin," *McGraw-Hill Modern Scientists and Engineers,* 1980, vol. 1: 180–81. Lois Soule kindly showed me around the labo-

ratory and discussed the significance of the circular design. The laboratory was funded by NSF and the Charles F. Kettering Foundation with matching funds from the state.

23. BMS Annual Report FY 1962, 10.

24. *NSF Annual Report FY 1963,* 196; Harlyn O. Halvorson, interview by author, 6 July 1989. Later NSF awards provided instrumentation for the laboratory. Wisconsin combined the NSF funds for a molecular biology facility ($600,000) with NIH funds for a biophysics facility ($500,000), which were matched by $1.1 million from the Wisconsin Alumni Research Fund. R. W. Fleming to Haworth, 22 March 1965, BMS, DSF 1965.

25. Herman Lewis, Annual Report for Genetic Biology FY 1964, 4, Prog. Ann. Repts, 1961–68; Minutes, BMS Divisional Committee, 29th Meeting, 28–29 April 1965, 6–8, HF. On molecular biologists at Cold Spring Harbor, see Judson, *Eighth Day of Creation*.

26. See BMS Annual Report FY 1960, 17.

27. Haworth to NSB, 4 January 1966; Carlson to Haworth, "Additional Information on Proposal B5-2385F, Research Laboratory, Marine Biological Laboratory, Woods Hole," 8 February 1966; Haworth to Randal M. Robertson, "Relative Priority of the MBL Proposal," 15 February 1966; Robertson to Haworth, "Relative Priority of the MBL Proposal," 2 March 1966; Haworth to Carlson, "Marine Biological Laboratory, 7 November 1966; and Carlson to Haworth, "Material for the Director's Special Book for the November 17–19, 1966 Meeting of the National Science Board," 9 November 1966, in BMS, DSF 1966.

28. Ray D. Owen to Carlson, 14 August 1968, copy supplied by Estelle ("Kepie") Engel.

29. BMS Annual Report FY 1959, 9–10; Kenton L. Chambers, Paul D. Hurd Jr., and William E. Sievers, Annual Report for Systematic Biology FY 1968, Prog. Ann. Repts, 1961–68.

30. Alan Leviton, California Academy of Sciences, interview by author, 29 December 1989.

31. David D. Keck, Annual Reports for Systematic Biology FY 1960 and FY 1961, Prog. Ann. Repts., 1958–60, 1961–68.

32. See BMS annual reports for figures. NSF had early hoped to support basic research by scientists in federal agencies, but aside from the Smithsonian, BMS made only a few such awards. When NSF funding was withdrawn, the Smithsonian persuaded Congress to support staff research through the Smithsonian Research Fund. Nevertheless, in 1968, Smithsonian Secretary S. Dillon Ripley attempted unsuccessfully to negotiate a reinstatement of eligibility for NSF awards. Communication from Pamela Henson, Smithsonian Archives; Carlson to Haworth, "Smithsonian-NSF Relations," 4 June 1968, BMS, DSF 1968.

33. "Report of the Committee on Systematic Biology of the American Insti-

tute of Biological Sciences," April 1957, Ad hoc Comm. Repts. BMS funded the committee through AIBS as a convenience.

34. Annual Reports for Systematic Biology: FY 1955–FY 1959, Prog. Ann. Repts, 1952–57 and 1958–60.

35. Jack T. Spencer, Annual Report for Facilities and Special Programs FY 1968, 6, Prog. Ann Repts, 1961–68.

36. *NSF Grants and Awards* FY 1965, 32–33; *NSF Grants and Awards* FY 1966, 35.

37. "Systematic Biology Research Training Grants," n.d., and [David B. Tyler], "(On) Research Training Grants," 5 May 1967, Tab C, BMS Advisory Committee Book, 4th Meeting, 11–12 May 1967, HF; Chambers, Hurd, and Sievers, Annual Report for Systematic Biology FY 1968.

38. Chambers, Hurd, and Sievers, Annual Report for Systematic Biology FY 1968, 1.

39. For example, the Genetic Biology panel was concerned in 1964 that "most proposals in quantitative genetics used the same methodology used for many years." In 1965, Lewis reported that quantitative genetics had been "virtually discontinued in terms of NSF support." Lewis, Annual Reports for Genetic Biology FY 1964, 3, and FY 1965, 2, Prog. Ann. Repts, 1961–68.

40. See David L. Hull, *Science as Process: An Evolutionary Account of the Social and Conceptual Development of Science* (Chicago: University of Chicago Press, 1988).

41. Chambers, Hurd, and Sievers, Annual Report for Systematic Biology FY 1968, 1. A breakdown of research grants by year and methodology is found on p. 21.

42. *NSF Annual Report FY 1965,* 83. In its section of the report, BMS featured comparisons of amino sequences in the cytochrome C of organisms ranging from yeast to human. NSF physical scientists, such as Assistant Director for Research Edward Creutz and his staff member Wayne Gruner, later delighted in publicizing this research on the Hill. *NSF Annual Report FY 1972,* 15–17; Wayne Gruner, interview by author, 2 and 9 April 1991. On numerical taxonomy, see Hull, *Science as Process,* and Keith Vernon, "The Founding of Numerical Taxonomy," *British Journal for the History of Science* 21 (1988): 143–59.

43. William Dawson, University of Michigan, interview by author, 20 June 1989; University of Michigan Research Institute, Museum of Zoology, Proposal to NSF: "A Regional Facility for Research in Biosystematics," 15 March 1961, copy provided by the Museum of Zoology, University of Michigan.

44. National Science Foundation, *Grants and Awards for the Fiscal Year Ended June 30, 1966* (Washington, D.C.: NSF, 1967), 114; interviews by author at the University of Florida: Walter Auffenberg, 29 October 1989, and J. C. Dickinson Jr., 30 October 1989.

45. Ernst Mayr, interview by author, 17 April 1991; *The Museum of Comparative Zoology and Its Role in the Harvard Community* (Cambridge, Mass.: MCZ, 1969), 28 pp.; NSF, *Grants and Awards* FY 1966, 130.

46. *NSF Annual Report FY 1954,* 88; *American Type Culture Collection: A History, 1964* (Rockville, Md.: ATCC, n.d.), copy in ASM Archives, University of Maryland–Baltimore County; NSF press release 61-125, "NSF Helps Defray Cost of New Building for American Type Culture Collection," 24 April 1961; Rita R. Colwell, ed., *The Role of Culture Collections in the Era of Molecular Biology, ATCC 50th Anniversary Symposium* (Washington, D.C.: American Society for Microbiology, 1976), esp. Harve J. Carlson, "Germ Plasm as a National Resource," 73–76. ATCC remained in Rockville until its recent move to Manassas, Virginia.

47. The resulting reports were: Conference of Directors of Systematic Collections, *The Systematic Biology Collections of the United States: An Essential Resource. Part 1. The Great Collections: Their Nature, Importance, Condition, and Future. A Report to the National Science Foundation* (January 1971); National Research Council, Division of Biology and Agriculture, *Systematics in Support of Biological Research* (Washington, D.C.: January 1970); Office of Science and Technology, Panel on Systematics and Taxonomy of the Federal Council for Science and Technology, "Systematic Biology: A Survey of Federal Programs and Needs," May 1969.

48. David M. Gates, "Report Concerning Systematics Collections of Museums and Herbaria" 24 April 1968, in Spencer, Annual Report, Facilities and Special Programs, 1968, Appendix D, 46–53; Spencer, Annual Report, Facilities and Special Programs, 1968, 3.

49. Data on plant scientists among staff members of BMS is based on NSF annual reports and *American Men and Women of Science,* various editions.

50. Paul J. Kramer, "Botany in a Changing World," *Plant Science Bulletin* 12(1) (March 1966): 1–4, esp. 3. William Stern, University of Florida, kindly lent his complete collection of this newsletter. On NSF funding of systematic biology, see William Campbell Steere, "An Appraisal of Present and Future Trends in Botany," *Plant Science Bulletin* 5(3) (July 1959): 1–6; and Steere, "Plant Taxonomy Today," in ibid. 6(2) (March 1960): 1, 3–5.

51. See NSF, *Federal Grants and Contracts for Unclassified Research in the Life Sciences, Fiscal Year 1958* (Washington, D.C.: NSF, 1961); National Institutes of Health, *Research Grants Index,* 2 vols. (Bethesda, Md., 1965). NIH, which listed grants by research topic, showed over 250 entries under the general heading "plants" and some 50 under "photosynthesis." On the AEC laboratory, see *Plant Science Bulletin* 11(1) (April 1965): 8.

52. See for example, Sydney S. Greenfield, "The Challenge to Botanists," *Plant Science Bulletin* 1(1) (January 1955): 1, 3–4, and Harriet Creighton, "Botanists of the World, Unite—and Get Going!" in ibid. 3(4) (October 1957): 1–4.

53. See Creighton, "Botanists of the World," 2; Robert W. Long, "Which Hat Shall I Wear?" *Plant Science Bulletin* 6(1) (January 1960): 8. Neither author recommended that the society change its name, but the issue of name and image has persisted. A later acrimonious debate led to the defeat of a proposal to change "botany"

to "plant biology." V. B. Smocovitis, "Disciplining Botany: A Taxonomic Problem," *Taxon* 41 (1992): 459–70. NSF used the label "plant sciences" in the 1970s and "plant biology" in the 1980s.

54. George S. Avery, "Botany in the Framework of Captive Education—A Search for Policy," *Plant Science Bulletin* 4(5) (November 1958): 1–4; Kramer, "Botany in a Changing World"; A. J. Sharp, "The Botanist as Scientist and Citizen," in ibid. 12(4) (December 1966): 1–3.

55. William L. Stern, "Quo Vadis, Botanicum?" *Plant Science Bulletin* 15(2) (June 1969): 1–4. See also Sydney S. Greenfield, "Botany in the Academic Jungle," in ibid. 16(4) (December 1970): 6–9, and Eshbaugh and Wilson, "Departments of Botany, Passé?"

56. "Notes from the Editor" [Bonner to Stern], *Plant Science Bulletin* 15(3) (October 1969): 6–7; See also Kramer, "Botany in a Changing World"; Arthur W. Galston, "Botany is Alive and Well and Living, Among Other Places, in New Haven," in ibid.: 7; Ralph W. Lewis, "Quo Vadis, Botanicum? Procede, Terge!" in ibid. 16(2) (June 1970): 5–6; and National Academy of Sciences–National Research Council, *The Plant Sciences Now and in the Coming Decade* (Washington, D.C.: NAS-NRC, 1966), 124.

57. See Creighton, "Botanists of the World, Unite," 3; "Report of Botanical Society of America Committee to Study the Role of Botany in American Education," *Plant Science Bulletin* 4(3) (May 1958): 1–3; "Education Committee Venture," in ibid. 2(2) (April 1956): 4.

58. See Ralph H. Wetmore, "Do We Need a National Center for Plant Sciences?" *Plant Science Bulletin* 7(2) (May 1961): 1–3, and minutes of BSA annual meetings in ibid. 6(4) (October 1960): 6; See also *Plant Science Bulletin* 7(3) (October 1961): 8 and 8(4) (December 1962): 6. BMS made the $12,000 feasibility study award through AIBS. See *NSF Annual Report FY 1961*. Another attempt to form a federation took place in the early 1970s. See Robert W. Long, "Editor's Notes," *Plant Science Bulletin* 17(1) (March 1971): 4, and minutes of BSA annual meetings in ibid. 17(1) (March 1971): 7, and 18(1) (March 1972): 8.

59. Adolph Hecht, "Report of the Committee on Education," *Plant Science Bulletin* 11(2) (October 1965): 4–5; Greenfield, "Botany in the Academic Jungle," 6.

60. "SEPEL. The Southeastern Plant Environment Laboratories of Duke University and North Carolina State University. A History of Their Planning and Financing . . . ," n.d., 2, Phytotron Records, 1955–78, Botany Department Records, Duke University Archives, Durham, N.C.; BMS Annual Report FY 1960, 2; ibid., FY 1961, 1.

61. John W. Purseglove, "Neotropical Botany Conference," *Plant Science Bulletin* 8(3) (September 1962): 4–7. The conference report stated: "In the course of the discussions, it became apparent that the field should not be limited to neotropical botany alone, but that consideration had to be given to the study of tropical biology

as a whole." "News and Notes," in ibid.: 9. William L. Stern, interview by author, September 1992.

62. Plant pathologist A. J. Riker, University of Wisconsin, chaired the section. "Some Opportunities and Challenges in the Plant Sciences," *BioScience* 12 (April 1962): 31–40, esp. 40.

63. NRC, *The Plant Sciences,* esp. 137; Perry R. Stout and C. A. Price, "Plant Research Strength in American Universities, 1965–1975," *BioScience* (July 1966): 456–63, esp. 462–63. On the effectiveness of the chemistry report, see Daniel S. Greenberg, *The Politics of Pure Science* (New York: Plume Books, 1967), 151–69.

64. After leaving NSF, Sprugel became successively chief of the National Park Service Division of Natural Sciences and chief of the Illinois Natural History Survey. He served a term as president of AIBS in 1974. "AIBS Officers for 1974," *Bio-Science* 23 (1973): 722; Josephine Doherty, interview by author, 10 April 1990.

65. George Sprugel, Annual Report for Environmental Biology FY 1957, esp. 1; ibid., FY 1955; ibid., FY 1956, esp. Appendix A, "Spring Meeting of the Advisory Committee for Environmental Biology, March 9, 10 and 11, 1956," 1–2, Prog. Ann. Repts. 1952–57.

66. Sprugel, Annual Report for Environmental Biology FY 1956, Appendix A, 1. Names were obtained from NSF annual reports and from *Federal Grants and Contracts in the Life Sciences, 1958.* A few, including Hutchinson and Sears, were supported by programs other than Environmental Biology.

67. BMS Annual Report FY 1959, 23; Doherty interview.

68. National Research Council, Committee to Evaluate the IBP, *An Evaluation of the International Biological Program* (Washington, D.C.: National Academy of Sciences, 1975), bound copy in NSF Library, 28. Other sources give somewhat variant figures. Chunglin Kwa, "Representations of Nature Mediating Between Ecology and Science Policy: The Case of the International Biological Programme," *Social Studies of Science* 17 (1987): 413–42, esp. 413–16.

69. See W. Frank Blair, *Big Biology: The US/IBP* (Stroudsburg, Pa.: Dowden, Hutchinson & Ross, 1977); E. B. Worthington, *The Evolution of IBP* (Cambridge: Cambridge University Press, 1975); Chunglin Kwa, "Mimicking Nature: The Development of Systems Ecology in the United States, 1950–1975" (Ph.D. diss., University of Amsterdam, 1989); Stephen Bocking, *Ecologists and Environmental Politics: A History of Contemporary Ecology* (New Haven: Yale University Press, 1987); and Joel B. Hagen, *An Entangled Bank: The Origins of Ecosystem Ecology* (New Brunswick, N.J.: Rutgers University Press, 1992), 164–88.

70. Sprugel, Annual Report for Environmental Biology FY 1956, Appendix A, 5.

71. On the early development of IBP, see G. Ledyard Stebbins, "International Horizons in the Life Sciences," *AIBS Bulletin* 12(6) (December 1962): 13–19; Stebbins, "Toward Better International Cooperation in the Life Sciences," *Plant Science*

Bulletin 8(3) (September 1962): 1–5; and Worthington, *Evolution of IBP,* 1–16. The 1963 document was reprinted as "International Biological Programme: Report of the Planning Committee, November 15th, 1963," *BioScience* 14(4) (April 1964): 43–49.

72. "International Biological Programme"; Stebbins, "International Horizons."

73. Waterman to Hiden T. Cox, 25 August 1961, ATW Notes.

74. According to the NRC evaluation committee, NSF had until this point "vacillated between reluctant support and opposition, its attitude apparently stemming from concern that research of unknown quality in the IBP was to be superimposed on ongoing research activities for which funding was already inadequate." NRC, *Evaluation of the IBP,* 11. See George Sprugel, "International Biological Program," [1963], and Carlson to NSB, "International Biological Program," 14 April 1964, Div. Staff Papers II. Negative reactions are detailed in NRC, *Evaluation of the IBP,* 6–7, and Blair, *Big Biology,* 1–26.

75. NRC, *Evaluation of the IBP,* 7–9; Minutes, BMS Divisional Committee, 28th Meeting, 2–3 October 1964, 3, HF.

76. Minutes, BMS Divisional Committee 26th Meeting, 11–12 October 1963, 1, HF; ibid. 27th Meeting, 6–7 March 1964, 4.

77. Minutes, BMS Divisional Committee 28th Meeting, 2–3 October 1964, 3–4, HF; ibid 29th Meeting, 28–29 April 1965, 1–6, esp. 4; NSB Minutes, 100th Meeting, 27–28 May 1965, 20–21.

78. Carlson interview; NRC, *Evaluation of the IBP,* 10; Blair, *Big Biology,* 22, 105. Blair wrote, "Carlson's strength within NSF and in 'arm twisting' representatives of other Federal agencies to support the USNC/IBP and its subcommittees with funding were to be a major factor in eventual success of the US/IBP" (p. 22).

79. NRC, *Evaluation of the IBP,* esp. 11–12, 17–18.

80. Members of the USNC/IBP had already been moving in this direction. NRC, *Evaluation of the IBP,* 12, 18.

81. See Kwa, "Mimicking Nature," and Hagen, *An Entangled Bank,* 130–36. Hagen discusses the problems of defining systems ecology in relation to ecosystem ecology. By one definition, systems ecology is a subdiscipline of ecosystem ecology concerned with ecosystem modeling.

82. See Blair, *Big Biology,* 24–25, 149.

83. NRC, *Evaluation of the IBP,* 13–14; House Committee on Science and Astronautics, Subcommittee on Science, Research and Development, *Hearings on the International Biological Program, H. Con. Res. 273,* 90th Cong., 1st sess., 1967, esp. 38–40, 147; Blair, *Big Biology,* 67, 91, 97, 102.

84. President's Science Advisory Committee, *Restoring the Quality of Our Environment* (Washington, D.C.: The White House, 1965).

85. Kwa, "Representations of Nature." See also Kwa, "Mimicking Nature," and Bocking, "Environmental Concerns and Ecological Research," 119–25.

86. Blair, *Big Biology,* 61–115.

87. Minutes, BMS Advisory Committee, 5th Meeting, 23–24 October 1967, 6, Adv. Comm.

88. NRC, *Evaluation of the IBP,* 19–21.

89. Haworth to NSB, "Proposed Project from Biological and Medical Sciences," 1 May 1968, esp. 2, BMS, DSF 1968; NSB Minutes, 80th Meeting, 16–17 May 1968, 9–10. The board insisted that approval "in no way commits the Foundation at this time to the support of the other five ecosystem projects." See also NSF press release, "NSF Grant Initiates Grasslands Study of International Biological Program (IBP)," 4 June 1968, 68–136, which refers to all six proposed biome studies.

90. Lewis, Annual Report for Genetic Biology, FY 1969, 7, Box 3, 76-172, NARA. Josephine Doherty recalled other members of the BMS staff slighting Johnson out of resentment of IBP. Doherty interview.

91. Minutes, BMS Advisory Committee, 7th Meeting, 14–15 November 1968, 2. For statistics, see Appendix D.

92. Lewis, Annual Report for Genetic Biology FY 1968, 1, 3, 4, Prog. Ann. Repts, 1951–68. Carlson, though partial to field biology, appreciated the problem, which he discussed sympathetically in the division's annual report. See BMS Annual Report FY 1968, 27–28.

93. NSF, *Report on Funding Trends and Balance of Activities: National Science Foundation, 1951–1988* (Washington D.C.: NSF, 1988), 1, 3.

94. Minutes, BMS Advisory Committee, 6th Meeting, 2–3 May 1968, esp. Appendix D, "Report of BMS Advisory Committee's Subcommittee on Institutional Support."

95. BMS Annual Report FY 1968, esp. Appendix E, "Meeting of the Advisory Committee . . . November 14–15, 1968."

96. Ray D. Owen, "Annual Report, 1968, Advisory Committee for Biological and Medical Sciences," 24 December 1968, Adv. Comm. Ann. Repts.

Chapter 9   Forging New Directions after the Golden Age, 1968–1972

1. John T. Wilson, interview by Frank Edmonson, 12 November 1980, HF, 25–26. NSF press release 68-123, "John T. Wilson, NSF Deputy Director, Appointed Vice President of University of Chicago," 23 March 1968; ibid. 68–153, "National Science Foundation Authority Broadened by Law Amending NSF Act," 24 July 1968; ibid. 69–151, "NSF Announces Reorganization," 5 November 1969, 69–151.

2. On science policy during this period, see Daniel J. Kevles, *The Physicists: The History of a Scientific Community in Modern America* (Cambridge, Mass.: Harvard University Press, 1987), 393–409, and Jeffrey K. Stine, *A History of Science Policy in the United States, 1940–1985* (Washington, D.C.: Government Printing Office, 1986), 57–68.

3. Harve J. Carlson, interview by author, 4 November 1988.

4. See William J. Hoff to H. Guyford Stever, "Some Key Events in the Development of NSF Programs," 5 June 1972, HGS Notes, HF.

5. "Grantees May Face Cutback Problem," *Science & Government Report* 3 (15 January 1973): 8.

6. Minutes, BMS Divisional Committee 28–29 April 1965, 11, HF. On NSF in the role of "balance-wheel," see House Committee on Science and Astronautics, Subcommittee on Science, Research and Development, *The National Science Foundation: Its Present and Future,* report, 89th Cong., 1st sess., 1966, 80–86, 88, 89. On "dropouts," see Senate Committee on Labor and Public Welfare, Special Subcommittee, *Hearings on National Science Foundation Authorization Act of 1972,* 92nd Cong., 1st sess. Appendix A provides a list of "dropouts."

7. Greenberg wrote, "McElroy has successfully tuned in to what Congress is all about: power, influence, and personal glorification of the membership, with the furtherance of the public well-being sometimes an acceptable ingredient." He was "now regarded as the shrewdest scientific operator to ascend the Hill since NIH's James Shannon went there some years back to coax out several odd billion dollars for a breakneck expansion of medical research and training." *Science & Government Report* 1(2) (1 February 1971): 3. See also "McElroy's Exit Dismays NSF," *Science & Government Report* 1(12) (1 September 1971): 4.

8. Wilson interview by Edmondson, 34, see also 27. Interviews by author: Estelle ("Kepie") Engel, 19 June, 13 July, and 14 November 1988; Josephine Doherty, 10 April 1990; and Wayne Gruner, 2 and 9 April 1991. NSF press release 69-141, "NSF Appoints New Assistant Director for Administration," 5 September 1969, and ibid., 69–153, "NSF Appoints Deputy Assistant Director for Administration," 10 November 1969.

9. John T. Wilson, interview by J. Merton England and Milton Lomask, 21 May 1974, 3, HF. Josephine Doherty and Randal Robertson, Associate Director for Research, were among those who welcomed the integration of basic and applied research in IRRPOS. Doherty and Gruner interviews; Randal M. Robertson, interview by J. Merton England, 21 June 1985, HF. *NSF Annual Report FY 1970,* 55–60; Office of the Director, Notice No. 24, 11 December 1969, filed with NSF press releases.

10. "RANN: Growth at NSF Stirs Concern, but . . . ," *Science & Government Report* 2(10) (15 July 1972): 1–4; Wilson, interview by England and Lomask, 29, see also 42; NSF, *Report on Funding Trends and Balance of Activities: 1951–1988* (Washington, D.C.: NSF, 1988), 21.

11. BMS Annual Report FY 1968, Appendix H, Tables II-A, II-B, and VI, HF. The 51.4 percent represents the ratio of awards to all actions, or 1300/2527. BMS funded 23.1 percent of all dollars requested, and 908 of 1056 renewal applications as compared to 450 of 1396 new requests.

12. NSF, *Science and Engineering Doctorates: 1960–1982* (Washington, D.C.: GPO, 1983), 18–19; House Committee on Science and Astronautics, Subcommittee on

Science, Research and Development, *1970 National Science Foundation Authorization Hearings,* 91st Cong., 1st sess., II: 154.

13. NSF Program Review Office, *Director's Program Review, Biological Sciences, 1971* (Washington, D.C.: NSF, 1971) 9–10. Copies of these internal printed reports are in HF. NIH had funded Arnon, discoverer of photophosphorylation, for fifteen years; NSF assumed his support. Daniel I. Arnon, interview by author, 26 July 1990; Arnon to William I. Gay [NIGMS], 5 December 1969, and Elijah B. Romanoff to McElroy, 20 January 1970, BMS, DSF 1970.

14. Among recipients from McElroy's reserve was the aging Nobelist Albert Szent-Gyorgyi. See William D. McElroy to Szent-Gyorgyi, 29 April 1971, and William V. Consolazio to McElroy, 13 April 1971, BMS, DSF 1971. This folder contains a number of complaints concerning loss of funding.

15. David M. Gates, "Twenty-five Years of NSF and the Future," *BioScience* 25 (1975): 295. John Mehl noted that BMS was funding about the same number of investigators in 1974 as 1964 at an annual rate that compensated for inflation but not for increased overhead rates. Mehl, letter to the editor, ibid., 480–81.

16. H. Guyford Stever to James D. Ebert, 31 July 1972, and attachments, BMS, DSF 1972; Stever to Ebert, 20 December 1974, and attachment, BMS, DSF 1974; James D. Ebert, interview by author, 27 March 1988.

17. NSF, *Grants and Awards* FY 1970 (Washington, D.C.: GPO, 1971), 73.

18. Minutes, BMS Advisory Committee, 10th Meeting, 30 April and 1 May 1970, 4, and 9th Meeting, 30–31 October 1969, 12, Adv. Comm. One of the arguments raised by the NSF hierarchy against the continuance of BMS training grants was that the program had never been formalized.

19. Minutes, BMS Advisory Committee, 11th Meeting, 29–30 April 1971, 16, Adv. Comm.; NSF, *Grants and Awards* FY 1972, 6–13.

20. Carlson to McElroy, "Biological Oceanography," 30 June 1970, Carlson and Mehl to McElroy, "The Decision-Making Process within the Foundation," 15 July 1970, BMS, DSF 1970. See Toby A. Appel, "Marine Biology/Biological Oceanography and the Federal Patron: The NSF Initiative in Biological Oceanography in the 1960s," in Keith R. Benson and Philip F. Rehbock, eds., *The Pacific and Beyond: Proceedings of the Fifth International History of Oceanography Meeting* (Seattle: University of Washington Press, in press).

21. The 135-foot schooner *Te Vega,* costly to operate, was replaced in 1971 by the more smaller, more efficient motor vessel *Proteus.* By 1973, neither Stanford nor NSF could afford the program and it was abandoned. Alan Baldridge (Librarian, Hopkins Marine Station), interview by author, 24 July 1990. Duke's *Eastward* terminated its training cruises but continued research cruises in biological and geological oceanography. See *Research Vessel Eastward: Fifteen Years of Service,* compiled by Charlene and Orin Pilkey (Durham, N.C.: Duke University, n.d., ca. 1978), copy in Duke University Archives.

22. Mehl to McElroy, "Alpha Helix," 15 August 1969, BMS, DSF 1969; Per F. Scholander to McElroy, 12 December 1969, BMS, DSF 1970; McElroy to Scholander, 16 February 1971, BMS, DSF 1971; Appel, "Marine Biology/Biological Oceanography." The Alpha Helix program ended in 1980.

23. William D. McElroy, "The Transitional Decade," *NSF Annual Report FY 1970,* 1–8, esp. 3.

24. "Long-Range Plan for Support of Biological Sciences, Biological Oceanography—Marine Biology, Specialized Biological Facilities and Equipment, International Biological Program," 20 April 1970, Div. Staff Papers II; Minutes, BMS Advisory Committee, 10th Meeting, 30 April and 1 May 1970, 16, Adv. Comm.

25. The Institute of Ecology (TIE), founded independently in 1971 after two BMS-supported feasibility studies, took over coordinating activities of the biome programs after U.S. IBP ended in 1974, supervised the NSF-funded publication of the US/IBP Synthesis Series, and carried out significant studies for federal agencies and other organizations. Overly grandiose initial goals, insufficient support from ecologists, organizational and management problems, and dependence on granting agencies contributed to its demise in 1984. See Josephine K. Doherty and Arthur W. Cooper, "The Short Life and Early Death of The Institute of Ecology: A Case Study in Institution Building," *Bulletin of the Ecological Society of America* 71 (1990): 6–17; W. Frank Blair, *Big Biology: The US/IBP* (Stroudsburg, Pa.: Dowden, Hutchinson & Ross, 1977), 151–57.

26. *Director's Program Review,* 1971.

27. David B. Tyler to McElroy, "Further Explorations of the Desirability for the NSF to Initiate a 'Crash' Program on the Basic Elements of Drug Dependence and Addiction," 4 February 1970, Edward P. Todd to McElroy, "Recommended Special Program in Basic Elements of Drug Dependence and Addiction," 25 February 1970, and Carlson to John Cohrsson, President's Advisory Council on Executive Organization, "Program on Drug Dependence and Addiction," 19 March 1970, BMS, DSF 1970.

28. NSF Program Review Office, *Director's Program Review,* 1971, 29–30; ibid., 1972, 3; ibid., 1973, 3. The document on collaboration with NIH has not been located.

29. *Director's Program Review,* 1971, 30; Eloise Clark and William Riemer, "FY 1972 Budget Emphases, Centers for the Study of Molecular Structure," 6 October 1970, attached to Carlson to McElroy, 23 April 1971, BMS, DSF 1971.

30. Carlson to McElroy, "Meeting with NIH and AEC Representatives, July 1, 1970 re - Merit of Establishing Centers for the Study of Macromolecular Structure," 13 August 1970, BMS, DSF 1970; Director's Program Reviews: 1972, 2, and 1973, 3. BMS did support the development of a Crewe Electron Microscope and improved access to facilities for neutron diffraction studies by visiting biologists to the Brookhaven National Laboratory.

31. Peter H. Raven to "Dear FNA Participant," 21 February 1973, files lent by James Rodman, NSF (hereafter, Rodman files).

32. National Museum of Natural History, Smithsonian Institution, "Proposal to the National Science Foundation: Flora North America," 30 June 1971, 1, 7. In 1969, the printed volumes still predominated over the data base. See John T. Mickel, "Introduction to the Symposium and the Flora North America Project," *BioScience* 19 (1969): 702–4, 707; "AIBS—Flora North America Project," n.d. [ca. 1969], Rodman files.

33. FNA Proposal, 5.

34. Memorandum to NSB, "Flora North America Program," n.d. [1972], and Carlson to Stever, "Agenda Item from the Division of Biological and Medical Sciences and the Office of Science Information Service—National Science Board Meeting, September 7–8, 1972," 7 August 1972, BMS, DSF 1972.

35. "Flora North America Directory," January 1973, Rodman files; FNA Proposal, ii.

36. Porter M. Kier to William E. Sievers, 12 February 1973, Rodman files.

37. Mehl to McElroy, "Flora North America," 15 February 1973, and Raven to "Dear FNA Participant," Rodman files.

38. Stanwyn Shetler to "Dear FNA Collaborator," 1 February 1973, Rodman files; John Walsh, "Flora North America: Project Nipped in the Bud," *Science* 179 (23 February 1973): 778; Edward P. Todd to Stever, "Potential Problem Relative to the Flora of North America Project," 21 February 1973, Rodman files. See also Howard S. Irwin, "Flora North America: Austerity Casualty?" *BioScience* 23 (1973): 215. It is noteworthy that the current revival of FNA is a modest affair, focusing on printed volumes and initially costing less per year, even without inflation, than the original FNA.

39. See Paolo Palladino, *Entomology, Ecology and Agriculture: The Making of Scientific Careers in North America, 1885–1985* (Amsterdam: Harwood, 1996). NIH had, for example, supported Stern and colleagues' pioneering work in integrated control of insect pests. See Vernon M. Stern, Ray F. Smith, Robert van den Bosch, and Kenneth S. Hagen, "The Integrated Control Concept," *Hilgardia* 29 (1959): 81–154, esp. 81n.

40. Rachel Carson, *Silent Spring* [1962] (Boston: Houghton Mifflin, 1987); *NSF Annual Report FY 1966*, 18–20; President's Science Advisory Committee, Environmental Pollution Panel, *Restoring the Quality of Our Environment* (Washington, D.C.: The White House, 1965), 30; NSF press release, "Education Project Aimed at Pest Population Control," 28 May 1969, 69–126. On the history of biological control of pests in America, see Richard Sawyer, *To Make a Spotless Orange: Biological Control in California* (Ames: Iowa State University Press, 1996).

41. John Brooks, David Tyler, and William Riemer, "FY 1972 Budget Emphases. Biological Control of Populations (Biological Control of Unnaturally

Abundant Plant and Animal Populations)," 12 August 1970, in Agenda Book, BMS Advisory Committee Meeting, 26–27 October 1970, HF; *Director's Program Review,* 1971, 26–27. The earlier formulation of the initiative was broader, including general research on population dynamics. Brown appears to have been giving a classic argument for biological control based on the notion of a balance in nature. See Palladino, *Entomology, Ecology, and Agriculture,* 75–76, 104–7, and Paolo Palladino, "Ecological Theory and Pest Control Practice: A Study of the Institutional and Conceptual Dimensions of a Scientific Debate," *Social Studies of Science* 20 (1990): 255–81.

42. John L. Brooks, "Biological Regulation of Pest Populations," in *Director's Program Review,* 1972, 39–50; NSF press release 71-106, "NSF Grant Supports Research on Hormonal Control of Insects," 4 February 1971, and ibid., 71–153, "Hormones and Viruses for Insect Control to be Tested for Possible Harmful Side Effects Under NSF Grants," 16 May 1971; *NSF Annual Report FY 1969,* 37–38; ibid., FY 1975, 15–16. See also Stever to John D. Baldeschwieler [OST], 14 April 1972 and attachment, BMS, DSF 1972.

43. Historian John H. Perkins called it "the largest coordinated research project ever launched for insect control." John H. Perkins, *Insects, Experts, and the Insecticide Crisis: The Quest for New Pest Management Strategies* (New York: Plenum, 1982), 61–95, 142–53, esp. 65; Palladino, *Entomology, Ecology, and Agriculture,* 138–42, 162–63, 166–67, 189; NSF press release 72-152, "250 Scientists to Begin Research Aimed at Biological Control of Crop Pests," 23 June 1972.

44. Perkins, *Insects, Experts,* 71–73, 145; NSF press release 72-152, 23 June 1972. Palladino argues that the Huffaker project used systems analysis very differently than the biome projects. For Huffaker and his colleagues, systems analysis was more of a tool than a theory; they continued to draw upon concepts from theoretical population ecology. Palladino, *Entomology, Ecology, and Agriculture.*

45. Carlson to McElroy, "Briefing on Biological Control of Pests, March 8, 1971," n.d., and Carlson to McElroy, "NSF-Agriculture Agreement for Cooperation in Research on Alternative Methods of Control of Insect Pests," 19 March 1971, BMS, DSF 1971; Brooks, "Biological Regulation of Pest Populations." Huffaker had already received in FY 1968 a two-year NSF award for "The Role of Natural Enemies in Suppressing Spider Mites in Important Food Crop Ecosystems." NSF, *Grants and Awards* FY 1968, 5.

46. Perkins, *Insects, Experts;* Palladino, *Entomology, Ecology, and Agriculture,* 136–37. Perkins regarded integrated pest management and the USDA's approach, which he labeled "total pest management," as competing paradigms.

47. Palladino, *Entomology, Ecology, and Agriculture,* 137–41; Perkins, *Insects, Experts,* 87–89; John Erlichmann to Stever, received 22 February 1972, and Stever to Russell Train, 7 April 1972, BMS, DSF 1972. I disagree in part with Palladino's claim that NSF did not want to support the Huffaker project. Certainly, BMS staff expected to fund it.

48. NSF press release 72–152, 23 June 1972; Palladino, *Entomology, Ecology, and Agriculture,* 13, 138; Perkins, *Insects, Experts,* 61, 65–66. On conflicts after 1972 between project managers and BMS staff, see Perkins, 142–53, and Palladino, *Entomology, Ecology, and Agriculture,* 162–63. As a complement to the Huffaker project, BMS funded a second "integrated" project by Robert L. Rabb at North Carolina State University, which studied the life cycle and quantitative population dynamics of the corn earworm and tobacco budworm in relation to all the crops on which they fed. See *Director's Program Review,* 1973, 13–21, and *NSF Annual Report FY 1973,* 16–17.

49. Minutes, BMS Advisory Committee, 12th Meeting, 25–26 February 1972, 3, Adv. Comm.

50. Herman Lewis, interview by author, 18 March 1991; Herman Lewis, interview by Charles Weiner, 10 November 1975, 23–38, Box 10, MC 100, Recombinant DNA History Collection, MIT Archives, Cambridge, Mass. (hereafter, Lewis oral history). Lewis recalled in 1991 that one of the products foreseen was a complete map of the chromosomes of the chosen cell (sequencing of DNA bases was not yet a possibility). Thus HCB can be viewed as a precursor of the current Human Genome Project.

51. Lewis interview; Lewis oral history, 25–35.

52. Lewis oral history; Murray Rosenberg, "Human Cell Biology," in *Director's Program Review,* 1972, 14–23, esp. 14–15.

53. Robert Haselkorn to McElroy, 22 May 1970, and attachment, "A Program in Molecular Biology of Human Cells," BMS, DSF 1970. See also *Director's Program Review,* 1971, 24.

54. John Mehl, Abraham Eisenstark, and Herman Lewis, "FY 1972 Budget Emphases. Proposed Program on the Molecular Biology of the Human Cell," 7 August 1970, 10, in Agenda Book, BMS Advisory Committee Meeting, 26–27 October 1970, HF; Lewis to McElroy, "Human Cell Program," 21 May 1971, BMS, DSF 1971. On National Cancer Act, see Stephen P. Strickland, *Politics, Science, and Dread Disease: A Short History of United States Medical Research Policy* (Cambridge, Mass.: Harvard University Press, 1972).

55. Lewis interview; Carlson to McElroy, "Proposed Program on the Molecular Biology of the Human Cell," 13 October 1970, BMS, DSF 1970; McElroy to Staff, "Establishment of the Biology of the Human Cell Program . . . ," 9 December 1971, BMS, DSF 1971; *Director's Program Review,* 1973, 2; Mehl, Eisenstark, and Lewis, "FY 1972 Budget Emphases"; NSF Budget to Congress, FY 1977, D-I-1, HF.

56. Lewis interview; Rosenberg, "Human Cell Biology"; NSF press release, "University of Alabama to Build Mammalian Cell Facility," 7 October 1974, 74-207.

57. Lewis interview; Lewis oral history, 50–59; Sheldon Krimsky, *Genetic Alchemy: The Social History of the Recombinant DNA Controversy* (Cambridge, Mass.:

MIT Press, 1982), 24–69; A. Hellman, M. N. Oxman, and R. Pollack, eds., *Biohazards in Biological Research* (Cold Spring Harbor, N.Y.: Cold Spring Harbor Laboratory, 1973). NSF requested assistance in funding from the National Cancer Institute and the American Cancer Society. Sociologist-historian Krimsky (who scarcely mentions NSF's role) saw Asilomar I as emerging out of earlier debates among biologists concerning the risks of recent or proposed experiments using tumor viruses to create hybrid cells.

58. Krimsky, *Genetic Alchemy,* 70–96, esp. 82–84; Lewis oral history, 70; Lewis to Stever, 17 July 1974, BMS, DSF 1974. As a federal employee, Lewis could not sign the statement.

59. Krimsky, *Genetic Alchemy,* 97–153, esp. 104; Lewis interview.

60. Krimsky, *Genetic Alchemy,* esp. 147–53; Lewis oral history, 100–101, 110–36. Lewis was the NSF liaison to the NIH advisory committee which drafted NIH guidelines.

61. Among those who contributed to the initial position paper on EHVIST in 1971 were Louis Levin, then NSF deputy director, and William Consolazio, then special assistant to the director. Lewis suggested an office to deal with the social implications of science as early as 1964. See Lewis oral history, 40–48; Herman Lewis, Annual Report, Genetic Biology Program, FY 1964, 6–8, Prog. Ann. Repts., 1961–68; and Cristine Russell, "A Program Whose Time Has Come," *BioScience* 24 (1974): 672–73. HCB was phased out after its originally designated ten-year period. One reason for termination was that the tissue culture centers were of much greater use to NIH's grantees than to NSF's. John S. Cook, interview by author, 3 April 1990.

62. Psychobiology Program, "Neurobiology Discussion," 26 October 1970, 4, in Agenda Book, BMS Advisory Committee Meeting, 26–27 October 1970, HF.

63. Lewis to Carlson, "Neurobiology Program," 19 November 1965, Tyler to Carlson, "Neurobiology," 11 April 1966, and John F. Hall to Carlson, "Comments on Memo of February 9, 1967," 24 February 1967, copies lent by James H. Brown, NSF.

64. See Robert W. Doty, "Neuroscience," in John R. Brobeck, Orr E. Reynolds, and Toby A. Appel, eds., *History of the American Physiological Society: The First Century, 1887–1987* (Bethesda, Md.: American Physiological Society, 1987), 427–34. Henry S. Odbert to Carlson, "Recommendation for the Establishment of a Section on Psychobiology," 5 November 1969, and James H. Brown and David Tyler, "Neurobiology," 9 July 1970, lent by Brown. See also Henry S. Odbert, James H. Brown, and Mary Ann Sestili, "The Status of Psychobiology at the National Science Foundation," 6 October 1970, in Agenda Book, BMS Advisory Committee Meeting, 26–27 October 1970, HF.

65. McElroy to "Organization," "Establishment of the Neurobiology Program in the Division of Biological and Medical Sciences," 23 March 1971, BMS, DSF 1971. In preparation, the Psychobiology Program had handled neurobiology pro-

posals in FY 1971 and increased its panel. James H. Brown and Henry S. Odbert, "Request for a Separate Neurobiology Panel," 20 August 1971, ibid.

66. James H. Brown, interviews by author, 17 and 24 January 1989. Information on grantees is taken from the annual NSF *Grants and Awards*.

67. Brown interview.

68. *Director's Program Review,* 1971, 27–28; James H. Brown, "Learning and Memory," in ibid. 1972, 51–60, esp. 59; ibid., 1973, 2–3.

69. Carlson to McElroy, 16 April 1970, BMS, DSF 1970; Report, BMS Advisory Committee, 10th Meeting, 30 April and 1 May 1970, 6; David M. Gates, interview by E. Vernice Anderson, 12 November 1981, transcript, 2, HF.

70. Conference of Directors of Systematic Collections, *Systematic Collections of the United States: An Essential Resource. Part I, The Great Collections: Their Nature, Importance, Condition, and Future* (January 1971), iv.

71. Ibid., esp. 32.

72. McElroy to David M. Gates, 5 January 1972, BMS, DSF 1972.

73. Walter H. Hodge, "Operational Support of Resource and Research Centers," in *Director's Program Review,* 1972, 24–38, esp. 24, 26.

74. Gates interview, 10–11. The Board adopted a resolution in favor of the initiative in January 1972.

75. Carlson to McElroy, 16 April 1970, BMS, DSF 1970.

76. American Institute of Biological Sciences, *Report to the National Science Foundation: The Role of Field Stations in Biological Education and Research* (February 1971), copy formerly in Division of Biotic Systems and Resources, BBS, NSF. The two most provocative participants were former BMS staff members Jack T. Spencer and John Cantlon. Cantlon suggested that to be relevant to national needs, stations must address "urban-industrial" and "agricultural systems" as well as "semi-wild systems" (p. 346). The report focused primarily on inland field stations, though it also contained information on marine stations.

77. James Edwards, "National Science Foundation, Biological Research Resources Program, Categorized Support from FY 72 to FY 87," 1987, copy furnished by Edwards.

78. Edward P. Todd to Assistant Director for Administration, "Approved Program in BMS," 28 September 1973, Edward P. Todd Chron File, HF.

79. *Director's Program Review,* 1973, 10–12; Edwards, "NSF Biological Research Resources Program"; James Edwards, interview by author, 6 May 1991.

80. See for example Kenneth V. Thimann, "Presidential Address at Opening Plenary Session, XI International Botanical Congress," *Plant Science Bulletin* 15(3) (October 1969): 1–3; Philip Abelson, [Editorial] "Opportunities in Plant Science," *Science* 172 (18 June 1971): 1195, reprinted in *Plant Science Bulletin* 17(3) (September 1971): 23.

81. Elijah Romanoff, interview by author, 25 May 1992.

82. Minutes, BMS Advisory Committee, 12th Meeting, 25–26 February 1972, 7, Adv. Comm.; Romanoff interview.

83. *NSF Annual Report FY 1972,* 17–18; ibid., *FY 1973,* 18–19. See also *Director's Program Review,* 1973, 35–43 on photosynthesis. In FY 1975, BMS highlighted "Energy and Plants." *NSF Annual Report FY 1975,* 14, 17.

84. Romanoff interview; NSF Budget to Congress, FY 1977 [looseleaf binder], D-I-1-2, D-I-9, HF; Interagency Committee on Plant Sciences, *Federal Support for Research in the Plant Sciences* (Washington, D.C.: NSF, 1979), 49. As a result of Romanoff's campaign, regulatory biology ceded all its plant research to metabolic biology. The other programs continued to support plant, animal and microbial research.

85. *Federal Support for Research in the Plant Sciences,* esp. iii, 16, 23–31, 49; "Federal Support for Basic Research in Plant Biology at Academic Institutions, Competitively Awarded Grants, FY 1978–1993," 20 February 1992, and Robert Rabin, "NSF Postdoctorate Fellowships in Plant Biology: A Five-Year Review," August 1987, copies lent by Mary Clutter; Mary Clutter, interview by author, 25 September 1992. The interagency committee estimated total federal funding for plant research in FY 1977 at $209 million, including applied, intramural, and agricultural experiment station research.

86. New York State College of Agriculture and Life Sciences, Cornell University, "Potential Increases in Food Supply through Research in Agriculture. Researchable Areas Which Have Potential for Increasing Crop Production: A Grant Report to the Science and Technology Policy Office of the National Science Foundation," February 1976, copy lent by Mary Clutter.

87. Rabin, "NSF Postdoctorate Fellowships in Plant Biology," 2; "Federal Support for Basic Research in Plant Biology"; E. M. Leeper, "New for Scientists: RFP's from USDA, EIS's from NIH," *BioScience* 27 (1977): 297–98. NSF estimated that in FY 1991 it spent $82 million on basic academic research in plant biology compared to $44 million by USDA, $34 million by DOE, $16 million by NIH, and $3.3 million by NASA.

88. Director's Program Reviews: 1971, 23–24; 1972, 6; Roger Mitchell, Ramona A. Mayer, and Jerry Downhower, "An Evaluation of Three Biome Programs," *Science* 192 (28 May 1976): 859–65, esp. 865. IBP was labeled "big biology" at the time. See Blair, *Big Biology;* "Big Biology," *Nature* 216 (2 December 1967): 842; and Philip M. Boffey, "International Biological Program: Was It Worth the Cost and Effort?" *Science* 193 (3 September 1976): 866–68.

89. *NSF Annual Report FY 1971,* 38–39; NSF press release 71-150, "NSF Expands Tundra Research," 14 May 1971, ibid., and 71-187, "NSF Continues Project Aimed at Better Scientific Basis of Land, Water Use," 28 September 1971.

90. Carlson to Haworth, "Administrative Reorganization of Environmental and Systematic Biology Section," 20 February 1969, Haworth to Staff, "Administrative Reorganization of Environmental and Systematic Biology . . . ," 22 April 1969, BMS, DSF 1969; National Research Council, Committee to Evaluate the IBP, *An*

Evaluation of the International Biological Program (Washington, D.C.: National Academy of Sciences, 1975), 16. Philip L. Johnson had been hired as a rotator the previous year to handle the IBP proposals.

91. NRC, *An Evaluation of the IBP,* 6, 14–16, 23–26, 30, 49, 61.

92. John M. Neuhold, "International Program" in *Director's Program Review,* 1972, 6–13; NSF press release 73–210, "Primitive Tribes Found to Have Chromosome Damage, High Mercury Levels," 28 November 1973. AEC contributed to Neel's project.

93. See Chunglin Kwa, "Mimicking Nature: The Development of Systems Ecology in the United States, 1950–1975" (Ph.D. diss., University of Amsterdam, 1989), 125–32, and Stephen Bocking, *Ecologists and Environmental Politics: A History of Contemporary Ecology* (New Haven: Yale University Press, 1997), 63–101, esp. 93–101. See also Battelle Columbus Laboratories (Kenneth M. Duke et al.), *Evaluation of Three of the Biome Studies Programs Funded Under the Foundation's International Biological Program (IBP)* (Columbus, Ohio: Battelle Columbus Laboratories, 1975).

94. See Kwa, "Mimicking Nature," 91–124; Joel B. Hagen, *An Entangled Bank: The Origins of Ecosystem Ecology* (New Brunswick, N.J.: Rutgers University Press, 1992), 176–77; and Battelle, *Evaluation of Three of the Biome Studies.* For data on NSF expenditure for biomes, see Mitchell et al., "Evaluation of Three Biome Programs," 861.

95. Carlson to Haworth, "Material for the Director's Special Book, National Science Board Meeting, May 15–16, 1969," 2 May 1969, and Haworth to NSB, "Memorandum to Members of the National Science Board," 9 May 1969, BMS, DSF 1969; McElroy to NSB, "Grant for Ecological Research on the Alaskan North Slope," 21 April 1970, BMS, DSF 1970. The Tundra biome was evaluated in Battelle, *Evaluation of Three of the Biome Studies.*

96. NSF press release 70-172, "NSF Supports Study of Hawaiian Evolutionary, Ecological Processes," 19 June 1970; Carlson to Haworth, "Material for the Director's Special Book, National Science Board Meeting, September 3–4, 1970," 24 August 1970, BMS, DSF 1970. NSF spent about $4.2 million on Solbrig and Mooney's cooperative project.

97. Doherty interview; James T. Callahan, interview by author, 2 July 1991. For Callahan, $55 million was a low price for this achievement. Blair, *Big Biology,* 162. For other highly favorable assessments by ecologist-historians, see Robert L. Burgess, "United States," in Edward J. Kormondy and J. Frank McCormick, eds., *Handbook of Contemporary Developments in World Ecology* (Westport, Conn.: Greenwood Press, 1981), 67–101; and Robert P. McIntosh, *The Background of Ecology: Concept and Theory* (Cambridge.: Cambridge University Press, 1985), 213–21.

98. Consolazio to McElroy, "Some Thoughts on Framing the 1974 Budget," 31 July 1972, Consolazio Memos—1972–73, HF.

99. Minutes, BMS Advisory Committee, 12th Meeting, 25–26 February 1972,

4, 9–10 Adv. Comm.; Philip M. Boffey, "'Boondoggle' Criticism Hits International Bio Program," *Science & Government Report* 2(18) (15 December 1972): 1–3, esp. 3; Nelson G. Hairston to Stever, 27 November 1972, BMS, DSF 1972. Acting NSF Director Raymond Bisplinghoff replied to Hairston that the biomes were "partly an experiment in methods of conducting research which we will have no hesitation in terminating when we are as convinced as you that it is relatively unproductive." He reminded Hairston that funding of IBP was supplementary to regular programs, not competitive. Bisplinghoff to Hairston, 12 December 1972, ibid.

100. NRC, *An Evaluation of the IBP,* 51, 53–54, 62. The Academy report was critically discussed in Boffey, "International Biological Program: Was It Worth the Cost and Effort?"

101. Battelle, *Evaluation of Three of the Biome Studies,* I-1-43, esp. I-41; Mitchell et al., "An Evaluation of Three Biome Programs," 865; "The Biome Programs" (letters from J. H. Gibson, Blair, and Downhower and Mayer), *Science* 195 (4 March 1977): 822–23; and Stanley I. Auerbach, Robert L. Burgess, and Robert V. O'Neill, "The Biome Programs: Evaluating an Experiment," ibid., 902–4. See NSF's own evaluation in *NSF Annual Report FY 1974,* 44–45.

102. See Bocking, *Ecologists and Environmental Politics,* 4–5, 116–47; Hagen, *An Entangled Bank,* 181–88; F. Herbert Bormann and Gene E. Likens, *Pattern and Process in a Forested Ecosystem: Disturbance, Development and the Steady State Based on the Hubbard Brook Ecosystem Study* (New York: Springer-Verlag, 1979); and James T. Callahan to Stever, 25 September 1974, BMS, DSF 1974. Bocking relates that in 1971, Charles Cooper told Bormann and Likens that their renewal proposal's strip cutting experiment, which simulated commercial logging practice, did not fall within NSF responsibility. Bormann and Likens had to appeal to McElroy to avoid restrictions on the grant. Bocking, op. cit., 132–33. This is another example of BMS's ambivalence toward applied research.

103. Hagen, *An Entangled Bank,* 178–79.

104. Bocking, *Ecologists and Environmental Politics,* 101–15, 179–205.

105. Kwa, "Mimicking Nature," 115, 117.

106. Hagen, *An Entangled Bank,* 175–88.

107. Callahan interview. The US/IBP Synthesis Series, which included Blair's *Big Biology,* was published by Dowden, Hutchinson & Ross.

108. James T. Callahan, "Long-Term Ecological Research," *BioScience* 34 (1984): 363–67.

109. *NSF Annual Report FY 1975,* 13–14, 17.

Chapter 10   End of an Era, 1972–1975

1. On Project Hindsight, see Roger L. Geiger, Research and Relevant Knowledge: American Research Universities Since World War II (New York: Oxford University Press, 1993), 191–93.

2. Melvin Kranzberg, "The Disunity of Science-Technology," *American Scientist* 56 (1968): 21–34. Edwin T. Layton Jr., "Mirror-image Twins: The Communities of Science and Technology in 19th-century America," *Technology and Culture* 12 (1971): 562–80.

3. William D. McElroy, "The Utility of Science," in James A. Shannon, ed., *Science and the Evolution of Public Policy* (New York: Rockefeller University Press, 1973), 19–29, esp. 23. The papers in the volume were based an earlier lecture-seminar series at the Rockefeller University partially funded by NSF.

4. For examples of criticism, see Jerome Fregeau to Edward Todd, "Information for Response to Dr. Guttman," 29 June 1972, Edward Todd Chron File, HF; Edward Creutz to Guyford Stever, 6 September 1974 [on response to Senator Edward Kennedy's criticism of NSF recruiting methods], "AD/R Assistant Director/ Research 1974," Box 7, 79–010, NARA; Stever to Marjorie C. Caserio, 27 April 1973, HGS Notes, HF. On peer review, see George T. Mazuzan, "'Good Science Gets Funded . . . ': The Historical Evolution of Grant Making at the National Science Foundation," *Knowledge: Creation, Diffusion, Utilization* 14 (1992): 63–90.

5. Minutes, BMS Advisory Committee, 11th Meeting, 29–30 April 1971, 19–20, Adv. Comm. In 1970 Carlson had discussed with the Advisory Committee expanding and broadening it to include a representative of a small college, of a grants-poor school, or of industry. Minutes, BMS Advisory Committee, 10th Meeting, 30 April and 1 May 1970, 23. On the beginnings of the women's movement in science and engineering, see Margaret W. Rossiter, *Women Scientists in America: Before Affirmative Action, 1940–1972* (Baltimore: Johns Hopkins University Press, 1995), 361–82.

6. NSF Program Review Office, *Director's Program Review, Biological Sciences,* 1971, 33–36, HF; Minutes, BMS Advisory Committee, 11th meeting, 29–30 April 1971, 13–14.

7. Committee on Research in the Life Sciences of the Committee on Science and Public Policy, National Academy of Sciences, *The Life Sciences* (Washington, D.C.: National Academy of Sciences, 1970); Philip Handler, ed., *Biology and the Future of Man* (New York: Oxford University Press, 1970).

8. See Minutes, BMS Advisory Committee, 1969–1972, esp. 12th Meeting, February 1972. On the termination of the committee, see Guyford Stever to Robert S. Bandurski, 10 November 1972, "BMS, Advisory Committee," Box 1, 77-080, NARA.

9. "Guidelines Issued to Thwart Open Advisory Committee Act," *Science & Government Report* 2 (15 December 1972), 8, and "NSF Committees Realigned to Defy Open Advisory Act," ibid, 3 (15 January 1973), 4.

10. Herman Lewis, Annual Report for Genetic Biology, FY 69, esp. 5, Box 3, 76-172, NARA.

11. See David M. Gates to Guyford Stever, 34 April 1972, BMS, DSF 1972.

12. Creutz to Stever, 6 September 1974.

13. Interviews by author of NSF staff.

14. See Joseph S. Murtaugh, "Biomedical Sciences," in James A. Shannon, ed., *Science and the Evolution of Public Policy* (New York: Rockefeller University Press, 1973), 157–85.

15. Directors Program Review, Biological Sciences, 1973, 4–5, 49–51, 56–58. The attempt to distinguish NSF funding from NIH is seen in *NSF Annual Report FY 1976*, 55–69

16. On Proxmire's ascent into this position, see "Notes on the New Congress: Proxmire, OTA and S.32," *Science & Government Report* 3 (15 January 1973), 7.

17. See, for example, Senate Committee on Appropriations, Subcommittee, Department of Housing and Urban Development, space, science, veterans, and certain other independent agencies appropriations for fiscal year 1975, 93rd Cong., 2nd sess., 1974, 388; and Senate Committee on Appropriations, Department of Housing and Urban Development—Independent Agencies Appropriation bill, 1988, report to accompany H.R. 2783, 370–74.

18. In 1996, for example, Yale spun off from the Department of Biology a new Department of Ecology and Evolutionary Biology. See Wade E. Roush, "Biology Departments Restructure," *Science* 275 (14 March 1997), 1556–58.

19. Some saw the reorganization as benefiting the social sciences. See "NSF Reorganization Elevates Social Sciences," *Science & Government Report* 5 (1 August 1975), 6–7. However, Larsen, a social scientist at NSF, viewed the reorganization as dropping the social sciences "one critical step down the decision-making hierarchy." Otto M. Larsen, *Milestones and Millstones: Social Science at the National Science Foundation, 1945–1991* (New Brunswick, N.J.: Transaction Publishers, 1992). See also Mark Solovey, "The Politics of Intellectual Identity and American Social Science, 1943–1970" (Ph.D. diss., University of Wisconsin–Madison, 1996).

20. For example, Harve Carlson, interview by author, 4 November 1988.

21. *Adapting to the Future: Report of the BBS Task Force Looking to the 21st Century* (Washington, D.C.: Directorate for Biological, Behavioral, and Social Sciences, NSF, 1991). *Adapting to the Future: Report of the BBS Task Force Looking to the 21st Century: Executive Summary* (Washington, D.C.: Directorate for Biological, Behavioral, and Social Sciences, NSF, 1991).

22. Jeffrey Mervis, "Keeping up with Rita Colwell," *Science* 279 (13 March 1998), 1622–23; Jeffrey Mervis, "The Biocomplex World of Rita Colwell," *Science* 281 (25 September 1998): 1944–45, 1947.

Note on NSF Primary Sources

The documentary record for researching biology at the National Science Foundation in its first two decades is an exceedingly rich one. This note is intended to serve as a brief guide to the chief types of NSF records consulted for this book.

Reports

The most basic source for NSF is the published annual reports beginning in 1952. For most of the period covered by this book, the official title for this report is *Annual Report for Fiscal Year . . .* (Washington, D.C.: NSF). Each volume contains a section on the Division of Biological and Medical Sciences (BMS) and a list of grants awarded by each program. Each entry includes name of grantee(s), institution, title of grant, duration of the award, and amount. Summary statistics compare amounts expended by each program to previous years. Also in the annual reports are lists of program officers, members of the BMS Divisional Committee, members of the program panels, and highlights of biological projects funded. Beginning in 1964, the list of grants was published in a separate annual volume titled *Grants and Awards for Fiscal Year Ended . . .* (Washington, D.C.: NSF).

NSF's *Federal Funds for Science* (later *Federal Funds for Research, Development and Other Scientific Activities*) (Washington, D.C.: GPO), which began with FY 1953, contains summary statistics on the support of science by federal agencies. These statistics are problematical for the life sciences, since each agency was allowed to categorize its own research as basic or applied, and biological, medical or agricultural (see chapter 5).

The Division of Biological and Medical Sciences printed for limited distribution its own series of reports on funding of the biological sciences by federal agency, *Federal Grants and Contracts for Unclassified Research in the Life Sciences*

(Washington, D.C.: NSF). These cover 1952 and FY 1954 through FY 1958 and were published from 1954 to 1961. Grants are listed by category (molecular biology, regulatory biology, etc.) and by individual. One can determine from which agencies a given individual was receiving funding, though not the total amount of funding an individual received. A set of these reports is available in the NSF Library.

From 1971 through 1973, BMS had printed for internal distribution a review of its activities and of its program "emphases" for future development. The three issues are *Director's Program Review, Biological Sciences,* 3 March 1971, 11 January 1972, and 16 January 1973. Copies are in the NSF Library and in the NSF Historian's Files.

The NSF Library has a full set of the printed congressional bills related to NSF from 1945 to 1950 as well as the hearings of the House and Senate subcommittees overseeing NSF appropriations.

Manuscript Sources at NSF

The bound minutes of the National Science Board are kept by the National Science Board at NSF. There is a detailed index to them, a copy of which is in the NSF Historian's Files (HF). Minutes of the closed meetings of the board are maintained in a separate series. The Office of Legislative and Public Affairs maintains a complete set of all of NSF's press releases.

National Archives and Records Service (NARA)

PROCESSED RECORDS

NSF records are found in Record Group 307. NARA has long held Alan T. Waterman's subject files. These records, arranged in eighty-three boxes with an index, are found under the heading Office of the Director Subject Files (ODSF). (Personal papers of Waterman are in the Manuscripts Division of the Library of Congress.)

UNPROCESSED RECORDS

NSF Historian's Files (HF)

The NSF Historian's Files formerly maintained by NSF Historians J. Merton England and George T. Mazuzan contain a gold mine of information. When I used them, they were located in the Office of Legislative and Pubic Affairs at NSF, but they have recently been given to NARA. They are now in Archives II. The arrangement is basically alphabetical by subject. Most useful are the annual reports of the Division of Biological and Medical Sciences (under BMS) which were submitted by the assistant director (later division director) for BMS to the director of NSF. These exist from 1952 through 1968. The reports discuss achievements, new directions, and policy issues. They contain more detailed summary statistics on awards than can be found in the *NSF Annual Reports* as well as fuller lists of staff members and names of divisional committee and panel members.

Minutes of meetings (sometimes called "Staff Notes") of the BMS Divisional Committee from 1953 through 1965 are also located in HF. In 1965 the BMS Divisional Committee became the BMS Advisory Committee. For the period 1965 until the committee's dissolution in 1972, an invaluable resource are the loose-leaf notebooks put together for members of the committee before each meeting. These contain minutes, reports of committees, discussion papers, annual reports of the division, and miscellaneous information copied for the benefit of members.

A very revealing resource are the chronological sets of Alan T. Waterman Diary Notes (ATW Diary Notes) and Waterman Notes (ATW Notes). These were created by the first NSF Historian by making copies from the "Chron File," formerly located in a storage closet at NSF. Originals of most of these materials are in the Office of the Director Subject Files at NARA (see below). The Diary Notes are Waterman's memos to files which record his impressions of conversations with visitors, telephone calls, etc. They were circulated among the senior staff. The series, ATW Notes, consists of copies and extracts of correspondence. Louis Levin, program director for regulatory biology, also penned a set of diary notes, 1954–58, located in HF. For the formative years, the series of minutes of senior staff meetings has been useful.

The two transcripts of previous interviews of John T. Wilson in 1974 and 1980, in HF, as well as transcripts of interviews of National Science Board members carried out by Vernice Anderson, added much insight and color to my story.

Materials Formerly in Records Storage

When I used these records, they were maintained in the Washington National Records Center in Suitland, Maryland, and were still legally owned by NSF. They were retrieved by record group (RG 307), accession number, and box (actually cardboard carton) number. The records through 1986 have recently been transferred to the Archives II building of NARA. They will eventually be processed and rearranged, at which time the locations given in the notes to this book may no longer be accurate.

Each program director submitted an annual report to the assistant (division) director to serve as a basis for the Divisional Annual Report. These very rich documents are complete from FY 1952 through FY 1968 and are spotty from FY 1969 to FY 1972. They are located in loose-leaf notebooks in Box 1 and Box 2, 76-172, RG 307, NARA.

The minutes of the Divisional Committee for Biological Sciences (1952–1953) and the Divisional Committee for Medical Research (1953) prior to the first meeting of Divisional Committee for BMS in 1954, are located in "Biological Sciences Div. Comm., 1st–7th Meetings (1952–53)," Box 20, 70A-2191, RG 307, NARA.

Reports (Minutes) of the BMS Advisory Committee, 1965–1972, and Annual Reports of the BMS Advisory Committee sent by the Chairman to the National Science Board are in Box 1, 76-172, RG 307, NARA.

Reports on various topics prepared by ad hoc committees of BMS staff and members of the BMS Divisional Committee in the 1950s and 1960s are found in loose-leaf notebooks "Ad hoc Committee Reports, 1954–61," "Division Staff Papers, Book I," and "Division Staff Papers, Book II," Box 3, 76-172, RG 307, NARA.

Information on the evaluation of the first series of proposals is in folder "Advisory Panel Meetings, January & April 1952," Box 20, 70A-2191, RG 307, NARA.

The individual folders for grants have also just been transferred from the Washington National Records Center to NARA. Boxes containing the folders for the earliest awards are in Accession 60A-262, RG 307, NARA. The grants folders were difficult to use in their unprocessed state, and I did not attempt to peruse them beyond the first few years of grants. Folders for proposals turned down have not been saved, although information is available on the earliest sets of unfunded proposals in folder "Advisory Panel Meetings, January & April 1952" (see above).

Assistant Director H. Burr Steinbach's informative "Diary Notes," 1953–54, are in Box 5, 72A-1808, RG 307, NARA.

For the period 1964–74, there is much useful material in the subject files of NSF Directors Leland O. Haworth (1963–68), William McElroy (1968–72), and H. Guyford Stever (1972–76). For each year until 1974, except for 1973 when files were lost, there is at least one folder for BMS as well as folders for major biological undertakings (see the list of abbreviations at the beginning of the notes for locations of BMS folders).

Index

Library of Congress Cataloging-in-Publication Data

Appel, Toby A., 1945–
    Shaping biology : the National Science Foundation and American biological
research, 1945–1975 / Toby A. Appel.
        p.   cm.
    Includes bibliographical references (p.) and index.
    ISBN 0-8018-6321-X (alk. paper)
    1. Biology—Research—United States—History.   2. National Science
Foundation (U.S.)—History.   I. Title.
QH319.A1 A66 2000
570'.7'2073—dc21   99-089620